# COMPRESSION for GREAT DIGITAL VIDEO

Power Tips, Techniques, and Common Sense

*Ben Waggoner*

**CMP*Books***

CMP Books
an imprint of CMP Media LLC
1601 West 23rd Street, Suite 200
Lawrence, Kansas 66046
USA
www.cmpbooks.com
email: books@cmp.com

Designations used by companies to distinguish their products are often claimed as trademarks. In all instances where CMP is aware of a trademark claim, the product name appears in initial capital letters, in all capital letters, or in accordance with the vendor's capitalization preference. Readers should contact the appropriate companies for more complete information on trademarks and trademark registrations. All trademarks and registered trademarks in this book are the property of their respective holders.

Copyright © 2002 by CMP Media LLC, except where noted otherwise. Published by CMP Books, CMP Media LLC. All rights reserved. Printed in the United States of America. No part of this publication may be reproduced or distributed in any form or by any means, or stored in a database or retrieval system, without the prior written permission of the publisher; with the exception that the program listings may be entered, stored, and executed in a computer system, but they may not be reproduced for publication.

The programs in this book are presented for instructional value. The programs have been carefully tested, but are not guaranteed for any particular purpose. The publisher does not offer any warranties and does not guarantee the accuracy, adequacy, or completeness of any information herein and is not responsible for any errors or omissions. The publisher assumes no liability for damages resulting from the use of the information in this book or for any infringement of the intellectual property rights of third parties that would result from the use of this information.

Technical editor: Dominic Milano
Acquisitions editor: Dorothy Cox
Proofreader: Robin Small
Managing editor and layout design: Michelle O'Neal
Graphic design: Justin Fulmer
Cover layout design: Damien Castaneda

Distributed in the U.S. by:
Publishers Group West
1700 Fourth Street
Berkeley, California 94710
1-800-788-3123
www.pgw.com

Distributed in Canada by:
Jaguar Book Group
100 Armstrong Avenue
Georgetown, Ontario M6K 3E7 Canada
905-877-4483

ISBN: 1-57820-111-X

**CMP Books**

# Table of Contents

**Foreword** .................................................................. xi

**Chapter 1     Introduction** ........................................... 1
    Who Should Read This Book..........................................3
    Book Organization..........................................................3
    Acknowledgments..........................................................4

**Chapter 2     Seeing and Hearing** ............................... 5
    Introduction....................................................................5
    Vision..............................................................................5
    Hearing........................................................................13
    For Further Reading....................................................18

**Chapter 3     Fundamentals of Compression** ............ 19
    Introduction..................................................................19
    Sampling and Quantization.........................................19
    Color Spaces...............................................................28
        *What Colors are Cb and Cr?*......................................*32*

Quantization Levels ..................................................................................33
    *The Delights of Y'CbCr Processing* ........................................................*36*
Quantization Errors ..................................................................................37
Compression Basics—an Introduction to Information Theory ................39
Data Compression ....................................................................................40
Spatial Compression Basics ....................................................................42
Temporal Compression ............................................................................49
Audio Compression ..................................................................................56
What Makes a Good Codec? ....................................................................58
For More Information ................................................................................61

# Chapter 4  The Digital Video Workflow .............................. 63
Introduction ...............................................................................................63
Planning ....................................................................................................63
Production .................................................................................................65
Postproduction ..........................................................................................66
Digitization ................................................................................................66
Preprocessing ...........................................................................................66
Compression .............................................................................................67
Delivery .....................................................................................................67

# Chapter 5  Producing and Editing for Optimal Compression .. 69
Introduction ...............................................................................................69
Preplanning ...............................................................................................70
Production .................................................................................................70
Postproduction ..........................................................................................74

# Chapter 6  Capture and Digitization ................................... 77
Broadcast Standards ................................................................................77
Tape Formats ............................................................................................80
Connections ..............................................................................................85
Audio Connections ....................................................................................87
Capture Resolution ...................................................................................89
Device Control and Timecode ...................................................................90
Data Rates ................................................................................................91
    *The 2GB Limit* .........................................................................................*93*

# Chapter 7  Preprocessing ................................................. 95

Introduction ................................................................................. 95
Cropping ..................................................................................... 95
Deinterlacing and Inverse Telecine ........................................... 101
Progressive Scan—Perfection Incarnate .................................. 104
Scaling ..................................................................................... 105
Aspect Ratio ............................................................................. 105
Pixel Shape .............................................................................. 108
Resolution Constraints ............................................................. 108
Noise Reduction ....................................................................... 110
Lowpass Filtering ..................................................................... 111
Sharpening ............................................................................... 111
Composite Noise ...................................................................... 111
Motion Tracking ....................................................................... 111
Luma Adjustment ..................................................................... 112
Remapping from ITU-R BT.601 ................................................ 112
Brightness ................................................................................ 114
Chroma Adjustment ................................................................. 116
Saturation ................................................................................. 117
Hue ........................................................................................... 117
Color Correction ....................................................................... 117
Frame Rate .............................................................................. 118
Audio Preprocessing ................................................................ 119

# Chapter 8  Video Codecs ................................................. 121

Introduction .............................................................................. 121
Codec Settings ........................................................................ 121
Data Rate ................................................................................. 126
Achieving Balanced Mediocrity with Video Compression ........ 134

# Chapter 9  Audio Codecs ................................................. 141

Introduction .............................................................................. 141
Audio Codec Settings .............................................................. 141
Trade-offs ................................................................................. 145

## Chapter 10  CD-ROM, DVD-ROM, and Kiosk Delivery ............ 147

Introduction ..................................................................................................... 147
Characteristics of Disk Playback ..................................................................... 147
Authoring for Disk Playback ............................................................................ 148
Formats for Disc Playback ............................................................................... 152
    *Director* ...................................................................................................... *153*

## Chapter 11  Web Video ........................................................ 155

Introduction ..................................................................................................... 155
Connection Speeds on the Web ...................................................................... 156
Four Kinds of Web Video ................................................................................. 157
Selecting the Right Transmission Mode ......................................................... 161
Integrating Web Video ..................................................................................... 162

## Chapter 12  Rich and Interactive Media ............................. 167

Introduction ..................................................................................................... 167
Interactive Video .............................................................................................. 167
Rich Media ....................................................................................................... 168
Authoring Rich and Interactive Media ............................................................ 169
Skins ................................................................................................................. 173

## Chapter 13  Choosing Platforms and File Formats ................ 175

Introduction ..................................................................................................... 175
OS Compatibility .............................................................................................. 175
Installed Base ................................................................................................... 178
Codec Options .................................................................................................. 180
Disc-based Playback ........................................................................................ 181
Progressive Download ..................................................................................... 183
Real-time Streaming ........................................................................................ 185
Rich Media Support ......................................................................................... 186

## Chapter 14  QuickTime ........................................................ 189

Introduction ..................................................................................................... 189
QuickTime Format ............................................................................................ 189
    *History of QuickTime* ................................................................................. *190*
QuickTime Tracks ............................................................................................. 190
Delivering Files in QuickTime .......................................................................... 195

*MPEG-4 in QuickTime* .................................................................................. 196
The Standard QuickTime Compression Dialog ........................................... 203
QuickTime Integration Tips ........................................................................ 208
QuickTime Alternate Movies ...................................................................... 209
Cool QuickTime Tricks ............................................................................... 211
QuickTime Delivery Codecs ....................................................................... 215
Legacy Delivery Codecs ............................................................................. 226
QuickTime Authoring Codecs .................................................................... 230
QuickTime Audio Codecs ........................................................................... 234
Legacy Audio Codecs ................................................................................. 237
For Further Reading ................................................................................... 238

# Chapter 15    Windows Media ........................................................ 239

The Advanced Streaming Format ............................................................... 239
Windows Media Player ............................................................................... 240
Windows Media DRM ................................................................................. 243
Windows Media Encoding Tools ................................................................. 246
Windows Media Video Codecs ................................................................... 251
Windows Media Audio Codecs ................................................................... 254

# Chapter 16    RealSystem ............................................................... 257

Introduction ................................................................................................ 257
RealMedia Format ...................................................................................... 258
RealOne Player .......................................................................................... 258
RealVideo Encoding Tools ......................................................................... 261
RealVideo Codecs ...................................................................................... 264
RealAudio Codecs ...................................................................................... 266

# Chapter 17    MPEG-1 ..................................................................... 271

Introduction ................................................................................................ 271
MPEG-1 Video ............................................................................................ 271
I, P, and B-Frames ..................................................................................... 272
GOP ............................................................................................................ 273
Video Buffer Verifier (VBV) ........................................................................ 273
MPEG-1 Aspect Ratios ............................................................................... 274
MPEG-1 Audio ............................................................................................ 274
MPEG-1 for VideoCD ................................................................................. 276
MPEG-1 Playback ...................................................................................... 279
MPEG-1 Authoring and Tools .................................................................... 280

## Chapter 18  MPEG-2 .................................................................. 285

Introduction ..........................................................................................285
The MPEG-2 Format .............................................................................285
MPEG-2 Video ......................................................................................285
MPEG-2 Audio .....................................................................................287
MPEG-2 for DVD Video ........................................................................289
MPEG-2 for DVD-ROM ........................................................................292
MPEG-2 for Digital Cable and Satellite ................................................293
MPEG-2 for CD-ROM ...........................................................................295
MPEG-2 for Authoring .........................................................................296
MPEG-2 Encoding Tools ......................................................................296
For Further Reading .............................................................................298

## Chapter 19  MPEG-4 .................................................................. 299

Introduction ..........................................................................................299
MPEG-4 Architecture ...........................................................................299
MPEG-4 Profiles and Levels .................................................................300
MPEG-4 File Format .............................................................................301
MPEG-4 Servers ...................................................................................301
MPEG-4 Players ...................................................................................301
MPEG-4 Video Codecs .........................................................................302
MPEG-4 Audio Codecs ........................................................................303
  *Rectangular Visual Profiles* .............................................................*304*
  *Arbitrarily Shaped Video* ................................................................*308*
ISMA Standards ...................................................................................309
Wireless Standards ...............................................................................311
DRM in MPEG-4 ...................................................................................311
  *Still Visual Profiles* ..........................................................................*312*
Applications .........................................................................................313
  *Audio Codecs and Profiles* .............................................................*314*
Where and How MPEG-4 Matters .......................................................315
MPEG-4 Authoring Tools .....................................................................317
  *Things Called "MPEG-4" That Aren't MPEG-4* ...............................*319*

## Chapter 20  AVI ......................................................................... 321

The AVI Format ....................................................................................321
AVI Interleave ......................................................................................322
Delivering Files in AVI .........................................................................322
AVI Architectures .................................................................................323

AVI Authoring Tools...........................................................................................323
AVI Delivery Codecs..........................................................................................325
Legacy Delivery Codecs.....................................................................................331
AVI Authoring Codecs........................................................................................333
AVI Audio Codecs..............................................................................................334
Legacy Audio Codecs........................................................................................335
For Further Reading..........................................................................................336

# Chapter 21  Flash MX .................................................................337

Introduction......................................................................................................337
Authoring Flash MX..........................................................................................338
Spark Video Codec...........................................................................................340
Audio Codecs in Flash MX................................................................................342

# Chapter 22  Miscellaneous Formats .........................................343

Bink..................................................................................................................343
Java Media Framework....................................................................................345
Kinoma.............................................................................................................348
MNG.................................................................................................................349
Animated GIF...................................................................................................350
Advanced Authoring Format (AAF)..................................................................351
MP3..................................................................................................................351
Ogg Vorbis........................................................................................................356
Ogg Tarkin........................................................................................................356
Theora..............................................................................................................356
Obsolete Formats.............................................................................................356

# Chapter 23  Compression Tools .................................................359

What to Look for in a Compression Tool.........................................................359
Cleaner 5.1.2....................................................................................................362
Canopus ProCoder 1.0.1..................................................................................366
Adobe Premiere 6.01.......................................................................................368
Adobe After Effects 5.5...................................................................................371
VirtualDub 1.4.10.............................................................................................374
Sorenson Squeeze 2.0.....................................................................................376
HipFlics 1.1......................................................................................................378
Helix Producer.................................................................................................380
Vegas Video 3.0a.............................................................................................381
Windows Media Encoder 8/7.1.......................................................................382

## Chapter 24    Workflow Optimization ............................................................. 385

    Why Workflow Matters ........................................................................................385
    Choosing an Encoding Platform .........................................................................385
    Intermediate Files ................................................................................................389
    Media Freshening ................................................................................................391
    Multi-machine Rendering ....................................................................................391
    Scripting ................................................................................................................393
    Workgroup Encoding Products ..........................................................................394

## Chapter 25    Tutorials ........................................................................................... 397

    Tutorial 1:  NASA Footage for Web Streaming ...............................................397
        *Fixing Sorenson B-frame Sync* ........................................................................ *410*
    Tutorial 2:  Music Video for Progressive Download .......................................415
    Tutorial 3:  Animation for Video Game Cut Scene .........................................424

## Glossary ............................................................................................................... 429

## Index ..................................................................................................................... 439

## What's on the CD-ROM? ....................................................................................... 452

# Foreword

Understanding compression is as important to producing digital video as understanding film is to producing a motion picture. The importance will only increase, as digital video is rapidly replacing analog, and will eventually replace film. And yet compression remains a "mysterious art" to many professionals working in digital video. Hopefully, this book will help to change that.

Every major video distribution technology introduced in the last few years has featured compressed digital audio or video. This includes not only streaming video, but also DVD, MP3, satellite radio, and others. There is a very good reason for this. The "digital revolution" packs more quality and versatility into a smaller amount of space than was ever possible before. And compression is a key element of this revolution.

Compression is sometimes viewed as a negative; the cause of nasty artifacts. But in reality, it is only the use of sophisticated compression technologies that make many applications possible at all. Uncompressed video would require a dozen DVDs to hold a single movie. Uncompressed audio could not be streamed over a modem. The challenge lies in using these technologies effectively. As with any tool, the first step is to understand the medium.

Just as the choice of f-stop, film type, shutter speed, etc., has a profound effect on the character of film, the choice of codec, bitrate, filtering, and so on has a profound effect on a streaming video experience. The tools are constantly improving—doing a better and better job of running on automatic. But there are good reasons a professional photographer doesn't always rely on even the best camera's auto mode and there are just as good reasons to fine tune the video compression process.

If your goal is to hand-tweak the absolute best possible streaming presentation of a movie, you'll want to know all the controls to which you have access. If your project requires you to run a high volume of totally automated compressions, there are still important choices to be made in setting up an efficient process.

It is getting easier to produce decent results. But as with any endeavor, the gap between "decent" and "first rate" can be a challenging one to bridge. It takes is a lot of trial and error—or a good teacher. Sometimes a bit of both.

Ben Waggoner has been a leading expert on the use of digital video on the desktop—from the days when CD-ROM was the cutting edge of delivery technology through the latest developments in streaming video and MPEG-4.

Ben has a unique passion for his subject. He can wax rhapsodic about color spaces, rant about the weaknesses of interlaced video, and extol the virtues of a truly cutting-edge codec. And he's always willing to share his knowledge.

If you've ever heard Ben speak, you know that if pressed, he could probably talk through the entire contents of this volume in a single hour. But very few of us can handle information coming in at that rate! So here we have the benefit of a large chunk of his knowledge in written form to assimilate at our own pace.

Of course, if this book was digitally compressed, it would be much more efficient...

Darren Giles
CTO, CustomFlix; and creator of Cleaner

# Chapter 1

# Introduction

It was the fall of 1989, and I was taking a computer animation class at UMass Amherst (I went to Hampshire College, another school in the Five Colleges consortium). We were using a number of long-gone tools like Paracomp Swivel 3D to create a video about robots. This was long before After Effects came out, so we were using Macromind Director 1.0 as a very basic compositing app (all straight luma keyed). We needed files that would play back faster on our 20MHz Macintosh IIcx machines so we could preview animation timing. I started experimenting with a utility called Director Accelerator, which would composite all the layers together of the Director project, and export them as a series of Run Length Encoded (RLE) images, a very early form of video compression. I was amazed how, sometimes, the resulting files were smaller than the original files. And so I started messing around with optimizing for RLE, and forgot about animation and robots.

I got a D in the class, while the TAs and some other students went on to create Infini-D, a pioneering 3D app for desktop computers. Infini-D lives on as Carrara from Eovia. Me, I didn't do any more animation after that, but darn if that compression stuff didn't stay interesting.

So interesting I've made a career of it, although it took me a while to recognize my destiny. Right after college, I spent a year writing and producing a comedy-horror mini-series. For reasons none of my collaborators understood, I wrote an overly long scene which hinged on the compression ratio of Apple Compact Video (later known as Cinepak).

Next, I co-founded a video postproduction company called Journeyman Post in 1994, which became Journeyman Digital in 1995. We started trying to be a standard NLE shop, using a Radius VideoVision Studio card in a PowerMac 8100/80, with a whopping 4GB RAID drive. However, due to some bugs in the system, we couldn't capture more than about 90 seconds in sync, making the whole system a boat anchor for video editing.

Then, one day, someone called us up and asked if we could do video for CD-ROM. Because they were willing to pay for it, we said "yes!" and started to figure out what the heck they were talking about. Turns out the VideoVision was perfect for short CD-ROM clips, which never had good sync anyway. That system cost $18,000 and was far less capable for editing and compression than my wife's two-year-old iBook.

For compression, we used Adobe Premiere 3.0 and Apple's MovieShop. The codecs of choice were Cinepak and Indeo 3.2, with 8-bit uncompressed audio. Data rates for a typical 240×180 15fps movie were around 120KBps, more than enough for full-screen Web video today. We targeted 80 minutes of encoding per minute of output, so there was a lot of time for "bartending."

Still, primitive as the technology was, it enabled some amazing things. For those who find the notion obvious that computers can play videos, it's hard to explain the thrill of seeing those first postage stamp videos play in a window. The video might have been small and blocky, but it was *interactive*. We created all kinds of very successful sales training discs, kiosks, and multimedia training projects for fun and profit.

By the late 1990s, video on the Web was all the rage. Better, faster, cheaper CPUs and Internet connections, as well as new Web technologies brought us to the point where Web video looked like something more than pixel soup. The future looked promising. New businesses were springing up like dandelions—everybody wanted to cash in on the opportunities presented by the coming of ubiquitous broadband connectivity. Companies like Akamai and Digital Island and many, many others set up Content Distribution Networks, and raised billions in their initial public stock offerings.

In 1999, I joined Terran Interactive, the original developers of Media Cleaner (the leading professional video compression application at the time). I founded and ran Terran's Consulting Division. My job was to reap services revenue, as well as to help out with marketing and product development. Media 100 acquired Terran, and I was laid off with most of the rest of the ex-Terrans in 2001.

While the Internet bubble has gone the way of the tulip frenzy, the number of outlets for compressed video has continued to increase. Folks still want to put video on the Net. They also want to stream it to mobile wireless devices, use it in Point of Purchase kiosks, send it over satellites to homes and businesses, display it within real-time 3D games, and burn it to DVD. The days of the get-rich-quick Internet scheme are long gone, but the need for skilled compressionists are only just beginning. While Web video didn't turn into a thriving industry, it is becoming a crucial part of the already enormous video industry. My clients today aren't trying to make money by selling $1 of advertising around a stream that costs $5 to delivery anymore, which is a relief. Instead, they are schools and businesses who are finding ways for compressed video to do things that traditional video can't—such as distance learning and training on demand. It might not be as flashy as the impossible dreams of 1999, but the technology is proving to be very useful and cost-effective in the right areas.

This book has been a long time in the making. I wrote the first outline for it back in 1998, back when CD-ROM video was all the rage and Web video a "someday" thing. While the subjects have grown to include things that weren't around back then, my overall goal hasn't changed a bit. This book is all about how to make good compressed video on the many delivery platforms available.

## Who Should Read This Book

This book is for compressionists, people who want to be compressionists, and people who on occasion need to pretend they're compressionists. Most of you will have come from the worlds of either video or Web authoring. Video people will be happy to hear that this stuff isn't as alien as it might initially seem. The core skills of producing good video are just as important when dealing with compression, although compression adds plenty of new twists and lots of new acronyms to keep track of. Web folks may find the video world intimidating—it's populated by folks with 20-plus years of experience, and comes festooned with enough jargon to fill a separate volume. Heck, you may also come to the subject with little or no background in either the Web or video.

No matter. I've tried to include definitions and explanations of this jargon as it comes up. If you read this book in nonlinear fashion and jump straight to the middle, there's a glossary that should help you decode some of the more common terms that crop up throughout.

## Book Organization

The chapters are organized in three main sections. The first section (Chapters 1–11) is about general principles of vision, compression, and how compressed video operates. While a fair amount of it will be old hat for practicing compressionists, there should be something of interest for everyone. Even if you skim the book the first time around, it'd be a fine idea to go back and give it a more careful reading sometime while you're bartending. I've tried to keep it readable—there's no math beyond basic algebra, and I kept it to details that matter for compression.

Chapters 12–24 cover specific video tools and technologies. Each chapter is self-contained, so it's fine to skip straight to the ones you care about.

The last chapter (Chapter 25) consists of three extensive tutorials that highlight many of the concepts, formats, and tools discussed earlier. The CD-ROM includes project files and sample media for the tutorials, so you can follow along. For the full media files that couldn't fit on the disc, you can order a DVD-ROM from CustomFlix with all the goodies; keep an eye on my Web site (www.benwaggoner.com) for information.

The world of compression changes fast enough to give you chronic whiplash. As this book was being written, I was constantly having to update chapters to keep up with the lightning pace of updates, acquisitions, and new releases. But we had to stop at some point. No matter what the current release of some tool or other is, the techniques described in these pages should prove useful. But it's a good idea to keep up-to-date with changes in the industry.

Check http://www.benwaggoner.com/bookupdates.html for updates, corrections, and other resources.

Ben Waggoner
Self-proclaimed "World's Greatest Compressionist"
June 2002

## Acknowledgments

A book like this requires a lot of work. Dominic Milano went far above and beyond the call of duty as technical editor, making sure it was readable, relevant, and right. Jim Feeley edited many articles for *DV* that were later incorporated here and continues to teach me an enormous amount about how to express complex technical ideas clearly. The long course of this book provided many opportunities to demonstrate the patience, talent, and good humor of Dorothy Cox, Paul Temme, and Michelle O'Neal from CMP Books.

Lots of people taught me the things I said in this book. The founders of Terran Interactive, Darren Giles, Dana LoPiccolo-Giles, and John Geyer answered many dumb questions from me, until I knew enough to ask some smart ones. I learned a lot about compression late at night.

During the gestation of this one book, my wife Sacha gestated Alexander and Aurora; a much harder job than mine! I'm looking forward to spending more time with them instead of in the basement.

My family helped me start a business back when I didn't know enough about what I was doing to even know what I didn't know. I'm sure they bit their lips more than they let on over the past few years of this industry. Hopefully, they're able to relax now.

And finally, none of this would have happened without Halstead York, who over the years has talked me into writing scripts, producing video, starting two companies, and learning compression. Without his infectious enthusiasm, vision, and drive, I'd probably still be writing banking software.

# Chapter 2

# Seeing and Hearing

## Introduction

While the rest of this book is about practical compression issues, it's important to understand how the human brain perceives images and sounds. Compression is the art of converting large images into small files. Compression algorithms accomplish that feat by describing various aspects of images or sounds in ways that efficiently describe detail so that those elements that are important to seeing and hearing are preserved, while elements that aren't important are discarded.

Understanding how your brain receives and interprets signals for your auditory and visual receptors (your ears and eyes) is half the battle in understanding how compression works and which compression algorithm will work best in a given situation.

## Vision

### What light is

Without getting too overly technical, light is composed of particles (loosely speaking) called photons that vibrate (loosely speaking) at various frequencies. Visible light is formed by photons vibrating within a range known as the visible light spectrum (faster photons are x-rays, slower photons are radio waves). This visible light spectrum can be seen, quite literally, in a rainbow. The higher the frequency of vibration, the further up the rainbow the light's color appears, from red at the bottom to violet at the top.

The colors of the visible light spectrum are known as black-body colors (Figure 2.1). That is, when an ideal black object is heated, it takes on the colors of the visible spectrum, turning red, then orange, then green, on through violet, as it gets hotter. These colors are measured in degrees

Kelvin, for example as 6500K. What about colors that don't appear in the rainbow like, say, slate green? They result from combining the pure colors of the rainbow.

### See also
See the color version of Figure 2.1 in the color section on page I.

The frequency of the light is measured in terms of wavelength. Visible light's wavelength is expressed in nanometers, or nm. Unless you're looking at something that radiates light, you're actually seeing a reflection of light generated elsewhere—by the sun, lights, computer monitors, television sets, and so on.

### See also
The range of colors from black body radiation at various temperatures are shown in Figure B on page I of the color section.

## What the eye does

So there's a lot of light bouncing around the universe. When some of that light hits the retina at the back of our eyes, we "see" an image of whatever that light has bounced off of. The classic metaphor for the eye is a camera. The light enters our eye through the pupil, and passes to the lens, which focuses the light. Like a camera, the lens has a focal plane, and objects nearer or farther away than that become increasingly blurry. The amount of light that enters is controlled by

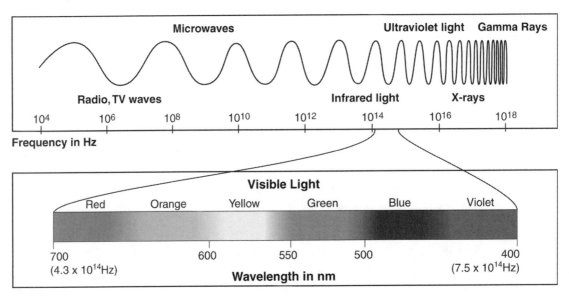

**2.1** The range of the electromagnetic spectrum. Visible light only takes up a very small portion.

the iris, which can expand or contract like a camera's aperture. The light is focused by the lens on the back of the retina, which turns the light into nerve impulses. The retina is metaphorically described as film, but it's closer to a CCD in a video camera, because it continually produces a signal. The real action takes place in the retina.

In high school biology, your teacher yammered on about two kinds of light-receptive cells that reside in the retina: rods and cones. We have about 100 million of these photoreceptors in each eye, 95 percent of which are rods. Rods are sensitive to low light and sharp motion, but they detect only luminance (brightness), not chrominance (color). Cones detect detail and chrominance, and come in three different varieties, those that detect blue, red, and green. Cones are much less sensitive to motion and detail than rods are. This explains why people don't see color well at night—rods work when it's too dark for cones to function at all.

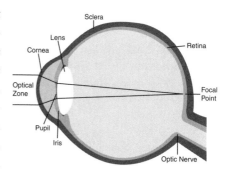

2.2  A technical drawing of the human eye.

The good news is you can safely forget about rods for the rest of this book. They only contribute to vision at low light (below indoor lighting levels), much lower levels than you'll get out of a computer monitor or television. At normal light levels, we see with just the cones.

Still, the rods make a remarkable contribution to vision. The ratio between the dimmest light we can see and the brightest light we can see without damaging our eyes is roughly one trillion to one! And the amazing thing is that, within the optimal range of our color vision, we aren't really very aware of how bright light in our environment is. We see relative brightness very well, but our visual system rapidly adjusts to huge swings in illumination. Try working on a laptop's LCD screen on a sunny summer day, and the screen will be almost unreadable even at maximum brightness. Using that same screen in a pitch-black room, even the lowest brightness setting can seem blinding.

 **Note**

We are most sensitive to green, less to red, and least of all to blue.

Folks with normal vision have three different types of cones, each specific to a particular wavelength of light, matching red, green, and blue (as in Figure 2.3). We're not equally sensitive to all colors however. We're most sensitive to green, less to red, and least of all to blue. That statement is repeated in the margin because it's important. Really. Trust me on this.

Also, not everyone has all three types of cones. About two percent of men have only two of the three cone types, causing red-green colorblindness, the most common type. Dogs and most other mammals also have only two cone types. Not that it's that much of a problem. Many people with

# 8 Chapter 2: **Seeing and Hearing**

**2.3** Relative sensitivity of the eye's receptors to the different colors. The black line is the rods, the white lines the three cones.

**See also**
The color version of Figure 2.3 is in the color section, page II.

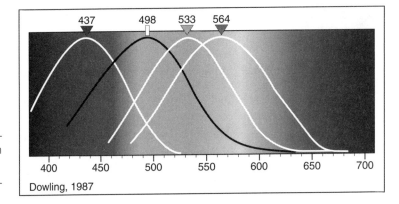

Dowling, 1987

red-green colorblindness don't even know they're colorblind until they get their vision tested for their driver's license. Red-green colorblindness is also why my dad wears more pink than he realizes.

Another surprising aspect of the retina is that most of the detail we perceive is picked up within a very small area of what the eye can see, in the very center of the visual field, called for *fovea*. The outer portion of the retina consists mainly of rods for low-light vision. The critical cones are mainly in the center of the retina. Try staring directly at a wall, and notice how small a region is actually in focus at any given time. That small region of clear vision is what the fovea sees. Fortunately, our eyes are able to flick around many times a second, pointing the fovea at anything interesting.

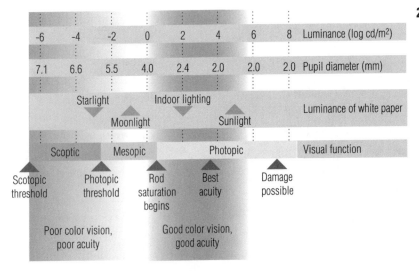

**2.4** Vision at different light levels. Note light level is measured on a logarithmic scale.

**Note**

***Late night debate topic answered once and for all!*** So, if you were prone to senseless philosophizing at late hours as I once was (having small children seems to have nicely displaced that niche in my schedule), you've no doubt heard speculation ad nauseam as to whether or not what one of us calls "red" could be what another perceives as "blue."

Kudos if you've never thought about this, because Figure 2.3 shows that what you see as red is what every other person with normal version perceives as red. In every culture in the world, in any language, if you ask what the most "red" color is and offer a bunch of samples, almost everyone picks the sample at 564nm. Thus what you see as red, I see as red, and everyone sees as red. I find it rather pleasing when something that once seemed abstract and unknowable turns out to have an extremely concrete definition like this. That is, unless you're talking to my dad, who's red-green colorblind and thus finds red and green look about the same...

## How the brain sees

It's hard to draw a real mind/body distinction with anything in neuroscience, and this is no less true of vision. What we call "seeing" takes place partially in the retina and partially in the brain, with the various elements interacting in complex ways. Of course, our sense of perception happens in the brain, but some basic processing like finding edges actually takes place in the retina. This area is being actively researched, but in general it's impossible to describe how vision works in the brain with anywhere near the specificity we can use in describing the function of the eye. It's clear that the visual system takes up about a quarter of the total brain. Humans easily have the most complex visual system of any animal.

Under ideal circumstances, humans can discriminate between about one million colors. Describing how might sound like advanced science, but the CIE (*Commission Internationale de l'Echairage*—French for International Commission on Illumination) had it mostly figured out back in 1931. The CIE determined that the whole visible light spectrum can be represented as mixtures of just red, green, and blue. Their map of this color space (Figure D, color section) shows a visible area that goes pretty far in green, some distance in red, and not far at all in blue. However, there are equal amounts of color in each direction; the threshold between a given color and black shows how much color must be present for us to notice. So we see green pretty well, red so-so, and blue not very well. Ever notice how hard it is to tell navy blue from black unless it's viewed in really bright light? The reason is right there in the CIE chart.

**See also**

The CIE color charts from 1931 and 1976 are on page II and page III of the color section.

Another important concept that the CIE introduced is "luminance"—how bright objects appear. Note for human vision in normal light, we don't actually see anything in black and white—even

# Chapter 2: Seeing and Hearing

monochrome images start with the three color receptors all being stimulated. But our brains process brightness differently than color, so it's very important to distinguish between them.

Computers normally turn RGB values into gray by averaging the three values. However, this does not match the human eye's sensitivity to color. Remember, we're most sensitive to green, less to red, and least to blue. Therefore, our perception of brightness is mainly determined by how much green we see in something, with a good contribution from red, and a little from blue.

For math geeks, here's the equation. We call luminance Y because… I really don't know why.

$Y = 0.299R + 0.587G + 0.114B$

Encoding images requires representing analog images digitally. Given our eyes' lack of sensitivity to some colors and our general lack of color resolution, an obvious way to optimize the data required to represent images digitally is to tailor color resolution to the limitations of the human eye.

Color vision is enormously cool stuff. I've had to resist the temptation to prattle on and on about the lateral geniculate nucleus and the parvocellular and magnocellular systems and all the neat stuff us neuropsychology majors got to learn about in college.

## Perceiving luminance

Even though our eye only works with color at typical light levels, our perception is much better tuned towards luminance than color. Specifically, we can see sharp edges and fine detail in luminance much better than we can in color. We can also pick out objects moving in a manner different from other objects around them a lot more quickly, which means we can also see things that are moving quickly a lot better than things that are standing still. In optimal conditions with optimal test material, someone with good vision is able to see about 1/60th of a degree with the center of their fovea—the point where vision is most acute. That's 0.07 inches at a distance of 20 feet.

**2.5** Kanizsa Figure. Note how the inner square appears brighter than the surrounding paper.

A lot of the way we see is, fundamentally, faking it. Or, more gracefully put, our visual system is very well attuned to the sort of things bipedal primates do to survive on planet Earth, such as finding good things to eat without being eaten ourselves. But given images that are radically different than what exists in nature, strange things can happen. Optical illusions take advantage of how our brains process information to give illusionary results—a neuropsychological hack, if you will. Check out Figure 2.5. Notice how you see a box that isn't there? And how the lines don't look quite straight, even when they are?

## Perceiving color

Compared to luminance, our color perception is blurry and slow. But that's really okay. If a tiger jumped out of the jungle, our ancestors didn't need to know what color stripes it had. Evolutionarily, color was likely important for picking out poisonous from non-poisonous fruits and vegetables, an activity that doesn't need to happen in a split second nor does it require hugely fine detail.

 ***See also***

Refer to Figure F in the color section.

For example, look at Figure F in the color section, page IV. It shows the same image, once without color, and once without luminance. Which can you make the best sense of? Bear in mind there is mathematically *twice* the amount of information in the color-only image, because it has two channels instead of one. But our brains just aren't up to the task of dealing with color-only images.

This difference in how we process brightness and color is profoundly important when compressing video.

## Perceiving white

So, given that we really see everything as color, what's white? We now know that the only way there can be an absence of color is if there is no light at all. But clearly white isn't black! When speaking of color as it's found in visible light, white is achieved when there are equal amounts of red, green, and blue. However, in different circumstances, what color we see as white varies quite a bit, especially by context. What "color" white is measured by the black-body temperature. As was mentioned previously, for outdoors on a bright sunny day, white is about 6500K. Videographers well know 6500K is the standard temperature of white professional lights.

***See also***

See Figure E on page III of the color section to see how one's perception of white can vary depending on its context or surroundings.

As you can see in Figure E in the color section, our perception of what white is varies a lot with context. It also varies a lot with the color of the light that you're reading this book in.

We aren't normally aware of the fact that white can vary so much. Our brains automatically calibrate our white perception, so that things look like the same color in different conditions, even though the relative frequencies of light bouncing off them and into our eyes vary widely. This is one reason why white balancing a camera manually is so difficult—our eyes are already doing it for us.

## Perceiving space

Now you understand how we see stuff on a flat plane, but how do we know how far things are away from us? There are several aspects to depth perception. For close-up objects we have binocular vision, because we have two eyes. The closer an object is to us, the more different the image we see with each eye, because they're looking at the object at different angles. But this only works up to 20 feet or so. And you can still get a darn good idea of where things are with one eye covered. Binocular vision is important, but it isn't the whole story.

Alas, we don't know all of the rest of the story. There are clearly a lot of elements in what we see that can trigger our perception of space. These include relative size, overlapping, perspective, light and shading, blue shading in faraway objects, and seeing different angles of the object as you move your head relative to the object. You don't need to worry about elements that contribute to our depth perception too much for compression, except when compression artifacts degrade an element so much that the depth cues are lost. This is mainly a problem with light and shading.

## Perceiving motion

Our ability to know how things are moving in our environment is critical, and it's crucial to staying alive. We have a huge number of neurons in the brain that look for different aspects of motion around us. These ensure we can notice and respond very quickly. These motion sensor cells are specifically attuned to noticing objects moving in one direction in front of a complex background (like, say, a saber-toothed tiger running out of foliage).

We're able to keep track of a single object among a mass of other moving objects—it isn't hard to pick out a woman with a red hat at a crowded train station. As long as we can see some unique attributes, like color or size or brightness, we can differentiate a single item from a huge number of things.

One important facet of motion for compression is persistence of vision. This refers to how many times a second something needs to move for us to perceive it as actually moving, and not just being a series of separate unrelated images. Depending on context and how you measure it, persistence of vision is typically achieved by playing back images that change 16 times per second, or in film-speak, 16fps.

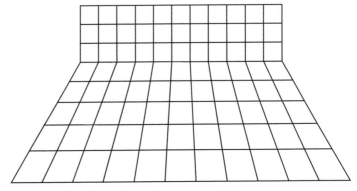

**2.6** This simple grid provides a powerful sense of perspective.

2.7 Paris Street: A Rainy Day by Gustave Caillebotte. This painting uses both shading and converging lines to convey perspective.

However, motion has to be quite a bit faster than that before it looks naturally smooth. Film running at 24fps shows each frame twice when the lamp in the film projector flashes twice for each film frame at 48Hz to present a smoother appearance. And most people can even tell a computer monitor set to refresh at 60Hz from one at 85Hz. A higher setting doesn't improve legibility so much, but a high refresh rate eventually makes the screen look as solid as paper (it takes at least 100Hz for me).

# Hearing

We know a lot less about how hearing works than vision, alas. But it's important not to forget its importance to compression. For most projects, the sound really is half the total experience.

## What sound is

Sound is an audible change in pressure. That's it. Sound is typically heard through the air, but can also be heard through water, Jell-o, rock, or anything elastic enough to vibrate and transmit those vibrations to your ear. There's no sound in a vacuum, hence the movie tag line "in space, no one can hear you scream." Of course, we can't hear all vibrations. Sound is produced when air pressure raises and lowers very rapidly. How much the air pressure changes determines loudness. How fast the pressure changes determines pitch. Like light, where you can have many different colors at the same time, you can have many different sounds happening simultaneously.

# Chapter 2: **Seeing and Hearing**

Different objects exhibit different audible characteristics, which we call *timbre* (pronounced "tam-ber"). For example, a note played on a piano sounds very different than the same note played on a saxophone. This is because their sound generating mechanisms vibrate in such a way that they produce different sets of overtones or harmonics.

Say what? When discussing pitch as it relates to Western musical scales, for example when someone asks to hear an *A* above middle *C*, they're asking to hear the fundamental pitch of that *A*, which vibrates at 440 cycles per second, or 440 Hertz (Hz). However, almost no musical instrument is capable of producing a tone that vibrates at just 440Hz. This is because the materials used to generate pitch in musical instruments (strings of various types, reeds, air columns, whatever) vibrate in very complex ways. These complex vibrations produce frequencies above that of the fundamental.

When these so-called overtones are whole number multiples of the fundamental frequency, they're called harmonics. The harmonics of A-440 appear at 880 (2×440), 1320 (3×440), 1760 (4×440), 2200 (5×440), and so on. The relative way the volumes of each harmonic change over time determines the timbre of an instrument. Overtones that are not simple whole number multiples of the fundamental are said to be enharmonic. Percussion instruments and percussive sounds such as explosions, door slams, and such, contain enharmonic overtones.

To achieve a "pure" fundamental, folks use electronic devices—oscillators—that produce distortionless sine waves. (In case you're curious, a distorted sine wave produces sounds with harmonics.) Clean sine waves look like this:

**2.8** The sine wave of a perfectly pure 440Hz tone.

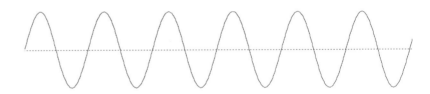

Of course, air doesn't know about overtones or notes or fundamentals or any of that. Sound at any given point in space and time is a change in atmospheric pressure. Here's what the same frequency looks like with overtones:

**2.9** This is the waveform of a single piano note at 440Hz.

When an electronic oscillator is used to produce a square wave (a sine wave with a distortion pattern that makes it look squared off hence its name), it produces a timbre containing the maximum possible amount of overtones. It looks like this and sounds pretty cool too.

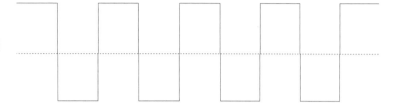

**2.10** A 440Hz square wave. Even though it looks very simple, acoustically it's the most loud and complex single note that can be produced.

When there are a lot of notes played simultaneously as in this chord, their waveform produces the following image. It may look like a mess, especially with overtones, but it sounds fine.

**2.11** A full piano A-major chord at 440Hz. Note how much more complex it is, but the 440Hz cycle is still very clear.

While tonal music is a wonderful thing—and a big part of what we compress—it's far from the only thing. Percussive sounds (drums, thunder, explosions) are made up of random noise with enharmonic spectra. Because there is so little pattern to them, percussive sounds prove a lot more difficult to compress, especially high-pitched percussives like cymbals.

**2.12** The waveform of a rim shot. Note how random the pattern seems.

## How the ear works

In the same way the eye is something like a camera, the ear is something like a microphone. First, air pressure differences cause the eardrum to vibrate. This vibration is carried by the three bones of the ear to the basilar membrane, where the real action is. On the membrane are many little hairs called cilia. These turn the vibration into electrical impulses, which go into the cochlea and then the brain. We only have 16,000 of these hairs, and they are damaged by excessively loud noises. This damage is the primary cause of hearing loss.

Like the rods and cones in our retina, cilia along the basilar membrane respond to different frequencies depending on where along the cochlea they reside. Thus, celia at specific locations respond to specific frequencies. The cilia are incredibly sensitive—moving them by as little as the width of one atom can produce a response (in proportion, that's like moving the top of the Eiffel tower one-half inch). They can also respond to vibrations up to 20,000Hz. And if that sounds impressive, bats and whales can hear up to 200,000Hz!

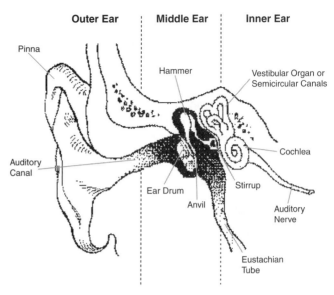

**2.13** The anatomy of the ear.

## What we hear

As Discovery Channel aficionados know, there is plenty of sound that humans can't hear. As with visible light, there is a range of vibrations that are audible. We measure the frequency of sound in Hertz (Hz), which means cycles per second. Someone with excellent hearing can hear the range of 20 to 20,000Hz. The maximum frequency that can be heard gets lower with age and hearing damage from loud noises. Attending a few impassioned rock concerts can dramatically lower your high-frequency hearing (wear your earplugs!). Below 30 to 20Hz, we start hearing sound as individual beats (20Hz is the same as 1,200 beats per minute—hardcore techno, for example, goes up to around 220bpm). Below 10Hz, we *feel* more than hear sound. We hear best in the range of 200 to 2000Hz. This happens to be pretty close to range of the human voice. Evolutionarily, it isn't clear whether our speech range adapted to match hearing or vise versa.

Sound levels (volume) are measured in decibels (dB). The range between the quietest sound we can hear (the threshold of hearing for those with good ears or 0dB) and the loudest sound that doesn't cause immediate hearing damage (120dB) is roughly 1 trillion to one (the same ratio as that between the dimmest light we can see and the brightest that doesn't cause damage to the eye, in fact).

Much less is known about how the brain processes audio than is known for vision. Still, it is clear our auditory system is capable of some pretty amazing things. Music, for example. Music is arguably the most abstract art form out there—there's nothing fundamental about why a minor chord sounds sad and a major chord doesn't. And the seemingly simple act of picking out instruments and notes in a piece of music is massively hard for computers. A few pieces of music software are capable of transcribing a perfectly clear, pure-tone monophonic (one note at a time) part played on a single instrument fairly well these days. But throw but throw chords or more than one instrument in the mix, and computers can't figure out much of what they're hearing.

We're also very capable of filtering out sounds we don't care about, like listening to a single conversation while ignoring the sounds in a crowded room. We're also able to filter out a lot of background noise. It's always amazing to me to work in an office full of computers for a day, and not be aware of the sound—keyboards clacking, fans whirring. But at the end of the day when things are shut off, the silence can be almost deafening. Even with all those things off, you can hear sounds that were completely masked by the others—the hum of a fluorescent light tube, birds twittering outside, and such.

We're able to sort things in time as well. For example, if someone is talking about you at a party, you'll often hear the whole sentence, even the part of it that preceded the mention of your name. This suggests to me that the brain is actually listening to all the conversations at once, at some level, and only brings things to our conscious attention under certain circumstances. This particular example is what got me to study neuroscience in the first place (although scientists still don't know how we're able to do it).

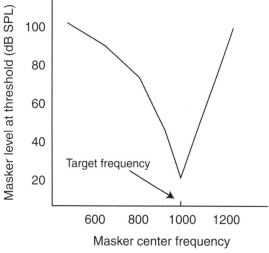

**2.14** Masking.

## Psychoacoustics

While our hearing system is enormously well suited to some tasks, we still aren't able to use all our sensitivity in all cases. This is critically important to doing audio compression, which is all about removing the parts of the music and sound we can't hear to spend bits on what we can. For example, two sounds of nearly the same pitch can sound just like the same note, but louder. In the center of our hearing range, around 200-2000kHz, we can detect very fine changes in pitch—for example, a 10Hz shift up or down in a 2kHz tone. Sounds can also be masked by louder tones of around the same frequency.

# Chapter 2: Seeing and Hearing

# For Further Reading

## Seeing

| | |
|---|---|
| http://webvision.med.utah.edu/Color.html | Peter Gouras created this fairly technical site, which brings together historical and current research on color vision. Good illustrations. |
| http://www.wendycarlos.com/colorvis/color.html | Wendy Carlos is best know for her pioneering work in electronic music (*Switch on Bach*, and many others), but she also provides a very readable site about some very interesting properties of color vision. |

## Hearing

| | |
|---|---|
| www.sfu.ca/sca/manuals/ZAAPf/d/decibels.html | Explains decibels, the range of human hearing, and so on. |
| www.anu.edu.au/ITA/drw/PpofM/INDEX.html | David Worrall's Physics and Psychoacoustics of Music course notes. Includes a wonderful overview of the physics of sound and psychoacoustics. |
| www-ccrma.Stanford.edu/marl/ | The Musical Acoustics Research Library of Stanford's Center for Computer Research in Music and Acoustics includes links to the John Backus Archive, the Catgut Acoustical Society Library, and more. |

# Chapter 3

# Fundamentals of Compression

## Introduction

This chapter is about how to turn light and sound into numbers and back into light and sound. Putting it that way makes it sound simple, but brace yourself. Compression involves some deep thought and some deep math. While it's not essential to know how codecs work in complete detail, having a working knowledge of what's going on under the hood will help you understand the result you're getting, spot problem areas, and plan strategies for overcoming the challenges they pose. I've tried to keep the chapter as readable and math-free as possible. I hope the many close encounters with the technical jargon of compression will help you get through the rest of the book with a minimum of confusion.

 **Note**

***So Far, So Fast.*** QuickTime turned 10 in the fall of 2001. To celebrate, some folks threw a party at which CD-ROMs were distributed to the attendees. Those CD-ROMs contained QuickTime movies, some dating back to the earliest QuickTime-compressed movies. Those early movies were typically 160x120 pixels—postage stamp-sized. They ran at 10 frames per second (fps), and had data rates of around 1300Kbps. Ten years later, those same files can be encoded with similar quality at around 40Kpbs. That's a 32× improvement in a decade!

## Sampling and Quantization

In the real world, light and sound exist as continuous analog values. Those values in visual terms make up an effectively infinite number of colors and detail; in audio terms, they represent an effectively infinite range of loudness (amplitude) and frequency. But digital doesn't do infinite. When analog signals are converted to digital form, that is, digitized, the infinite continuous scales of analog signals must be reduced to a finite range of discrete bits and bytes. This is accomplished via processes known as *sampling* and *quantization*.

# Chapter 3: Fundamentals of Compression

## Sampling space

Nature doesn't have a resolution. Take a standard analog photograph (one printed from a negative on photographic paper, not one that was mass-produced on newsprint or magazine paper stock via a printing press). When you scan that photo with a scanner, you need to specify a resolution in dots per inch (dpi). But given a good enough scanning device, there is no real maximum to how much detail you could go into. If you had an electron microscope, you could scan at 1,000,000 dpi or more, increasing resolution past where the individual molecules of the pigments are visible. Of course, beyond a certain point, you won't get any more visual detail from the image.

Sampling is the process of breaking up an image into discrete pieces, or samples, each of which represents a single point in 2D space. Imagine spreading a sheet of transparent graph paper over a picture. A sample is one square on the graph paper. The smaller the squares, the more samples. The number of squares is the resolution of the image. Each square is a picture element or *pixel* (or in MPEG-speak, a *pel*).

Most delivery codecs deal with square pixels, that is the height and width of each sample are equal. However, capture formats such as DV and ITU-R BT.601 often use non-square pixels. More on that later.

## Sampling time

The above describes how sampling works for a single frame of video. But video doesn't have just a single frame, so you need to sample temporally as well. Temporal sampling is normally described in terms of frames per second (fps) or Hertz (Hz, aka cycles per second).

## Sampling sound

Digitized audio, like video, is sampled. Fortunately, audio data rates are lower, and so it's a lot easier to store audio in a high-quality digital format. Thus, essentially all professional audio is uncompressed.

While video has two spatial dimensions, audio has only one—loudness (the air pressure at any given moment). So audio is stored as a series of measurements of loudness. In electrical terms, those changes are changes in voltage. The frequency at which voltage changes are sampled is called the *sampling rate*. For CD-quality audio, that rate is 44.1kHz (44,100 times per second). Other common sampling rates include 22.05kHz, 32kHz, 48kHz, 96kHz, and 192kHz. Note 96kHz and 192kHz so far are used for authoring, not for delivery.

But in digital audio, the sampling rate is only half the story. The other half is the sampling resolution. Resolution is measured in *bits*—a computer term derived from Binary digiT. A bit can have one of two values—one or zero. Each additional bit doubles the sampling resolution. For example, a two-bit sample can represent four possible values; three bits gives eight possible values; eight bits provides 256 possible values. The more bits, the better the sampling resolution.

# Sampling and Quantization 21

**3.1** The same source image with three different sampling grids.

## Chapter 3: Fundamentals of Compression

Each bit of resolution equals six decibels (dB) of dynamic range (the difference between the loudest and softest sounds that can be reproduced). The greater the dynamic range, the better the reproduction of very quiet sounds, and the better the signal-to-noise ratio.

Early digital audio sampling devices operated at 8-bit resolution. Audio CD players operate at 16-bit resolution, and provide 65,536 possible values. Contemporary professional digital audio recording devices operate at 20- and 24-bit resolution, and provide 1,048,576 and 16,777,216 values, respectively.

## Nyquist frequency

A concept crucial to all sampling systems is the *Nyquist theorem*. Harry Nyquist was an engineer at Bell Labs. Back in the 1920s, he proved that you need a sampling rate at least twice as high as the frequency of any signal to reproduce that signal accurately. The frequency and signal can be spatial, temporal, or audible.

The classic example of the Nyquist theorem is train wheels. I'm sure you've seen old movies where wheels on a train look like they're going backwards. As the train speeds up, the wheels go faster and faster, but after a certain point, they appear to be going backwards. If you step through footage of a moving train frame by frame, you'll discover the critical frequency at which the wheel appears to start going backwards is at 12Hz (12 full rotations)—half the speed of the film's 24fps.

The following images illustrate what's happening. Assume a 24fps film recording of that train wheel. I've put a big dot on the wheel so you can more easily see where in the rotation it is. In the first figure, the wheel is rotating at 6Hz. At 24fps, this means we have four frames of each rotation. Let's increase the wheel speed to 11Hz. We have a little more than two frames per rotation, but motion is still quite clear. Let's go up to 13Hz. With less than two frames between rotations,

**3.2** Two files scaled down 400%. The one on the right used filtering to weed out frequencies that would violate Nyquist, the one on the left didn't.

**3.3** The infamous wagon wheel example of the Nyquist limit. With a recording at 24Hz, wheels spinning at under 12Hz have a clear motion of direction. But above 12Hz, motion can't be accurately determined.

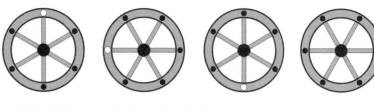

6Hz (90° per rotation): motion is clear

11Hz (165° per rotation): can still see direction of motion
per rotation): motion forward again, but very slowly

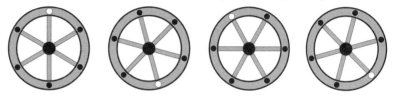

13Hz (195° per rotation): motion appears backwards

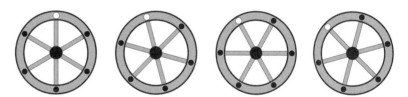

25 Hz (375°
(more than one full rotation between frames)

the wheel actually looks like it's going backwards! Once you get past 24Hz, the wheels look like they're going forward again, but very slowly.

Alas, there is no way to correct for a sampling rate that's too low to accommodate the Nyquist frequency after the fact. If you must have train wheels that appear to be going forward, you need slow wheels or a fast camera!

Another way to think about this is with a sine wave. If you sample a sine wave, as long as the source frequency is half or less of the sampling frequency, you can easily draw a new sine wave between the samples. However, if the sample rate is less than twice the frequency of the sine wave, what looks correct turns out to be completely erroneous.

**3.4** We're starting with two images, one within the Nyquist limit, and the other without. Each is scaled down, then back up. The one within the limit is jumbled, but still indicates something of the pattern of the original. The one where the source was beyond the limit shows no indication of the original frequency.

You'll encounter the same issue when working with images—you can't show spatial information that changes faster than half the special resolution. So, if you have a 640-pixel–wide screen, you can have at most 320 variations in the image without getting errors. Figure 3.4 (page 24) shows what happens to increasing numbers of vertical lines.

This problem comes up in scaling operations in which you decrease the resolution of the image, because reducing the resolution means frequencies that worked in the source won't work in the output. Good scaling algorithms automatically filter out frequencies that are too high for the destination output. Algorithms that don't filter out frequencies that are too high can cause serious quality problems. Such quality problems can sometimes be seen on Web sites where thumbnail-sized pictures seem unnaturally sharp and blocky.

# Aliasing

With audio, problems occur when frequencies that fall above the Nyquist frequency get picked up during the sampling process. When this occurs, digital recording combines the signal with the sampling rate. This new creates non-harmonic frequencies, that is, frequencies that are not whole number multiples of the frequencies in the main portion of the signal. These new frequencies—*aliases* of the errant high frequency plus the sampling rate—are called *aliasing noise*—very high pitched squeaks and chirps that are quite unpleasant to hear.

Antialiasing filters are used to reduce these noises, but the perfect antialiasing filter has yet to be created. An alternate way to eliminate audio aliasing is to use professional gear that oversamples. Oversampling is a technique that uses extremely high-frequency sampling rates that can be down-converted to the desired sampling rate while gentle digital filtering techniques are applied to reduce high-frequency distortion and aliasing.

For people coming to compression from the video world, one way to ensure that your audio won't suffer from aliasing is to avoid using your NLE's built-in audio inputs and outputs and instead use a professional external digital-to-analog converter that oversamples.

In general, you shouldn't have to worry about Nyquist frequencies (good tools handle the filtering behind the scenes), but understanding these concepts can help explain the limitations of lower resolutions and sample rates.

## Quantization

We've learned that sampling converts analog sounds and images into discrete digital values. The process of assigning discrete numeric values to theoretically infinite possible values is called *quantization*.

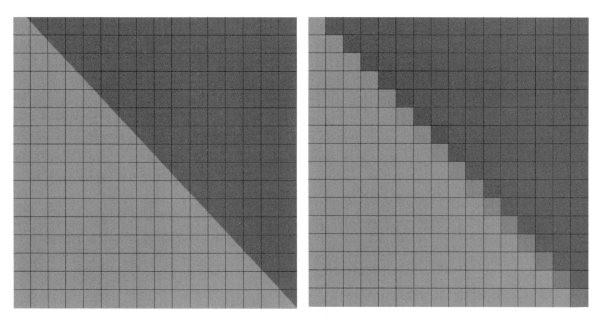

a) The source, with sampling grid overlaid.

b) After sampling. Note the purple squares, although there was no purple in source.

**3.5** When sampling, colors can exist in the output that weren't in the input, due to averaging.

 ***See also***

The color version of Figure 3.5 is in the color section (page V).

Continuing our graph paper metaphor, consider a single square (representing a single sample). If it's a solid color before sampling, it's easy to represent that solid color as a single number. But what if that square isn't a single solid color? What if it's partly red and partly blue? Because a sample is a single point in space, by definition, it can have only one color. The general solution is to average all the colors in the square and use that average as the sample's color. In this case, the sample will be purple, even if no purple appeared in the original image. Turn to page V in the color section, then prop the book up on your desk and look at the image from across the room. Looks perfect with a little distance, right?

The amount of quantization detail is determined by bit depth. Most video codecs use 8 bits per channel. Just as with 8 bit audio systems, 8 bit video systems provide a total of 256 levels of brightness from black to white, expressed in values between 0 and 255. Note, however, that Y'CbCr color space video systems (more on what that means in a moment) typically only use the range from 16 to 235 for luminance, while RGB color space computer systems operate across the full 0 to 255 range.

In the previous chapter we discussed the enormous range of brightness the human eye is capable of perceiving—the ratio between the dimmest and brightest light we can see is about one trillion to one. Fortunately, we don't need to code for that entire range. To begin with, the computer monitors and televisions compressed video is displayed on aren't capable of producing brightness levels high enough to damage vision or deliver accurate detail for dim images. Plus, when light is bright enough for the cones in our eyes to produce color vision, we won't see any detail in areas dark enough to require rods for viewing. In a dark room, such as a movie theater, we can get up to about a 100:1 ratio in absolute light intensity between the brightest light and the darkest light that is distinguishable from black. In a more typical environment containing more ambient light, this ratio will be less. We're able to discriminate intensity differences down to about 1 percent for most content (with a few exceptions). This means at least 100 discrete steps are required between black and white.

Note when I say we can distinguish differences of about 1 percent, I'm referring to differences between two light intensity levels. So, if we say there are 1,000 units of light, we can tell the difference between a value of 1,000 and 990 (1 percent less) and 1,010 (1 percent more). But if there are 100 units of light we can distinguish between values of 100 and 99 and 101. It's the ratio between the light levels that's important, not the actual amount of light itself. So, the brighter the light, the bigger a jump in absolute light levels we need to perceive a difference.

We take advantage of this by making the scale we use to encode brightness perceptually uniform. With a perceptually uniform scale, the percentage (and thus the perceived) jump in luminance is the same from 20 to 21 to 1 as from 200 to 201 (and not 210). This means that there is an exponential increase in the actual light emitted by each jump. But it makes the math a lot simpler, and enables us to get good results in our 256 values between 0 and 255. 256 is a magic number for computers, since one byte equals 8 bits, and 8 bits can describe 256 discrete levels. Most CPUs can process 256 numbers twice as fast as they can process even 257 numbers.

The exponential increase used to make the scale perceptually uniform is called *gamma*, after the Greek letter (γ). Things get complicated, because different systems use different gamma values. More (much more) on this later.

With infinite resolution, a gradient that starts as total black and moves to total white should appear completely smooth. However, as we've learned, digital systems do not enjoy infinite resolution. It's possible to perceive banding in 8-bit per channel systems, especially when looking at very gradual gradients. For example, going from Y=20 to Y=25 across an entire screen, you can typically see the seams where the values jump from one value to another. Note you aren't actually seeing the difference between the two colors; cover the seam with your hand, and the two colors look the same. What you're seeing is the seam—revealed by our visual system's excellent ability to detect edges.

This is why high-end professional video systems use 10 bits per channel. While this resolution only increases the data rate by 25 percent, it quadruples the number of gradations from 256 to 1,024 (and makes processing a lot more computationally intensive). There are no delivery codecs that use 10-bit resolution.

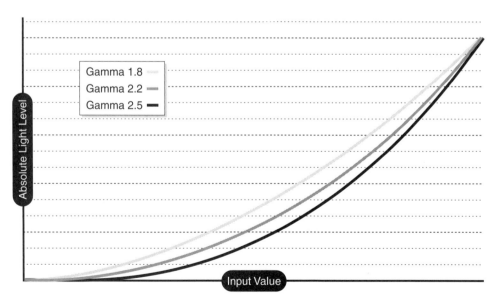

**3.6** In video gamma, the perceptually uniform spacing between 0–255 values means exponential increases in actual light output from the monitor. The amount varies depending on the gamma, with 2.2 for video, 1.8 for Mac.

**3.7** Our edge detection ability can see the boundaries between the regions of this gradient, but holding our hand over the seam, each side looks identical because the difference is below our threshold of perception for luma differences.

## Color Spaces

So far, we've been describing luminance, but haven't dealt with color. Converting real-world light to digital form requires good mathematical ways to describe the colors. There are a number of different ways to describe colors mathematically, called *color spaces*.

As we discussed in the last chapter, all visible colors are made up of varying amounts of red, green, and blue. Thus, red, green, and blue can be sampled separately and combined to produce any visible color. Three different values are required for each sample to store color mathematically.

These values fall into *channels*, which can be regarded as independent layers, or planes, of the image. Although the CIE (Commission Internationale de l'Eclairage) determined that these planes could be red, green, and blue, other combinations of three colors can achieve similar results. And importantly, combinations of two color channels plus varying degrees of luminance work as well. Because our visual system is so much more sensitive to brightness than color, by having those elements of the video separate, we're able to use less bandwidth for color and use more bandwidth for luminance, making more efficient use of available bits.

Let's examine all the major color spaces one by one and see how they apply to real-world digital image processing.

### RGB color space

RGB (Red, Green, Blue) is the native color space of the human eye. Red, green, and blue are the three basic colors the eye perceives. RGB is an additive color space, meaning that white is obtained by mixing the maximum value of each color together, and gray is obtained by having equal values of each color.

All video display devices are fundamentally RGB devices. The CCDs in cameras are RGB, and displays use red, green, and blue phosphors in CRTs, planes in LCDs, or light beams in projection displays. Computer graphics are almost always generated in RGB. All digital video starts and ends as RGB, even if it isn't ever stored as that.

This is because RGB is not ideal for image processing. Luminance is a critical aspect of video, and RGB doesn't encode luminance directly. Rather, luminance is an emergent property of RGB values, meaning that luminance arises from the interplay of the three channels. This complicates calculations enormously because color and brightness can't be adjusted independently. Increasing contrast requires changing the value of each channel. A simple hue shift requires decoding and re-encoding all three channels via substantial mathematics. Nor is RGB conducive to compression because luma and chroma are mixed in together. Theoretically, you might allocate fewer samples to blue or maybe red, but in practice this doesn't produce enough efficiency to be worth the quality hit.

## Y'CbCr color space

While digital video is displayed in RGB, it's nearly always stored in Y'CbCr or a variant thereof. Y stands for luminance, Cb for blue minus luminance, and Cr for red minus luminance.

Confusing? Yes, it is. Where did green go? Green is implicit in the luminance channel!

Let's expand the luminance equation from the previous chapter into the full formula for Y'CbCr.

```
Y  =  0.299R + 0.587G + 0.114B
Cb = -0.147R - 0.289G + 0.436B
Cr =  0.615R - 0.515G - 0.100B
```

Notice that Y mainly consists of G, some R, and a little B. This matches the sensitivities of the human eye. Another helpful aspect of Y'CbCr is that, because it separates luminance from chrominance, different compression techniques can be applied to each. We know that perception of chroma is much more limited than perception of luma, which suggests that two of the three channels can be compressed drastically without degrading image quality too much. This wouldn't work in RGB.

The simplest way to reduce the data rate of a Y'CbCr signal is to reduce the resolution of the Cb and Cr channels. Going back to the graph paper analogy, you might sample, say, Cb and Cr once every two squares, while sampling Y for every square. Other approaches are often used—one Cb and Cr sample for every four Ys, one Cb sample for every two Crs and Ys, and so on.

Formulas for Y'CbCr sampling are notated in the format x:y:z. The first number establishes the relative number of luma samples. The second number indicates how many chroma samples there are relative to the number of luma samples on every other line starting from the top, and the third represents the number of chroma samples relative to luma samples on the remaining lines. So, 4:2:0 means that for every four luma samples, there are two chroma samples on every other line from the top, and none on the remaining lines. This means that chroma is stored in 2×2 blocks, so for a 640×480 image, there will be a 640×480 channel for Y, and 320×240 channels for each of Cb and Cr.

## Chapter 3: Fundamentals of Compression

The quantization for Y runs as normal—0 to 255, perceptually uniform. Color is different. It's still 8-bit, but uses the range of –127 to 128. If both Cb and Cr are 0, the image is monochrome. The higher or lower the values are from 0, the more saturated the color. The actual hue of the color itself is determined by the relative amounts of Cb versus Cr. In the same way that the range of Y in video is 16 to 235, the range of Cb and Cr in video are limited to the range of –112 to 112.

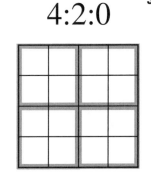

**3.8** The structure of the 4:2:0 colorspace. The black lines represent the pixels, and hence luma resolution, and the gray lines the chroma resolution. In this case, there is one color sample per 2×2 block of pixels.

### See also
See Figure I in the color section (page VI) showing the range of colors available in Cb and Cr when Y is at the middle of the range (127 in 8-bit).

### Note
***Terminology.*** In deference to Charles M. Poynton, who has written much more and much better about color than I, this book uses the term Y'CbCr for what other sources call YUV. YUV really only refers to the native mode of NTSC television, with its specific gamma and other attributes. Y'U'V' is also used in some cases. When used properly, U and V refer to the subcarrier modulation axes in NTSC color coding, but YUV has become shorthand for any luma-plus-subsampled chroma color spaces. Most of the documentation you'll encounter uses YUV in this way. It's a lot easier to type, so I was tempted to use it in this book, but thought better of it in the end.

*4:4:4 sampling*   This is the most basic Y'CbCr, and probably the least seen. It has one Cb and one Cr for each Y value. While not generally used as a storage format, it's used for transmission and internal processing. In comparison RGB is always 4:4:4.

Photoshop's delightful Lab color mode is a 4:4:4 Y'CbCr variant.

*4:2:2 sampling*   This is the subsampling scheme most commonly used in professional video authoring: one Cb and one Cr for every two Ys horizontally. 4:2:2 is used in tape formats such as D1, Digital Betacam, Digital-S, and D9. It's also the native format of Motion-JPEG editing systems. It looks good, but it's only a third smaller than 4:4:4.

So, if D1 is 4:2:2 and 4:2:2 results in a compression ratio of 1.5:1, why is D1 called "uncompressed"? Using this word to describe D1 doesn't take into account data reduction inherent in the color space—just as audio sampled at 44.1kHz isn't said to be "compressed" compared to audio sampled at 48kHz. In contrast, M-JPEG takes the subsampled 4:2:2 and then applies additional compression.

*4:2:0 sampling*   Here, we have one pixel per Y sample and four pixels (a 2×2 block) per Cb and Cr sample. This yields a 2:1 reduction in data before further compression. 4:2:0 is the ideal color space for compressing progressive scan video, and is used in all modern Web codecs, as well as the MPEG standards (except for the 4:2:2 HP@ML variant used in high-end editing systems). It is also used in PAL DV, except for PAL DVCPRO, which uses 4:1:1.

*4:1:1 sampling*   This is the color space of NTSC DV25 formats, including DV, DVC, DVCPRO, and DVCAM, as well as DVCPRO 4:1:1 in PAL. 4:1:1 uses Cb and Cr samples four pixels wide and one pixel tall. So, it has the same number of chroma samples as 4:2:0, but in a different (non-square) shape.

So why use 4:1:1 instead of 4:2:0? Because of the way standard analog video is put together. Each video frame is organized into two fields which can contain different temporal information—more about this later. The two fields are generally compressed separately, so a 720×480 frame is stored as two 720×240 fields. If DV were to use 4:2:0, only one color sample would encode every two lines in a field, so one sample would be seen spread over three lines in the final display (with a line from the other field in between). This will cause some quality problems when the video is later edited, especially if the video went through multiple generations.

4:1:1 turns out to be much better for interlaced content. It, too, yields only one color sample for every four pixels, but each line in the display has its own unique chroma samples.

*YUV-9 sampling*   A break from the 4:x:x nomenclature, YUV-9 denotes an average of nine bits per pixel. (By the same token, 4:2:0 has been called YUV-12). YUV-9 is only found in multimedia codecs and describes color in 4×4 pixel blocks. This isn't enough color information for professional content development or delivery, since sharp edges in highly saturated imagery in YUV-9 can look extremely blocky.

I spent a surprising amount of my professional life trying to deal with YUV-9 and yelling at codec vendors for using it. It was the Achilles heel of too many otherwise great products, including Indeo and Sorenson Video 2. Fortunately, Sorenson 3 and all the other modern codecs have adopted 4:2:0.

## CMYK color space

CMYK isn't used in digital video, but it's the other color space you're likely to have heard of. CMYK stands for cyan, magenta, yellow, and black (yup, that final letter is for blacK).

## What Colors are Cb and Cr?

So, what do Cb and Cr actually look like? On one level it's difficult to define. So, let's figure out how to describe colors in terms of an RGB computer display, which uses an 8-bit number between 0 and 255 to represent each color. To find Red, Green, and Blue, you take all of one color and none of the other two. Not so easy with Y, Cb, and Cr, as anything with zero Y is black, no matter what Cb and Cr are. So, let's assign Y=127, dead center in the luma range. Also, Cb and Cr are measured in −128 to 127, so we need to set our target color to that and the other color to −128, not just zero. Now, let's take a look at Cb=−126, Cr=+127, and Cb=+127, Cr=−126, with Y=127 in both cases.

From our previous equation, we can derive this:

```
R = Y + 1.4020Cr
G = Y - 0.3441Cb - 0.7139Cr
B = Y + 1.7718Cb
```

And with a little fiddling, we discover

```
All Cb is R=0,   G=210, B=352
All Cr is R=305, G=178, B=-96
```

Yipes! Some of those numbers are outside of the 0 to 255 range of the RGB channels!

That's actually to be expected. Y'CbCr -> RGB conversion often requires *clamping*, where values below 0 or above 225 are set to those limits (this is one of the reasons why converting from Y'CbCr to RGB isn't ideal). Clamping results in:

```
Cb is R=0,   G=210, B=255
Cr is R=255, G=178, B=0
```

(See Figure J on color page VI.)

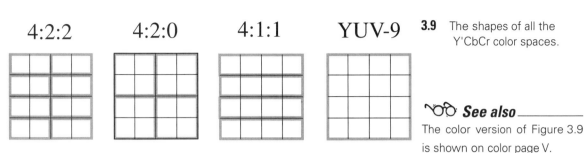

The four Y'CbCr color spaces we see in the wild.

**3.9** The shapes of all the Y'CbCr color spaces.

**See also**
The color version of Figure 3.9 is shown on color page V.

CMYK is used mostly in printing, and controls the placement and layering of inks on a page, not the firing of phosphors. Consequently, CMYK has some unique characteristics. For one, C+M+Y comes out a muddy brown, so a separate black channel is necessary.

In contrast to additive color spaces like RGB, white is obtained by removing all colors (which leaves white paper).

## Quantization Levels

Now that we've got a good handle on the many varieties of sampling, let's get back to quantization. There are a number of sampling options out in the world, although only 24-bit applies to the codecs we use most of the time.

*1-bit (Black and white)* In a black and white file, every pixel is either black or white. Any shading must be done through patterns of pixels. One-bit quantization isn't used in video at all. For the right kind of content such as line art 1-bit can compress quite small, of course. I mainly see 1-bit quantization in GIF files on the PBSkids.org Web site when I'm printing out "Bob the Builder" coloring paper for my two-year-old son.

*8-bit color (indexed color)* 8-bit is unique among color depths in that it doesn't use its eight bits to define a value for each of a pixel's three channels. Rather, the bits are used to define 256 discrete 24-bit RGB colors. The selection of discrete colors is called a palette, CLUT (Color Look Up Table), or index; thus 8-bit color is often called *indexed color*.

8-bit is never used in video production, and it is becoming increasingly rare for it to be used on computers either. Back in the Heroic Age of Multimedia (the mid-90s), much of your media mojo came from the ability to do clever things with palettes. At industry parties, swaggering pixel monkeys would make extravagant claims for their dithering talents. Now we live in less exciting times, with less need for heroes, and can do most authoring in 24-bit color depth. But 8-bit color depth still comes up from time to time. Eight-bit color's most common use remains in animated and still GIF files. Eight-bit color is also still used for compression captured screen animations for software tutorials.

Depending on the content, 8-bit can look better or worse than 16-bit. With 8-bit, you can be very precise in your color selection assuming there isn't a lot of variation in the source. Sixteen-bit can give you a lot more banding (only 32 levels between black and white on Mac), but can use all the colors in different combinations.

Making a good 8-bit image is tricky. First, the ideal 8-bit palette must be generated (typically with a tool such as Equilibrium's DeBabelizer). Then the colors in the image must be reduced to only those appearing in that 8-bit palette. This is traditionally done with dithering, where when the precise color isn't available, two different colors are randomly distributed to make up the missing color. This can yield surprisingly good results, but makes files that are very hard to compress. Alternatively, the color can be flattened—each color is just set to the nearest color in the 8-bit palette without any dithering. This looks worse for photographs, but better for synthetic images such as screen shots. Not dithering also improves video quality.

Many formats can support indexed colors at lower bit depths, like 4-color (2-bit) or 16-color (4-bit). These operate in the same way except with fewer colors. (Refer also to Figure 3.10.)

*8-bit grayscale*   A cool variant of 8-bit is 8-bit grayscale. This doesn't use an index; it's just an 8-bit Y channel without chroma channels. I used to use 8-bit grayscale compression for black and white video all the time—Cinepak has a great-looking 8-bit grayscale mode that can run 640×480 on an Intel 486 processor. You can also force an 8-bit index codec to do 8-bit grayscale by giving it an index that's just 0 to 255 luma with no chroma.

Alas, with the exception of JPEG, modern codecs don't really support 8-bit grayscale.

*16-bit color (High Color/Thousands of Colors)*   There are actually two different flavors of 16-bit color. Windows machines have six bits in Green and five bits each in Red and Blue. The Mac's Thousands mode 16-bit color is really 15-bit color consisting of three 5-bit channels, with 1-bit reserved for a rarely used alpha channel. 16-bit color is an improvement, in most cases, over 8-bit. Until a few years ago, most computers shipped in 16-bit mode by default, but 24-bit or 32-bit color is now standard.

16-bit can have trouble reproducing subtle color gradients. Given a 6-bit green (which is most of luminance) channel, only 64 gradations exist between black and white, which can cause visible banding in low-contrast images. Thus, sometimes even a well-tuned 8-bit palette can look better than 16-bit color depth. This becomes even more of a problem on a Mac OS system with only 32 levels between black and white.

16-bit should only ever be used as a delivery format. Multiple passes of image processing in 16-bit color generally result in unacceptable artifacts. No modern video codecs are natively 16-bit,

## Quantization Levels 35

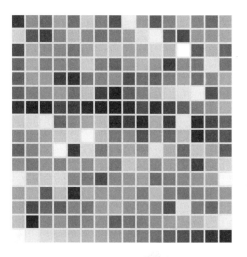

**3.10** Same image, converted to the shown 8-bit palette. The top image shows typical dithering, and the bottom shows flattening. The flat image is about 15% smaller when losslessly compressed.

**See also**
The color version of Figure 3.10 is on page VIII of the color section.

although they can display to the screen in 16-bit. The ancient video codecs of Apple Video (aka Road Pizza) and Microsoft Video 1 were 16-bit. But hey, what do you expect? They didn't have data rate control either.

*24-bit color (Millions of Colors)* This is the default depth for use in compression, for both RGB and YUV color space. 24-bit color allocates eight bits per channel. 24-bit color is adequate for most uses, although high-end prepress people working in CMYK color sometimes find it insufficient. 24-bit is used in most digital video hardware.

## The Delights of Y'CbCr Processing

Even though video formats are natively Y'CbCr, most video processing tools and their filters run in RGB. There isn't any fundamental reason for this beyond industry inertia. Otherwise excellent tools like After Effects are RGB-only. Apple's Final Cut Pro has the best implementation of a complete Y'CbCr processing mode I've seen to date, and it nicely proves that Y'CbCr is the no-compromises right way to go for video applications. There are three reasons why Y'CbCr processing rules over RGB.

- Speed

Compared to 4:2:0 or 4:1:1, there are twice as many pixels to process in RGB compared to Y'CbCr. For example, a 640x480 RGB needs three 640x480 8-bit channels per frame for a total of 7,372,800 bits per frame. 640x480 4:2:0 needs one 640x480 and two 320x240 for a total of 3,686,400 bits per frame, exactly half as many. That means only half as many pixels to move around in memory, run through filters, and so on.

With common filters, Y'CbCr's advantages get even stronger. Filters that directly affect luminance, like contrast, gamma, and brightness, only need to touch a single 640x480 channel in Y'CbCr, compared to all three in RGB, making Y'CbCr three times faster than RGB. Chroma-only filters like saturation only need to do two 320x240s instead of three 640x480s, making Y'CbCr *six* times faster.

And for operations like hue, RGB must be converted into a format like Y'CbCr, then processed, and converted back to RGB, which is slower yet. Many tools, even when converting from a Y'CbCr source file to a Y'CbCr output, will convert to RGB and back again, even when not doing any filtering!

- Higher quality

The gamut of RGB is smaller than Y'CbCr, which means that there are colors in Y'CbCr that you don't get back when converting into RGB and back to Y'CbCr. Also, every time this is done, some rounding errors are added, which increases noise by a bit (pun intended).

- Channel specific filters

One of my favorite ways to clean up bad video only works in Y'CbCr mode. Most ugly video, especially coming from analog sources like VHS and Umatic, are much uglier in the chroma channels than in luma. You can often get great results by doing an aggressive noise reduction on just the chroma channels. Lots of noise reduction adds a lot of blur to the image, which looks terrible in RGB mode, because it's equally applied to all colors. But when the chroma channels get blurry, the image can remain looking nice and sharp, while composite artifacts seem to just vanish.

(Refer to Figure K on page VII of the color section.)

You'll often hear people refer to 32-bit video. This is 24-bit color plus an 8-bit alpha channel for compositing, called *Millions+* on Mac OS. The images themselves are still stored as 24-bit color.

*30-bit color (10-bit per channel)*   30-bit allocates 10 bits to each channel per pixel, yielding 1,024 levels of gradation. While no delivery codecs use 30-bit color, it's the standard for high-end video authoring. 30-bit is commonly used in high-end studios for uncompressed digital video transmission via the Serial Digital Interface (SDI) standard between Digital Betacam and other high-end decks. Some SDI devices are still only 8 bits per channel, so don't assume using SDI automatically means you have 10 bits per channel.

I'd love to see 10 bits per channel become a minimum for processing within video software. It breaks my heart to know those bits are in the file, but are immediately discarded. Even if you're delivering 24-bit color, the extra two bits per channel can make a real difference in reducing banding after multiple passes of effects processing.

After Effects 5 Production Bundle can now process video in 16 bits per channel RGB, and hence can take advantage of the extra bits from 30-bit captures.

Note 30-bit color in video is generally referred to as 10-bit (10 bits × three channels = 30 bits) color.

*48-bit color depth*   My dream bit depth for processing, 48-bit, gives you 16 bits per channel. This is more than necessary in nearly all cases—which is why I like it. The main place we see 48-bit color now is in Photoshop's (RGB or Lab) and After Effects' (RGB only) 16-bit per channel modes. Quantization errors are effectively nonexistent. (12 bits per channel would be just as good in most cases, but 16 bits per channel is the next logical step up from eight for computers.)

The new Microcosm codec from Theory LLC is a losslessly compressed RGB codec that can do both 32-bit and 64-bit color (48-bit with a 16-bit alpha).

# Quantization Errors

Quantization errors arise from having to choose the nearest equivalent to a given number from among a limited set of choices. Digitally encoding colors inevitably introduces this kind of error, since there's never an exact match between an analog color and its nearest digital equivalent. The more gradations available in each channel—that is, the higher the color depth—the closer the best choice is likely to be to the original color.

For 8-bit channels that don't need much processing, quantization error doesn't create problems. But quant errors add up during image processing. Every time you change individual channel values, they get rounded to the nearest value, adding a little bit of error. With each layer of processing, the image tends to drift farther away from the best digital equivalent of the original color.

Imagine a Y value of 101. If you divide brightness in half, you wind up with 51 (50.5 rounded up). Later, if you double brightness, the new value is 102. Look at the histogram below before and after doing some typical gamma and contrast filtering. Note how some colors simply don't exist in the output, due to rounding errors, doubling the amount of a color in an adjacent value.

 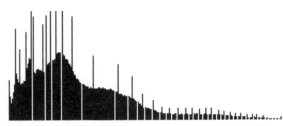

**3.11** A histogram before and after a contrast increase and contrast decrease. Note that even though the filters reversed each other, colors are missing in the output.

Quant errors also arise from color space conversion. For example, most capture formats represent color as Y'CbCr 4:1:1 or 4:2:2, but After Effects filters operate in RGB. Transferring from tape to tape via AE generally involves converting from Y'CbCr to RGB and back to Y'CbCr—at the very least. What's more, filters and effects often convert color spaces internally. For instance, using Hue, Saturation, and Lightness (HSL) filters involves a trip through the HSL color space.

The trouble with color space conversions is that similar values don't always exist in all color spaces. For example, imagine a sepia-tone movie in Y'CbCr. The tint is consistent, so the Cb and Cr channels have fixed values; only the Y signal varies. The Y value might range from full black to full white, where the chroma channels have no effect regardless of the information they hold. Obviously, the chroma channels in a sepia-tone movie should have constant values across the frame, yielding excellent compression. This is because the Cb and Cr channels can hold color information even when the luma channel makes the image all black or all white. However, upon conversion to RGB, the information in the Cb and Cr channels disappears—black in RGB means no R, G, or B; no color information at all. If the RGB version is converted back to Y'CbCr, the values in the Cb and Cr channels are no longer fixed. As Y gets near black or white, the Cb and Cr values approach zero.

In this case, Y'CbCr-to-RGB conversion definitely isn't reversible. This state of affairs isn't a huge problem for still images, but it can hamper interframe video codecs. Since they generally work by encoding each plane independently, these unnecessary and unseen perturbations in the color channels make compression less efficient.

## Gamma

Gamma controls how the intensity of the displayed image increases with higher values. A gamma curve describes how graphics hardware interprets luminance values to display brightness. Gamma issues occur when you're trading image files among platforms with different gammas.

Looking at the same image on a Mac and PC side by side, RGB=0 and Y=0 look equally black, RGB=255 and Y=255 look equally white, but at Y=127 the PC looks darker. The difference is the gamma. On most PCs, gamma is that of the CRT, and can be between 2.0 and 2.5 on different machines (with 2.2 or 2.5 considered the default in different circumstances). On Macs, the default is 1.8, although the Mac OS ColorSync control lets you change this. Mac gamma gives you a lot more detail in the whites at the expense of some detail in blacks. This is better for print projects, but less ideal for video.

Converting from one gamma to another is mathematically trivial as long as you do it before you apply compression. PC video gamma-corrected for Mac looks about as good as if it had been created on a Mac in the first place, although it doesn't look as good as the original, and vice-versa.

# Compression Basics—an Introduction to Information Theory

While most science is based on the patient accumulation of details collected by many people over many years, the basics of information theory were thought up almost entirely by one guy—Bell Labs Researcher Claude E. Shannon, whose "A Mathematical Theory of Communication" remains the cornerstone of the entire field. Shannon's paper was published in the *Bell System Journal* in 1948. The paper was meant to answer the then tricky question, "How much information can we send over all these telephone and telegraph lines anyway?" and went on to define information mathematically, including its limits. Shannon might not be a household name like Einstein or Edison, but the field he gave birth to is probably just as important to our modern world as the light bulb and relativity.

Tempted as I am to go off on a book-length discourse about the wonders of information theory, I'm just going to summarize a few critical insights that have deep relevance for video and audio compression.

## Any number can be turned into bits

This is an obvious point, but bears repeating. The bit, a single 0 or 1, is the basis of all math on computers, and of information theory itself. Any number can be turned into a series of bits. This means we can use bits as the basis for measuring everything that can be sampled and quantized.

## The more redundancy in the content, the more it can be compressed

English in plain text is rather predictable. For example, I'm much more likely to use the word "codec" at the end of this sentence than XYZZY. The distribution of letters in English isn't even. For example, the letter "e" appears a lot more often than "Q," so it is more efficient to design a system where "e" uses only one bit instead of eight, but Q takes more bits. The table that translates from the original characters to shorter symbols is called a *codebook*. Using a process called

*Huffman encoding*, an optimal codebook can be generated, using shorter codes for more common symbols and longer codes for less common ones. Different codebooks work better or worse for different kinds of content. A codebook for English won't work well for Polish, which has a very different distribution of letters. For an English codebook, the Polish distribution of letters would be very unexpected, meaning it would require more bits to encode, possibly even more than the original source would have taken up. The more you know about what you're trying to compress, the better a codebook can be generated.

Taken to the next level, the order of letters is also not random. For example, "ee" isn't as common a sequence in English as "th," even though "t" and "h" individually appear less often than "e." So a coding scheme can be expanded to take account of the probability of a given letter based on previous letters, and to assign symbols to a series of letters instead of a single one.

Codebooks can be agreed on in advance, so the decoder knows how to best decode the content in question. Multiple codebooks can be used, with a code at the start of the file or even changing throughout the file to indicate which codebook is to be used. Or, if there isn't an existing codebook that's adequate, a codebook tuned to that file can be dynamically created while the file is being compressed. This is the technique used by modern lossless compression applications like PKZip, Stuffit, and gzip.

The converse of "more redundant content compresses better" is that less redundant content doesn't compress as well. Truly random data won't compress at all, and sometimes can be larger after compression than before. For this reason, modern compressors have a flag that says "this file isn't compressed," so the output file won't be meaningfully larger.

### The more efficient the communication, the more random it looks

Using a codebook makes the file smaller by reducing redundancy. Because there is less redundancy, there is by definition less of a pattern to the data itself, and hence the data itself looks random. You can look at the first few dozen characters of a text file, and immediately see what language it's in. Look at the first few dozen characters of a compressed file, and you'll have no idea what it is. This means you can't usefully compress already compressed content because it is already essentially random.

Another surprising result is that you can feed random data into a decompressor, and often get something out on the other end that actually seems reasonable. This is why error correction is so critically important with all compressed media; getting one bit wrong might produce output that seems normal if you don't know what it was supposed to be like.

## Data Compression

Data compression is compression that works on arbitrary content, like computer files, without having to know any details in advance about their contents.

There have been many different compression algorithms used over the past few decades. Ones that are currently available use different techniques, but they share similar properties.

The best known formats for data compression are .zip from PKZip and many other tools that use the same format, .sit from Aladdin Systems' Stuffit products, and .gz from the open-source GZIP application as well as the many tools that use its libraries.

For example, the current draft of this chapter takes up 42,154 bytes as a text file. Compressed with GZIP at its highest compression mode, the chapter comes out as 16,136 bytes, or about a 61 percent reduction.

- **Already compressed data doesn't compress**

    As we said above, an efficiently encoded file looks pretty much random. This means there isn't any room for further compression, assuming the initial compression was done well. All the modern codecs already apply data compression as part of the overall compression process, so compressed formats like JPEG, QuickTime, RealVideo, and Windows Media typically won't compress much if at all. The one exception is with authoring formats, especially uncompressed formats. You might get up to a 2:1 compression when making a .ZIP or .SIT from an uncompressed video file, or the very simple run-length encoding of the Apple Animation codec.

    A time-honored Internet scam is a claim there is a compression technique that can compress random data and even its own output. While a wonderful fantasy, any such system is easily disproved with compression theory. Such systems are all bunk.

- **General purpose compression isn't ideal**

    Generating a codebook for each compressed file is time consuming and expands the size of the file and the time it takes to compress. Ideally, a compression technology will know more about the structure of the data, and will be able to do compression specific to that structure. This is why lossless still image compression will typically make the file somewhat smaller than doing data compression on the same uncompressed source file, and will be a lot faster doing it.

 **Note**

***Lossy and Lossless Compression.*** Lossless compression codecs preserve all of the information contained within the original file. Lossy compression codecs, on the other hand, discard some data contained in the original file during compression. Some codecs such as PNG and TIFF are always lossless. Others are always lossy. Others still may or may not be lossy depending on how you set their quality and data rate options. Lossless algorithms, by definition, might not be able to compress the file any smaller than it started. Lossy codecs generally let you specify a target data rate, and discard enough information to hit that data rate target.

- Small increases in compression require a lot of slowdown

    There is a fundamental limit to how small a given file can be compressed, called the *Shannon limit*. For random data, the limit is the same as the size of the source file. For highly redundant data, the limit can be tiny. A file that consists of the pattern "01010101" repeated infinitely should be able to be compressed down to only a few bytes. However, real-world applications don't get all the way to the Shannon limit, since it requires an enormous amount of computer horsepower, especially as the files get larger. Most compression applications have a value that controls some kind of trade-off between speed and compression efficiency. In essence, these controls expand the amount of the file that is being examined at any given moment. However, doubling compression time doesn't cut file size in half! Doubling compression time might only get you a few percentages closer to the Shannon limit for the file. Getting a file 10 percent smaller might take more than 10 times the processing time.

# Spatial Compression Basics

I'm going to spend a lot of time on spatial compression, because it forms the basis of all video compression.

Spatial compression starts with uncompressed source in the native color space, typically 4:2:0 at 8-bits per channel. Each channel is typically compressed by itself, so you can really think about it as three independent 8-bit bitmaps being compressed in parallel.

## Spatial compression methods

There are a ton of spatial compression algorithms and refinements to those algorithms. In the following, I briefly touch on the most common and best known. Most (but not all) modern codecs use Discrete Cosine Transformation (DCT) for images, so we'll focus on its description.

### Run-length encoding

The simplest way to compress data is with Run-Length Encoding (RLE). In essence, the data is stored as a series of pairs of numbers, the first giving the color of the next sequence and the second indicating how many pixels long that line should be. This works very well for content like animations of computer screens, since they have long lines of the same value. It can also work well for text, since a sharp edge isn't a problem, although a soft, antialiased edge can require several samples. However, RLE is terrible at handling natural images such as photographs. Just a slight amount of random noise can cause the image to be as large as an uncompressed file, because if each pixel is even slightly different from its neighbor, there aren't any horizontal lines of the same value. In the example I gave earlier of the "01010101" repeating pattern, RLE wouldn't be able to compress the file at all, since there is never more than one identical value in a row.

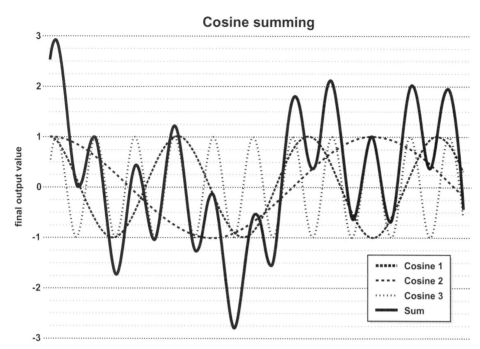

**3.12** Combining a few cosine waves can yield a very complex output.

### Discrete cosine transformation

Discrete Cosine Transformations, or DCTs, have been around for well over a decade, and we've often been told that they'd soon fade into irrelevancy. But there's a tremendous amount known about DCTs, so they continue to be enhanced. They work very well for temporal compression (more on that later). The description below closely follows how the older (pre-JPEG2000) JPEG works.

The basic idea behind DCT was developed by Joseph Fourier a couple of centuries ago. Fourier proved that any series of numbers can be produced by a sufficiently complex equation. In the modern era, the Fast Fourier Transform (FFT) provides a very easy way to take a series of numbers and turn them into the coefficients for a series of cosines that will produce the same numbers when added together. See Figure 3.12 for a very simple case for how things could work.

Mathematically, you get one coefficient on output for each sample pixel, so putting 64 pixels of source video into the DCT will output 64 coefficients. Because the coefficients are stored with more than eight bits each, just doing this would make a larger file, not a smaller one.

However, cosines are typically much simpler for expressing real-world information. A simple cosine would be *very* easy to encode, since a single coefficient would do the job. More complex images are more difficult to compress, because they require more coefficients. Rapid changes in values, as with a sharp edge, are quite a bit harder yet, with the most difficult thing to encode

## 44 Chapter 3: Fundamentals of Compression

**3.13** This grayscale 8×8 block is turned into a linear 1×64 sequence, using the standard zigzag pattern.

with DCT being an immediate change from the maximum or minimum to the other extreme. One of the limitations of DCT is that it doesn't handle text very well, because text if full of sharp edges. Our "01010101" example would be extremely hard to compress.

DCT becomes more efficient if a large number of samples can be processed at once. While the obvious solution would be to just encode a big chunk of a single line of video at a time, this proves to not be optimal, because a given line can show a lot of variation. In comparison, a square block of video tends to have a lot more in common. But the DCT itself only encodes a line—how do you encode a line of video into a block? See the following zigzag pattern. Note how much simpler the image looks when it is unwrapped than if a simple horizontal line is taken out of the video.

The search pattern is pretty simple for progressive scan images. However, the zigzag isn't optimal for interlaced content, where the video is really processed as two different 8×8 blocks in which every other line is interleaved to make one 8×16 block. Because it's most efficient to have a search pattern where left and right motion is spatially the same as up and down, the search pattern must move up and down by two pixels for every left and right motion (since each sample covers two horizontal lines). In a classic example of compressionist humor, this pattern is known as the "Yeltsin Walk" after the famously intoxicated Russian President. It is clearly more complex than the zigzag. It's one of the reasons interlaced video requires more bits than progressive to achieve the same quality. (See Figure 3.14.)

**3.14** The standard zigzag pattern used for progressive coding, and the "Yeltsin Walk" used for interlaced coding. It's clear why zigzag and hence progressive coding is more efficient.

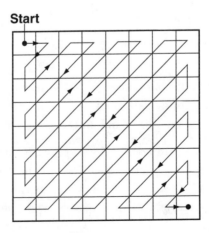
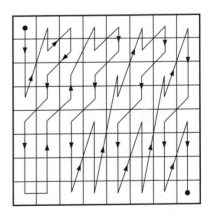

Early codecs experimented with other block sizes such as 4×4 and 16×16. However, 8×8 turned out to be better in almost all situations—although the forthcoming H.26L/MPEG-4 Part 10 codec will likely include support for 4×4 and other block sizes. An entire image can be broken down into these 8×8 blocks. However, because the color channels are subsampled, in 4:2:0 their 8×8 blocks actually cover a 16×16 area of the original image. This is called a *macroblock*, which is a combination of four 8×8 luma blocks and the corresponding 8×8 chroma blocks, one each for Cb and Cr.

This macroblock structure is why MPEG and other formats require resolution to be divisible by 16. Other formats that

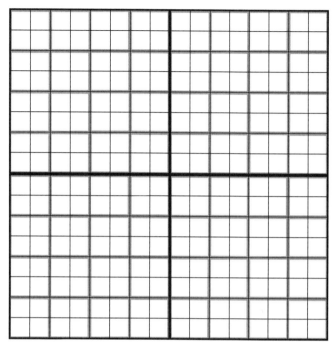

3.15   Macroblock structure. Black lines indicate edges of (luma) pixels, gray lines indicate edges of chroma samples, and thick black lines indicate edges of the 8×8 luma blocks. The whole 16×16 macroblock is a single chroma block.

allow arbitrary resolutions typically encode full macroblocks, and just pad the pixels at the edge with blank data. So, if you make a 240×180 file, you're actually making a 240×192 file where the bottom 12 pixels are filled with bogus data on compression, and then not drawn on playback. This is rarely a significant problem in real-world projects, but you do waste a little bandwidth and time. Ideally, keep resolution divisible by 16 to squeeze out the maximum bang for the bit.

We now take our source block of video, turn it into a linear sequence of 64 pixels, and then convert it to 64 coefficients via DCT. What does this give us? Something very cool, due to the properties of natural images transferred to DCTs. The structure of the coefficients is that big changes in the image become big coefficients early in the sequence, and that small changes become small coefficients. And, for reasonably normal images, we get a good deal of coefficents that equal zero at the end of the sequence. This is important because a long sequence of zeros can be easily compressed. Unlike RLE, where even a simple gradient won't compress at all because each pixel is different from its neighbor, DCT can produce great lossless compression of content consisting of slow, predictable changes. Text and other elements with sharp edges are a lot more difficult, and may require all 64 coefficients to losslessly encode the image.

Of course, in an image, each 8×8 block is going to share a lot of properties with its neighbors. So previously encoded blocks are used as a basis for new blocks, which means each additional block after the first is more easily compressed. The more blocks that contain similar content, the smaller the resulting file size.

But lossless compression isn't what's needed, since DCT in and of itself offers no control over the file size. You want to specify either a minimum quality level or file size, and get the best possible image within those constraints.

Almost all JPEG implementations feature a quality control, typically with a range of 0 to 100, with 100 equaling lossless and 0 as gawd-awful ugly, but tiny. So, how does this quality control work? In a surprisingly simple manner—it specifies a cutoff point below which coefficients are dropped. Since small coefficient values don't have much effect on the image, dropping them doesn't have much effect on the image. With a value of 100, no coefficients are dropped. With lower values, the cutoff becomes higher and higher, resulting in a smaller file (more zeroes) but lower quality (greater differences between the source and the output). Typically, anything above 60 won't look obviously compressed. Depending on how complex the image is, the file size can vary radically. A simple shot of clouds in the sky (very easy to compress) can come out at a fraction of the size of an image of something very difficult (say, text and foliage).

This quality value is also called a *quantization value*.

So what if you want to specify a file size? The problem is to figure out what quality value will produce a file of the desired size. At a fixed size, easy to compress content will wind up with a higher quality setting, and hence a higher quality, than a difficult to compress image.

Poor quality in DCT typically results in *ringing* and *blocking*. Ringing is errors around those sharp edges that are so difficult for DCT to encode. Blocking is where different macroblocks wind up with different errors, so the edge between them becomes visible. While JPEG has no feature for removing these artifacts, modern DCT video codecs like MPEG-4 support post-processing, which attempts to dynamically remove these artifacts on playback.

### See also

Two different files—as losslessly compressed source—encoded with fixed quality and encoded with fixed data rates are shown in Figure M in the color section (page IX).

## Wavelet compression

Wavelet compression represents a different model that's coming on strong for still image compression. Wavelets are the basis of JPEG2000 and MPEG-4 still image codecs. The fundamental idea behind wavelets is similar to DCT—turn a sequence of samples into an equation, but that's where the similarities stop.

# Spatial Compression Basics    47

**3.16** Screen shot from ICT Group's VCDemo. The top left shows the source, bottom left the quality of all the bands, and bottom right just the lowest band. Upper right shows the additional data added by each band, with the base band in the upper left.

Wavelet images are built out of a series of images, starting small and doubling in height and width until the final image is built at the file's resolution. These successively more detailed images are called *bands*. The space savings from this technique can be tremendous—each band is predicted from the previous band, so only those areas in which the band is different need to be encoded. Thus for gradually changing content, like clouds or ripples in water, the later bands need to add very little data. Conversely, more detailed images, such as faces in a crowd scene, might require a lot of detail in each new band.

One very intriguing aspect of these bands is you can display an image without the final bands. Wavelet files tend to present the base band first, so you can display a lower resolution version of the file when only a little bit of it has been transmitted. As more arrives, the resolution improves. The experience is similar to that of progressive JPEG or GIF.

Another wonderful thing about wavelets is that they don't have nearly the problems encoding with text as DCT codecs do, so at similar data rates, wavelets suffer less from ringing. They also block much less severely. This means that at lower data rates, wavelets tend to have much less objectionable artifacts, though the images still get quite soft. Even though there might not be any more actual information in the image, DCT artifacts look like false information (detail that wasn't part of the original) while wavelets are just missing information.

Wavelets are entering a second phase. Their initial use in video was with the old IMMIX Cube NLE systems, which used wavelets to great affect in providing decent quality video at much lower data rates than the Motion-JPEG cards they were competing against. However, the system had other issues, and the more open standard M-JPEG cards won out.

However, wavelets are proving to be the ideal format for today's still image codecs, and they're the basis of JPEG2000. And there is a corresponding Motion-JPEG2000, which hopes to bring wavelet compression back to video editing. Wavelets were used in video codecs, most notably in Indeo v4 and v5.

Alas, wavelets aren't nearly so promising for use in interframe codecs, because wavelets actually make temporal compression more difficult than DCT. More on this follows.

### Vector quantization

Vector Quantization, or VQ, was used in a number of older formats such as GIF, Cinepak, and Sorenson Video v1 and v2. Compared to other codecs, VQ requires a lot of CPU horsepower to encode, but its requirements for decoding are pretty reasonable.

Mathematically, VQ can be thought of as a superset of DCT. Given infinite computer power, VQ codecs should be able to beat DCT in compression efficiency. In practical terms, the processor power required means that VQ doesn't have any real advantages over modern DCT-based systems. Existing VQ codecs typically aren't capable of lossless or even visually lossless compression, although it would be possible with enough processing power.

Of course, Moore's Law is always giving us faster computers, so once we run out of ways to enhance DCT, VQ may yet have its day in the sun.

### Fractal compression

Fractals were the rotary engine of the 90s—the technology that promised to make everything wonderful, but never actually turned into mainstream products. The idea behind fractals is well-known to anyone who has read James Gleick's, *Chaos: Making a New Science* (Viking Penguin, 1987). In its classic definition a fractal image is able to reuse parts of itself, transformed in many ways, in other parts of the image. This concept is based on self-similar objects, so in fractal compression, things like ferns or trees can compress very well. Fractal compression isn't so useful with people—our fingers don't have smaller fingers hanging off their ends. Fractal compression can be thought of as extending motion search to rotation, scaling, and other kinds of transformations. The mathematical system was called *Iterated Function System*, which Iterated Systems, the company that tried to bring the technology to market, is named after.

The problem is that actually compressing a fractal image is hugely difficult. A fractal codec needs to see whether one part of an image matches another after being rotated, scaled, flipped, and otherwise looked at from every which way. All of which requires an insane amount of CPU power, and led to the Graduate Student Algorithm, where a grad student is locked in a room with a workstation for a few days to find the best compression for the image; the human visual system is a lot more capable of fast pattern recognition than software. Needless to say, the Grad Student Algorithm wasn't very practical in the real world.

In the end, the products that shipped as fractal compressors used what was called the *fractal transform*, which arguably wasn't really a fractal in the conventional sense, but a kind of VQ algorithm. The real advantage of these fractal compressors was that they offered resolution independence (the image could be decompressed at a higher resolution with fewer artifacts than other algorithms), and that a lower quality image could be generated from the beginning of the file, à la wavelets.

While Iterated Systems did ship a fractal video codec called ClearVideo, it was never particularly successful, and was eclipsed by Sorenson Video. Iterated's still image fractal technology, FIF, has gone through a couple of corporate owners, and is now mainly pitched as an algorithm that offers better scaling.

It is possible that Moore's Law will eventually give us, as it gave us the second coming of wavelets, enough CPU power to use some kind of fractal compression. For the time being, fractals aren't part of mainstream of video codec development.

# Temporal Compression

Okay, spatial compression is important. But spatial compression alone doesn't get us even close to common compression goals. Modern video codecs are able to produce 640×480, 30fps images at 600 Kilobits (Kbps) per second, or 75 KiloBytes per second. That would require encoding each frame at 2.5KiloBytes (KBps) per second in a still image codec, which no still technology is capable of doing with any quality.

All spatial compression works, fundamentally, by reducing redundancy within a frame. Temporal compression does the same between frames. Look at some video frame by frame. Note that, most of the time, most of the image looks like the frame before it. Redundancy! Information theory tells us we can remove that redundant information, and make the file smaller.

A very simple method of interframe encoding is to subtract the value of the previous frame. For video where the camera isn't moving at all, this can yield substantial savings. However, even a

# Chapter 3: Fundamentals of Compression

little camera motion can throw everything off. While subtracting the value of the previous frame worked for very early videoconferencing systems, modern codecs need something a lot more powerful. (See Figure 3.17.)

## Motion estimation and compensation

This brings us to motion estimation, the power behind modern codecs, and the major consumer of CPU cycles during compression. The basic idea is simple: Find out if and where parts of the previous frame have moved to in the new frame. If the codec knows this, it can use the previous information to predict the new frame, and so will need less information.

In practice, doing this right is hugely complicated and time consuming, and motion search algorithms are of a major focus of ongoing compression research.

To demonstrate how hard motion estimation is to do correctly, I'm going to describe the brute force method. If you had infinite CPU power and minimal time to write software, you'd simply take each block of the previous frame, and compare it to every pixel in the new frame, to find the closest match, if any. With a 640×480 image, this would mean comparing each of 64 pixels in 4800 blocks by the 307,200 pixels in the frame. Repeating that process for each and every frame. Riiight.

What's more, modern codecs can do motion estimation at half- or even quarter-pixel resolution, which requires four and 16 times as many comparisons, respectively. That's just not going to happen even with the fastest CPUs currently on the market.

So, many smart people have spent *a lot* of time figuring out ways to do motion estimation faster while still getting close to an optimal match, with impressive results. The details of the algorithms aren't particularly important to understand (although they're mind-blowingly clever). The important thing is that they provide results that come close to those a brute force approach can accomplish, but orders of magnitude faster. And the algorithms are still getting better and faster.

## Types of frames

All compressed video is made up of frames. Depending on the type, a compressed video will have one, two, or three of the major frame types.

### Keyframes (I-frames)

*Keyframes* (called I-frames in MPEG) are self-contained images, like a still frame. Other frames may be based on them, but keyframes aren't based on any other frames. Because of this, keyframes are typically several times larger than other kinds of frames providing the same quality.

However, keyframes are critical. The first frame of any video sequence needs to be a keyframe, so the following frames have something to reference.

**3.17** The difference between frames can be dramatic with different kinds of images. These images show the source frames and the differences between them. The black parts of the difference frame show where there was no difference.

## Chapter 3: Fundamentals of Compression

Keyframes are typically inserted automatically by a codec if it determines a frame is significantly different from the preceding frame. Content that will usually produce a lot of keyframes features lots of motion or a lot of cuts between scenes. Most codecs allow you to set keyframes at specific intervals and allow you to force the creation of keyframes in situations where a new keyframe hasn't been created within a certain number of previous frames.

Most editing formats like DV, M-JPEG, and I-frame-only MPEG make every frame a keyframe. This is because keyframe-only files offer great quality, and very fast random access, since such formats, unlike formats in which most frames only exist in reference to other frames [see the following section on "Delta frames (P-frames)"] don't need any other data to play a particular frame. Of course, such keyframe-only file formats require much higher data rates.

### Note

**3.18** "The picture worth a thousand words."

***Is a Picture Worth a Thousand Words?*** We've all heard the old saying, "A picture is worth a thousand words." And now, with the power of information theory, we check to see how good a picture we can make with the information in a thousand words. First, I'm going to take the first thousand words of this chapter, and save it as a text file. To make it a fair comparison, I'm going to take the redundancy out of the file by running it through Gzip at maximum compression. This gives a file of 2,593 bytes. Not a lot to play with, but we can definitely make some kind of image with it.

There aren't any good JPEG2000 tools available at the moment, so we'll make an old school grayscale-optimized JPEG instead. (I'm using grayscale because "a picture is worth a thousand words" was generally associated with newspapers back in the day when newspapers were black-and-white.)

While the image is recognizable, it ain't pretty. JPEG2000 would do a lot better at this file size, but still, a picture that's worth a thousand words really isn't much of a picture.

### Delta frames (P-frames)

Delta frames (P-frames in MPEG) are the workhorses of the modern codec. They are defined as "I am like the frame before me, but with these changes." This includes all the motion estimation goodness that a codec can supply. If there is little or no change from the frame before, or if the changes are very predictable (as in a camera dolly, where the entire frame shifts over a few pixels), a delta frame can provide the same quality as a keyframe at a fraction of the size.

The biggest limitation of delta frames is they only exist in reference to the previous frame. And if the previous frame is a delta frame, it only exists relative to the frame before that, and so on to the previous keyframe. This makes random access to a delta frame quite slow, because the codec must go back to the previous keyframe and decode every delta frame until the delta frame to be displayed is reached. Problems occur when playing back video on a slow computer or off a slow CD-ROM drive. If the video stutters, dropping a frame on which other frames are based, playback will stop until the codec decodes the next keyframe.

The bulk of any compressed video file will either be delta or in the case of MPEG files B-frames, with, "as few keyframes as possible, but no fewer," to paraphrase Albert Einstein.

 **Note**

***Recompression.*** So, what about recompressing already compressed video? This is done all the time with authoring formats such as DV—any time you change a frame, it needs to be recompressed when going back to a file or tape. Also, formats sometimes need to be converted from one to another.

The problem with recompression is that it can take your existing compression artifacts and treat them as data to be encoded. This can lead to generation loss, in which each successive round of compression creates more compression artifacts and echoes of past compression artifacts, leaving less of the original image data behind.

Recompression with the same codec and format doesn't tend to cause much trouble. Because the quantization of DCT or vectors or whatever follows the same rules, recompressing an image in which nothing has changed, information the codec would have thrown away has already been thrown away. However, even a simple change like shifting the video a few pixels to the left or right will cause the macroblocks to not line up in the same way, resulting in a very different quantization.

Generally, using the exact same codec, and for formats like MPEG using the same implementation from the same vendor, yields best results. Sticking to the same color space and mathematical model (DCT versus VQ versus wavelet) helps. And, of course, the fewer artifacts in the source, the fewer problems you'll have. Going between lossless formats is lossless.

So, going from DV to DV or from MPEG to MPEG can work. But going from a compressed delivery format to another, as in transcoding Cinepak to Windows Media, can yield horrible, horrible results. In such cases, always try to return to the original source and start over.

## Bi-directional frames (B-frames)

Bi-directional frames (B-frames in MPEG) are a twist on the delta frame—bi-directional frames are based on the keyframe or delta frame before or after them, but no frames are based on B-frames themselves at all. Because B-frames can draw on the motion-estimated source of two different frames, they typically compress quite a bit better than even P-frames. B-frames occupy less space yet produce the same image quality, although how much varies with codec and content. Nothing can be based on B-frames because such referencing entails some impossible dependencies; if a frame were based on the frame after it, which was in turn based on the frame before it, there wouldn't be any base frame to start from.

While the fact that nothing can be based on B-frames could be considered a drawback, it also becomes one of the B-frame's biggest advantages—you can drop them. With a file consisting of only keyframes and delta frames, any dropped frame means that playback stops until the next keyframe. But a B-frame can be dropped without any disturbance to later frames. This is very useful in when playback is constrained for some reason, be it in bandwidth or CPU decode performance.

B-frames are used in some codecs, notably the various MPEG formats, including Sorenson Video 3 and Indeo 4 and later versions. Windows Media Video 8, for example, doesn't use B-frames. B-frames can cause some very heated discussions. They definitely suffer some significant drawbacks. First, they're slower to encode and decode than normal P-frames, since they have to gather motion estimation from two frames instead of one. B-frames also add latency to the file, since they can't be displayed until the following P-frame is available. Lastly, for some kinds of rapidly moving content, B-frames can actually hurt quality, since they increase the difference between P-frames. In practice, this doesn't become a problem often enough to be of significant concern.

The biggest philosophical problem with B-frames is that they don't actually add any predictive value to later frames. In essence, all the data they provide is immediately thrown away. I don't find this argument compelling. If a file with B-frames looks better than one without them, what data is thrown away doesn't seem material. After all, as each keyframe is decoded and played back, all the previous data is no longer used.

Depending on the codec, you'll have at most one or two B-frames in a row. The pattern is generally described with IBP nomenclature, indicating the appearance of I-frames, B-frames, and B-frames. So, a stream with two B-frames for every delta frame would be described as IBBP. The actual pattern of frames would look something like this: IBBPBBPBBPBBPBBIBBPBBPBBPBBP

## Data rate control

Data rate control doesn't get the same attention as other aspects of video compression, but it is critically important to producing good results in the real world.

Armed with the descriptions presented in this chapter of different compression modes, you have the basis for understanding how frames can be compressed to a desired size. However, you should not compress every frame to the same number of bits as every other frame. In the simplest case, this would mean that keyframes would be much lower quality than the following delta frames, since keyframes need a lot more bits to achieve the same quality. Conversely, you don't want the keyframes to have so much more bandwidth than a delta frame that keyframes look a lot better than the frames that follow them. Both conditions were common occurrences with

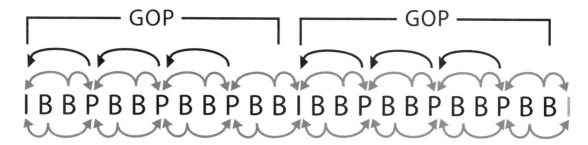

**3.19** The interrelationship between frames with B-frames. This is a IBBP sequence with a GOP size of 12, using open GOPs. The arrows from each frame indicate which frames that frame may be based on. Note that B-frames are based on two other frames, but no frame is based on them. P-frames are only based on the previous P-frame or I-frame. I-frames are self contained.

older codecs, and showed up as *keyframe flashing* where the visual quality of the video changed noticeably whenever a keyframe was displayed. You want to distribute bits so the visual quality of keyframes and delta frames is constant. There isn't any fixed ratio of sizes that will accomplish this, of course, because the more change in the video, the smaller the gap between the relative sizes of a keyframe and a delta frame of the same quality.

You also want to have all frames be of the same relative video quality. Inevitably, you'll encounter sequences of footage featuring wildly different levels of image complexity. In such cases, you'll need different data rates to achieve similar quality levels between scenes with disparate levels of complexity. Your codec should, ideally, allow for momentary data rate spikes. Of course, what goes up must come down, so for every bit above the average data rate, some other frame needs to be a bit below the average data rate to keep the average constant. This is actually harder than it sounds. Conventional one-pass codecs don't know what frames that have yet to be encoded look like. So, if the codec hits an increase in video stream's complexity, the codec won't know whether that complexity will go back down in a few frames, stay steady for a while, or become even more complex. With a streaming codec, the average data rate for each few seconds of video needs to be the same, so if the codec spends extra bits on a complex section, it might have to actually drop the bandwidth after a few seconds, even if faced with an even more complex video sequence. A good heuristic method that can respond to such situations intelligently is important to all codecs.

There are a variety of techniques different codecs use for data rate control such as 1-pass, 2-pass, buffered, and unbuffered. These are described in Chapter 8, "Video Codecs."

## Frame dropping

With sufficiently aggressive compression settings and complex source, the codec may simply not be able to produce a video frame of adequate spatial quality. This can be a problem should a frame have fine details, such as writing on a chalkboard, that need to be seen. Because of this,

# Chapter 3: Fundamentals of Compression

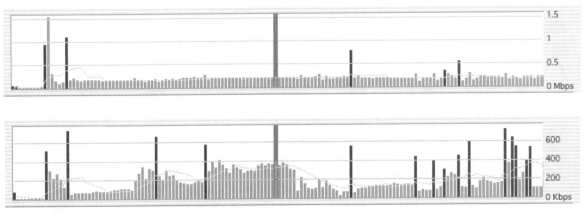

**3.20** The same source and data rate, encoded with 1-pass and 2-pass (Unbuffered) modes. The 2-pass is able to dramatically vary the data rate per frame to match the complexity, improving overall quality.

many codecs let you set a minimum quality threshold, below which a frame's quality can't drop. Thus, an overly difficult frame will be quite a bit larger than the bitrate and the frame rate of the file would support. After encoding that large frame, the codec will drop one or more frames to maintain the target data rate. So if a frame is three times the size of the desired average frame, two frames will be dropped before the next frame is encoded. In real-time compression, frames might also be dropped if the compressor can't keep up with encoding the frames as they stream.

The net effect is that the file will play normally for the easily encoded portions, but the frame rate can drop, sometimes to the point of becoming a slide show, in more difficult to encode sections. For most projects I find sporadic dropped frames more distracting to the user than simply using a lower frame rate in the first place.

Note different codecs may only support frame dropping in some of the encode modes. Neither Windows Media nor Sorenson Video support frame dropping in 2-pass whole file mode.

## Audio Compression

Audio compression uses a lot of the same principles as video compression, but with twists, mainly due to the differences in how we hear instead of see.

The earliest kind of audio "compression" was simply resampling the audio to lower sampling rates and bit depths. We discussed these techniques earlier in the chapter, and I don't really consider them compression per se. They're more like reducing the resolution of a video file, while leaving the pixels themselves uncompressed.

# Audio compression techniques

*ADPCM*   The earliest popular audio codec on personal computers, the IMA codec from the Interactive Multimedia Association, was based on Adaptive Differential Pulse Code Modulation (ADPCM). ADPCM is a simple technique that offers decent quality and excellent performance, but only limited compression.

The idea behind ADPCM is that, in most audio, the differences from one sample to the next aren't very large. With a loud, low frequency sound, the waves span across the entire dynamic range, but they don't change very quickly from sample to sample. With quiet but high frequency sounds, the cycles are short, but the change is never very great. The only exception to this is loud, high frequency sounds, which aren't nearly as common in most audio content.

ADPCM works by encoding the difference between samples, not the value of the voltage itself. This can result in distortion in high-pitched, loud sounds, but otherwise the quality is good. Note, however, that "high frequency" is relative. ADPCM sounds great at 44.1kHz, but an 8kHz ADPCM will distort frequencies above 2kHz. Such distortion is quite audible.

ADPCM codecs are really only used for legacy applications, and for times where absolute maximum decode performance is critical, as with high-bandwidth CD-ROM.

*Sub-band compression*   In the same way that DCT video compression turns samples into frequencies, and removes frequencies that are redundant or hard to see, sub-band codecs convert samples to frequencies, and use psychoacoustic models to find and remove frequencies that are redundant or hard to hear.

First the audio is broken up into discrete "frames" of about 20 to 500 fps. (These don't have any relationship to video frames the audio might be attached to.) Then, each frame is turned into a series of from two to over 500 frequency bands. Usually, the range of frequency bands is from 32 to 128. Because different bands have varying complexity, and lower complexity bands take up less space, and compression efficiency will be improved.

Psychoacoustic models can then be applied to remove any masked sounds. As we discussed in the previous chapter, you can't hear a quiet sound next to a loud sound of the same frequency. Nor will you hear a quiet sound of the same frequency as a loud sound that precedes it in time. Using these psychoacoustic techniques, the codec is able to remove unnecessary information.

Another important technique is used with stereo audio. Left and right channels almost always contain redundant information, so it would be inefficient to encode them separately. Instead the left channel is encoded, but what would be the right channel's data contains an encoding of only the differences between left and right channels. This substantially reduces the data rate requirements for stereo audio.

There are a lot of variations, tweaks, and flavors of these techniques. Most general-purpose audio codecs are based on them.

Of course, beyond a certain point, audio compression degradation will become audible. At that point, the codec tries to provide the best possible quality given the available bits.

Percussive sounds, which have relatively random waveforms, and hence flip between sub-bands with rapidly changing values, are the hardest for most codecs to compress. Cymbals and hi-hats are typically the most difficult to encode. Sometimes very clean, well-produced audio is difficult to compress as well, since even slight distortions are audible.

*Speech codecs*   Speech codecs are very different animals from the general-purpose codecs based on sub-band or ADPCM compression. Different codecs use different fundamental technologies, but they're all based on transmitting intelligible speech at very low data rates. The quality is rarely objectively good, and they're terrible at doing anything other than speech, but the data rates can go exceptionally low. Some of the newest cutting-edge speech codecs run at 1Kbps.

Speech codecs often mathematically model aspects of the human vocal system, so choral music and other sounds a human voice can make survive the compression process, but nothing else will. Speech codecs are generally limited to 8kHz and sometimes 16kHz sampling rates, and are mono.

## Audio data rate control

Unlike video codecs, most audio codecs are CBR—each frame takes up the same amount of bandwidth. Some codecs like MP3 allow a *bit-buffer* where extra bits from previous frames can be used in later ones.

There are some VBR 1-pass audio codecs now. MP3 and Ogg Vorbis are probably the best known of these. The VBR of MP3 audio means the data rate of each frame can vary within the allowed range of 8Kbps and 320Kbps. So, when the audio is silent or extremely simple in nature, the frame will be 8Kbps, and rising as the content becomes more complex. MP3 also allows a quality limited mode, with no explicit audio data rate target.

Early versions of the Ogg Vorbis codec only supported a vague VBR mode, although from RC3 (release candidate 3), Ogg Vorbis also includes a true CBR option.

# What Makes a Good Codec?

Given the complexity of all this, you've probably realized the answer to "what's the best codec?" doesn't have any simple answers. A lot of different parameters weigh on whether a codec is right for a particular task.

- **Compression efficiency**

    The most fundamental measure of a codec is compression efficiency, or "bang for the bit." Compression efficiency measures how good a playback experience the codec can provide at a given bitrate. In head-to-head comparisons, this is typically done by looking at two clips at the same bitrate, and by comparing the bitrates needed to achieve the same quality. Typically, when someone says, "Our new codec is 20 percent better than Brand X," they mean it can produce the same quality as X at data rate that's 20 percent lower. Of course, quality in itself is subjective. Comparing a codec that offers higher frame rates and lower spatial quality versus one with higher spatial quality and lower frame rates won't yield an obvious winner.

    Compression efficiency isn't a constant. Different codecs may be great at some data rates, or with some types of content, but much less competitive at others. For example, speech codecs have tremendous compression efficiency with speech content at low bit rates, but can't do music at all.

- **Maximum quality**

    Maximum quality is how good an experience the codec can provide when data rate isn't an object. Some codecs, especially VQ codecs like Cinepak and Sorenson Video 1 and 2, have a fairly low quality ceiling beyond which increasing the data rate doesn't help quality much. Other codecs can go up to artifact-free, where the video doesn't suffer any artifacts although it may not look identical to the source. The next step is visually or audibly lossless, where the compressed video can't be distinguished from the source. Lastly, there is mathematically lossless, where you can get out the exact bits you put in. Most codecs can't do mathematically lossless, due to rounding errors. True mathematically lossless codecs like JPEG-LS are different beasts, with different algorithms.

    While maximum quality is rarely a concern when using compressed video for delivery on the Web or on disc, it's very important when compressing files for later recompression.

- **Bitrate accuracy**

    How well the codec hits its target data rate is called *bitrate accuracy*. Most codecs come close to the target data rate with real-world content, but I've seen plenty fail by producing substantially higher data rates than requested when given extremely grainy or high-motion video.

    Another bitrate accuracy issue is whether the codec supports the bitrate mode needed for a particular project. If a codec can't do a buffered VBR, it's not a good choice for real-time streaming.

- **Encode performance**

    Encode performance is how fast the codec works. This is generally proportional to the frame rate and number of pixels; the content itself doesn't matter much. Depending on the motion search algorithm, some codecs actually get slower by a factor of the image area squared. For example, a

640×480 area might be 16 times slower than a 320×240 area, instead of the four times you'd expect.

Some codecs provide two or more speed versus quality modes. These are normally there to support real-time compression systems, which can drop a lot of frames if they don't have sufficient performance.

Encode performance can vary quite a bit depending on chip architecture. For example, the Sorenson codecs are typically a lot faster on a G4 processor than any Intel-compatible CPU. Some codecs are optimized for different SIMD architectures. An SSE-optimized codec might be faster on an AMD Athlon XP, but one with SSE2 optimizations would be much faster on an Intel Pentium 4. Some codecs are also optimized for multiple processors (though a maximum of two or four CPUs are typically supported); others only use a single processor. For example, Windows Media Video 8 can use up to four processors, while Sorenson Video v3.1 Pro can use up to two, but Sorenson Video 3 Basic can only use one.

- **Decode performance**

    Decode performance is how well the file plays back. This is typically in proportion to the number of pixels, the frame rate, and the data rate. And again, codecs optimized for different architectures will play back on very different types of computers.

    Many modern codes also have post-processing quality improvements they accomplish with surplus CPU power, so instead of dropping frames when running on a slower machine, the quality is scaled down. B-frame codecs may also drop frames on playback.

    These techniques are important for adding slack to playback requirements. Having insufficient decode power can be catastrophic for playback. Remember, once a frame is dropped, typically no new frames of video are seen until the next keyframe is decoded. Just a few dropped frames here and there can make the video unwatchable.

    With audio codecs, decode performance is typically not a problem, unless you're trying to do something on a very old machine (such as playing MP3 files on my 33MHz NeXTStation Color Turbo!). When that happens, there's usually a gap between the audio frames as they play back, so the audio stutters.

- **Post-processing**

    Post-processing is stuff a video codec does to improve quality on decode, such as deblocking or deringing, as described previously. Post-processing can improve quality substantially when playing back a video file on a machine a lot more powerful than needed. Post-processing turns off on slower machines—a good thing because higher frame rates that produce a lot of blocky frames look better than lower frame rates and less unblocky frames.

The RealVideo 8 codec is unique in that it can use post-processing to increase the frame rate. More on this in Chapter 16, "RealSystem."

- Compatibility

    Balancing quality versus compatibility—whether the codec will play on your target platforms—is always a struggle because older codecs are less efficient but are more commonly installed than newfangled codecs (especially when that new codec has just been released).

    The more compelling your content, the more likely users are to upgrade the codec to see it. Modern architectures make codec updating a much more elegant, automated experience, but the user still has to click "okay." When you're targeting older legacy systems that might not even have the codec or the required architecture available, such considerations become important. That's why MPEG-1 lives on.

## For More Information

| | |
|---|---|
| http://www.faqs.org/faqs/compression-faq/part1/preamble.html | The FAQ from comp.compression. |
| http://www.lucent.com/minds/infotheory/ | Lucent has this readable multimedia introduction to information theory. |
| http://cm.bell-labs.com/cm/ms/what/shannonday/shannon1948.pdf | Claude Shannon's seminal paper. |

# Chapter 4

# The Digital Video Workflow

## Introduction

We've talked about the fundamentals of compression. Now we're going to dive into how to use this stuff in the real world for real projects. While all the issues covered in this chapter will be gone over in loving detail throughout the rest of this book, it's important to establish a general understanding of the digital video workflow, to help you see how the different processes fit together.

## Planning

The first phase of any production is planning. Don't underestimate its importance. The best way to achieve good results is to know what you want before you start. Even if you're in charge of just a small portion of the overall project, knowing how it fits into the overall structure and having a solid understanding of what you're after will help you realize your goal and stay on budget.

Three basic elements to define as early as possible: content, communication goals, and audience. If you have a good grip on these three, the process will go much more smoothly.

### Content

What is the nature of your content? Music video? Movie trailer? Movie? Training video? Annual report? Will it be produced on film or video? Letterboxed or full frame? PAL or NTSC? How long is it?

Ideally, you'll design content to match your communication goals and audience, but as a compressionist, very often you'll be called upon to compress existing content produced for other purposes. Understanding the source as well as possible is critical, as you'll need to finesse the process to play to your content's strengths and avoid its weaknesses. The right compression settings vary radically with the content, even when targeting the same format and data rate.

## Communication goals

When compressing video, the compression is being done with a communication goal. Knowing that goal is essential. However, it's common for folks to explicitly not define what they're trying to achieve with a project.

How you would encode a piece of content can vary radically with different communication goals. Take a music video, for example. If you were a member of the band and wanted the video you were starring in to sell albums, you'd put a relatively large amount of the bandwidth in the audio, so the music would sound great, even if that meant reducing the video quality. Conversely, if you were the cinematographer of the video, and you were putting the video on your site for promotional purposes, you'd emphasize video quality at the expense of audio.

## Audience

It's beyond the scope of this book to discuss all the possible permutations you'll encounter when matching your content, your message, and your audience. So we'll leave those details aside and focus on the importance of understanding your audience as it relates to the task of delivering compressed video.

The single most important compression consideration with regard to your audience is knowing the extent of the bandwidth available to them. Of near-equal importance, is having some estimation of their playback platform configuration.

If each member of your target audience is sitting in front of one type of computer with a standardized amount of RAM, a specific type of monitor driven by identical graphics cards, all connected via a corporate LAN, start counting your blessings. Not only will your choice of media player be an easy one, you'll have a pretty darned good idea of how much excess bandwidth if any you have to burn. Likewise, if your audience is anyone who accesses a general interest Web site via the Internet, your choices though many are fairly clear-cut—you'll have to prepare your files for delivery to both dialup and broadband connections, decided whether those files should be available as progressive downloads or through true streaming, and target either all three major media players or some subset of them or settle on just one player. Of course, your decisions will likely be dictated as much by budget (bandwidth after all is expensive) as by your audience's preferences. If your content targets graphics or video professionals, you'll likely want to deliver QuickTime files. If your content targets corporate CFOs or IT folks, Windows Media will be your file format of choice.

If you choose an unusual format, or a brand new media player as your delivery vehicle, you'll want to take measure of the likelihood your audience will download a new player or new codec before proceeding.

## Balanced mediocrity

So, how do you balance content, communication goals, and audience? You're trying to achieve what I call, tongue-only-partially-in-cheek, balanced mediocrity.

For video on the web, don't assume you have as much bandwidth, as much resolution, as high a frame rate, or fast enough end-user computers as what you're developing your content on. Finding the right balance of elements means making them all seem mediocre, so you're not spending bits making one thing perfect at the expense of everything else getting worse. Balanced means "in the right proportion." So, given our music video example, you want to make the ratio between video and audio data rates be such that each seems about as good as the other, *relative to your goals*. The relative part there is critical—you'll want to weigh the quality difference differently between clips for the cinematographer and clips for the band.

A good test for balanced mediocrity is if changing any value up or down diminishes the overall experience. For example, when raising the data rate hurts more by slowing download times than it would help by improving video quality, and reducing data rate would hurt video quality more than it would help by reducing download times. Or increasing resolution would hurt more by adding artifacts than it would help by increasing readability, and reducing resolution would hurt readability more than it would help by reducing artifacts. The data rate and video quality or resolution and artifacts may all be mediocre, but if they're in the optimum combination for your project, you'll have done what you can.

# Production

Production is the actual shooting of the film or video, creating the animation, or otherwise creating the raw assets of the project. Production is often the most expensive and most difficult part of any project, so getting it right is critical.

Professional video and film production is much, much harder than video compression, and requires an experienced crew. One of the biggest mistakes many companies and individuals make in creating Web video is thinking that, because the content is going to be delivered on the Web, production standards don't matter. This is completely untrue. Amateur video on the Web looks at least as bad as amateur video on television.

There are many important, irrevocable decisions made during production. For video production, the content needs to be shot as progressive or interlaced, at 4:3 or 16:9 aspect ratios. In black and white or in color; at a certain frame rate; with soap opera-style lighting or with *X-Files* style lighting or under DOGME 95 rules. Of course, there is no one correct answer to such questions. It all depends on the nature of each project.

Production is covered in "Producing and Editing for Optimal Compression," Chapter 5.

## Postproduction

Postproduction is when assets created in production are turned into a final product. Postproduction includes editing, compositing, audio sweetening, and so on.

Much of the style of a production comes during post. Different editing styles can have dramatic effects on the final results, both in terms of the uncompressed project and in terms of the compressed version—MTV-style fast cutting yields a totally different viewing experience than long, slow cuts. Likewise, MTV-style cutting is enormously more difficult to compress than more conventional, sedate editing.

Production and Post compression issues are discussed in "Producing and Editing for Optimal Compression," Chapter 5.

## Digitization

Digitization is the process of getting video into a form where it can be compressed. If you're doing postproduction and compression, a compressible file can just be exported from the NLE software, and digitization isn't required.

However, digitization *is* required when all you have access to is a completed analog video tape, be it on VHS, S-VHS, or on Betacam SP. Digitization isn't particularly difficult, but it does require the right equipment. Getting DV format video off a miniDV or DVCAM tape is trivially easy with modern computers. High-end digital formats like Digital Betacam require access to expensive tape decks and capture cards, but they're easy to use correctly and there are plenty of facilities that can be hired to handle such transfers for you. Things get more complicated with analog formats. Cheap solutions are available, but they offer a lot lower quality than professional gear.

Digitization is covered in "Capture and Digitization," Chapter 6.

## Preprocessing

Preprocessing is everything done between the source video and audio, and what is finally handed off to the codecs. This includes cropping, scaling, deinterlacing, image adjustment, noise reduction, and audio normalization. The goal of preprocessing is to take video optimized for display on a television or other target, and optimize it for compression instead. It is the most artistic facet of compression.

Preprocessing is covered in "Preprocessing," Chapter 7.

## Compression

If you cared enough to buy this lovely book, you already know compression is the Big Kahuna. In the compression process, the postprocessed video is run through codecs, becoming the final compressed file(s).

With many tools, preprocessing and compression occur simultaneously. Still, it's useful to treat them as different steps conceptually.

In compression, many decisions need to be made, including choosing formats, codecs, data rates, and frame rates. While picking the right compression settings seems daunting, it becomes much easier when you understand what's driving our troika: content, communication goal, and audience.

Compression is covered in the bulk of this book. Make sure to read the chapters specific to the format you're using.

## Delivery

Delivery is how the files get to the user. This can be on a DVD, on a CD-ROM, downloaded from the Web, streamed through a streaming server, and in a number of other ways. The files can be standalone or integrated into a larger project such as a Web site or computer game.

More complex delivery schemes often have their own, more stringent issues. Delivery issues with particular technologies are covered within the chapters about those technologies.

color page — 1

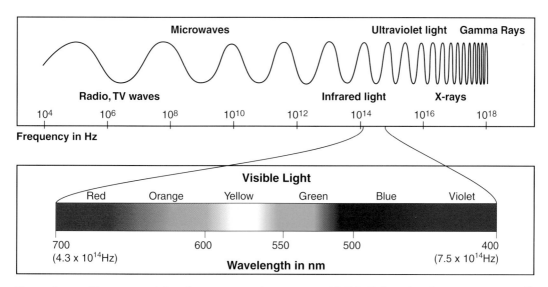

Figure A   The range of the electromagnetic spectrum. Visible light only takes up a very small portion (Figure 2.1).

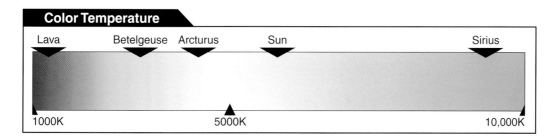

Figure B   The range of colors from black body radiation at various temperatures.

II — color page

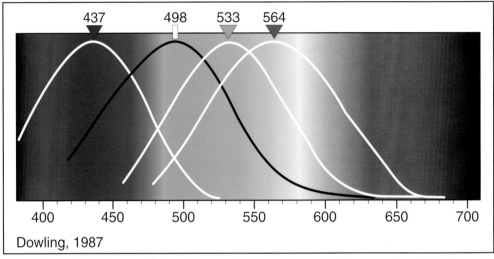

Dowling, 1987

Figure C    Relative sensitivity of the eye's receptors to the different colors. The black line is the rods, the white lines the three cones (Figure 2.3).

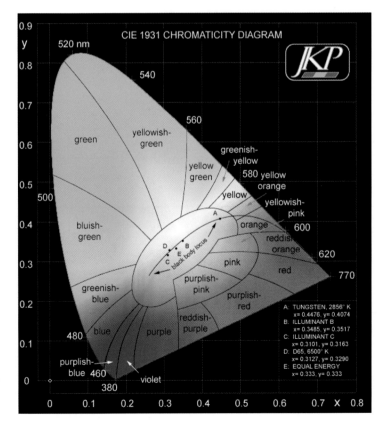

Figure D    The CIE color charts from 1931 and 1976. The black dots denote the color temperature of various kinds of lighting. The outside edge denotes the wavelength. Note the 1976 version renames the axes U and V, as in YUV *(see facing page as well)*.

Image courtesy of Joe Kane Productions.

color page — III

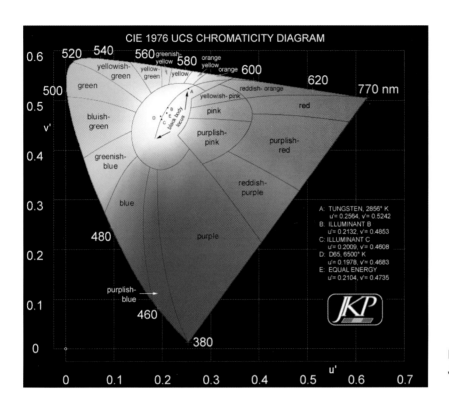

Image courtesy of Joe Kane Productions.

Figure E  Note how the surrounding color changes the appearance inside the square.

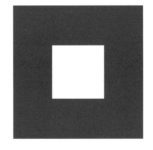

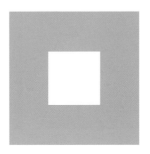
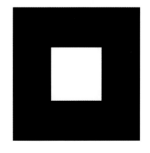

## IV — color page

Figure F   The same image as source *(left)*, with only luma *(bottom left)*, and only chroma *(bottom right)*. The chroma-only picture has twice as much information mathematically as the luma-only, but our visual system is much, much better able to take advantage of the luma information.

color page — V

Figure G  When sampling, colors can exist in the output that weren't in the input, due to averaging (Figure 3.5).

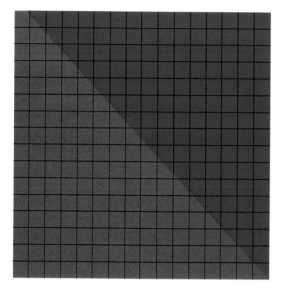

The source, with sampling grid overlaid.

After sampling. Note the purple squares, alghough there was no purple in source.

Figure H  The shapes of all the Y'CbCr color spaces (Figure 3.9).

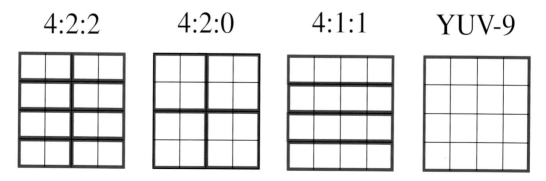

The four Y'CbCr color spaces we see in the wild.

# VI — color page

Figure I  The range of colors available in Cb and Cr when Y is at the middle of the range (127 in 8-bit).

Figure J  What Cb and Cr actually look like.

Cb                              Cr

Figure K  This process shows the power of doing noise reduction on just the chroma channels using Photoshop's Lab mode (analogous to YUV). Source has typical chroma noise. In this case, a 2.5 pixel Gaussian blur was applied to just the chroma channels. In Lab mode, color artifacts went away, while the image remained sharp. When the same value was applied in RGB, where all channels carry color, the image became unusable.

# VIII — color page

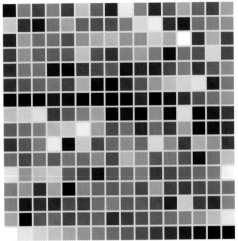

Figure L  Same image, converted to the shown 8-bit palette. The top image shows typical dithering, and the second shows flattening. The flat image is about 15% smaller when losslessly compressed (Figure 3.10).

Figure M  Two different files, as losslessly compressed source *(on the top)*, encoded with fixed quality *(middle),* and encoded with fixed data rates *(bottom)*. There would be even bigger differences at higher resolutions.

**X** — color page

Figure N   A clip with both tearing and edge blanking. The minimal cropping you'd want to pick is shown in Cleaner 5 (Figure 7.1).

Figure O    The same frame of video as interlaced source *(the two frames on the left)*, deinterlaced *(the two on the right)*, uncompressed on the top, and and then both compressed to progressive scan JPEG at the same file size. Note how much worse the quality of the compression is with the interlaced file.

## XII — color page

Figure P — The horror of a blended deinterlace.

Figure Q — This close-up shows typical composite noise artifacts at the edges of the curtain.

color page — **XIII**

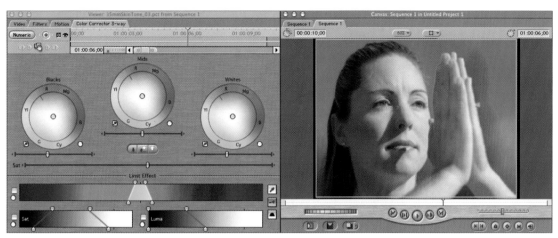

Figure R    The Color Correction Three Way interface of Final Cut Pro 3. It is fully Y'CbCr native, offering great performance and quality (Figure 7.10).

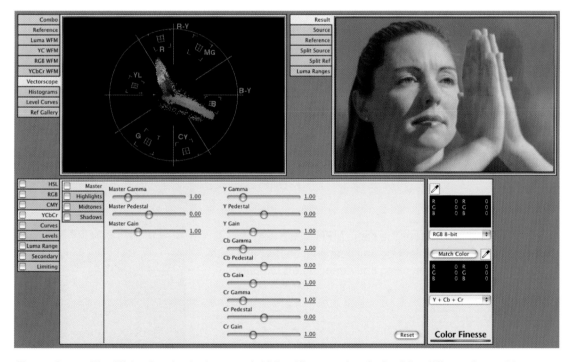

Figure S    The UI for Synthetic Aperture's Video Finesse plug-in for After Effects. It provides unrivaled control of the color correction process for desktop software. Note how the flesh tones are clustered along the flesh tone axis in the vector scope (Figure 7.11).

**XIV** — color page

Figure T   Doing hue adjustment to fix skin tone in the Final Cut Pro 3 Basic Color Corrector (Figure 7.16).

Figure U   *(right)* The same MPEG-4 file with deblocking and deringing on *(top)* and with deblocking and deringing off *(bottom)*.

color page — **XV**

Figure V   This frame and other archival shots have thicker black borders at the top and bottom of the screen (Figure 25.3).

Figure W   A progressive frame of the source video, showing some composite noise artifacts (Figure 25.22).

# XVI — color page

Figure X    Preprocessing in Squeeze on a Mac for Windows delivery (Figure 25.23).

Figure Y    A sample frame from the Myst trailer (Figure 25.29).

(Official Credits: Myst 3: Exile trailer produced by Presto Studios, Inc. published by Ubi Soft Entertainment, S.A. © 2001 Ubi Soft Entertainment, S.A. & Cyan, Inc. All rights reserved.)

# Chapter 5

# Producing and Editing for Optimal Compression

## Introduction

Shooting compression-friendly video is more difficult than shooting video that will be delivered in more traditional, high-bitrate forms. "It's only going in a small window, quality won't matter" was a pernicious assumption that lasted through the CD-ROM craze of the early 90s on to the dot-com boom of the late 90s. The assumption is far from true. In any video destined for viewing in some compressed form, be it online or on low-bitrate wireless devices or elsewhere, all the tenets of traditional product must be adhered to while addressing additional constraints imposed by compression.

## Mr. Rogers good, MTV bad

As we learned in Chapter 3, "Fundamentals of Compression," the two things that use up the most bits in compressed video are detail and motion. The more complex your image and the more it moves, the more difficult it is for a codec to reproduce that image well. At lower data rates, this difficulty manifests as increasingly poor image quality.

As a rule of thumb, classic, sedate shooting and editing compress well. Jump cuts, wild motion graphics, and handheld (shakycam) camera work are difficult or impossible to compress well at lower data rates. "Mr. Rogers good, MTV bad," says it all.

When designing for low data rate delivery, avoid using elements that aren't essential to what you're trying to communicate. If a static logo works, don't use a moving one. The lower the target data rate, the simpler the video needs to be. For 56k modem delivery, aim for Spartan simplicity. But if your delivery platform is a DVD player, you'll be able to shoot just about anything you would for broadcast delivery.

# Chapter 5: Producing and Editing for Optimal Compression

## Preplanning

As with any production, proper planning is critical to shooting compression-friendly video. Beyond the usual issues of location, equipment, cast, crew, script, and so on, with compressed video, you need to be aware of where your content is destined to be viewed. Is it going to DVD? CD-ROM? Broadband Web video? 56k dialup modem connection? Wireless handheld devices? All of the above? Knowing your delivery platform helps you understand what you can and can't get away with in terms of aspect ratios, backgrounds, framing, and so on. When targeting lower data rates, the whole production should be designed to work as a filmstrip, with low quality audio and slideshow-like video.

It's always better to plan ahead; to know in advance what you're trying to do and how you're going to do it. Even if you're making a documentary in which you'll find the story as you edit your footage, knowing your target delivery platform will dictate how you'll light and shoot your interview subjects, capture your audio, select archival footage, and so on.

Things that should be decided and communicated to your crew in advance of your shoot include: what aspect ratio you'll be shooting, whether you'll shoot progressive or interlaced video, your shooting style (locked down cameras or handhelds), whether you'll only use tight close-ups for talking heads or opt for a mix of close-ups and wide angle shots.

### Preplanning tips

If your crew aren't familiar with authoring for compressed video, it can be extremely helpful to expose them to what stuff looks like after it's compressed. Try taking their personal reel and compressing it according to the specs of the project at hand. It can be quite an eye opener to see your masterworks on the other end of an encoder. Seeing what works and what doesn't will definitely focus your sights on the production techniques required to produce acceptable results.

## Production

Production is generally the most expensive phase of any project, and hence the most expensive to get wrong. All the standard rules of producing good video apply to shooting video that will compress well: good lighting, quality cameras, and professional microphones are always required.

Doing professional video production is hard. (I know, I've said that already, but it bears repeating.) Professional video production is hard. A lot of Web content is shot with handycams by novice users who don't know enough to know they don't know what they're doing. I don't mean to sound elitist. Despite amazing advances in camera designs, which have brought costs down while bringing so-called broadcast quality images to consumer camcorders, video shot by hobbyists still looks like video shot by hobbyists. Such video played on TV usually looks bad. When it's compressed, it looks even worse. In general, it takes several years of experience working in a particular production discipline before you can call yourself a pro—it's not something you pick up over a weekend after reading a book, even one as fabulous as this.

Production decisions have a large impact on the compressed deliverable. If you're on a shoot and someone says, "We'll just fix it in post," be afraid. As someone who's been on both ends of that conversation, I can tell you that phrase contains one of the least accurate uses of the word "just." Repairing production problems in post is invariably more expensive than getting it right the first time, and the resulting quality is never as good as it would be if the shot had been done right the first time.

## Production tips

There are a number of simple and not so simple things you can do to improve the quality of your video. Optimizing it based on a single target delivery medium is relatively simple. However, we live in a world in which "write once, deliver everywhere" has become a business necessity. Shooting for multiple delivery platforms forces you to make choices, but there are things that shouldn't be optional, no matter what your target delivery platform.

- **Use pro gear.**

    If you want professional quality, use gear that's of a caliber you'd use to make content for non-compressed delivery. Where you draw the line of gear that's good enough depends on your budget and expectations. For some folks, the Canon GL-1 is a fully acceptable, low-cost DV camera. DV has revolutionized the price/performance of cameras that provide adequate industrial quality at a fraction of what you'd have to pay for Betacam SP gear. On the high-end, Digital Betacam rigs costing many times as much as sub-$5K cameras are the norm. Bear in mind tape format isn't all that matters in a camera. The quality of the lens, size and number of the CCDs, and many other features separate high-end from low-end cameras.

    Don't forget the importance of quality microphones. Avoid using the built-in microphone on video cameras. Use a high-quality external mic. A shotgun or boom is great in the hands of a skilled operator. A lavaliere (clip-on) microphone is more trouble-free, and will record voice well. Remember, audio is often the only part of your production that has a decent chance of making it through the compression process with most of its initial quality, so getting the sound right will really pay off.

- **Don't worry about NTSC/PAL safe colors.**

    If you're creating content exclusively for Web delivery, you can use a broader range of colors than you'd be able to use in the NTSC/PAL world. Because your video will be displayed on a computer monitor, which supports a much broader range of colors, the display will be able to handle just about anything your video codecs throw its way. This is generally true for content delivered on DVDs. Because many consumers use S-Video or Component connectors, they can get a much broader range of colors than traditional composite NTSC can support.

- **Don't worry about safe area.**

    If you're creating content exclusively for Web delivery, you don't need to account for the safe area. You can use the entire video frame and crop less or not at all on final rendering. This is particularly useful when targeting higher output resolutions, and you need every pixel you can get.

- **Produce in the right aspect ratio.**

    If you have content that benefits from a widescreen experience and are using a camera that supports shooting in 16:9, go for it. Most editing tools work well in widescreen mode, and you can easily deploy video on the Web at any aspect ratio. You can even make tall, skinny video, if you're doing a rock climbing clip.

    If you're doing a hybrid broadcast/Web delivery piece, you can always create a letterboxed version of the clip for playback on 4:3 video devices.

- **Use soft lighting.**

    Lighting plays an enormous role in determining how easily your video will compress. Soft, diffuse, even lighting is much easier to compress than lighting that produces deep shadows and hard edges. A three-light kit and some diffusers will usually be enough for a simple on-location shoot. For outdoors, bounce cards can be used to reduce shadows.

- **Record DV audio in 48kHz 16-bit.**

    DV supports two audio recording modes: 4-channel 32kHz 12-bit and 2-channel 48kHz 16-bit. Always use the 48kHz mode. Even though broadcast television's audio standard is 32kHz, as a rule, you should always start with the highest possible quality to achieve optimal compression results. This is especially true of audio, as high-quality audio can cover a multitude of compressed video evils. 12-bit audio is simply not enough quality to start with. If you need to record more than two channels on location, there are a few options available to you. You can use a multitrack audio recorder such as a Tascam DA-88 or Alesis (now Newmark) ADAT.

- **Produce in progressive scan.**

    Most modern cameras support progressive scan mode, which doesn't introduce interlacing. Progressive scan is great for content that will be viewed on computer screens, especially at resolutions above 320x240. By not worrying about deinterlacing your content before compressing it, you can double vertical resolution.

    However, progressive scan video can look very strange when displayed on a traditional television screen. If you're creating a single project for both broadcast and compressed delivery, it's best to go with interlaced footage. If the compressed video is quality-critical, you'll want to deinterlace before compression. One promising new format is 24p. With 24p, you can apply 3:2 pull-down to get film-like NTSC, and speed up 4% to get 25fps progressive PAL.

- **Control depth of field.**

    Compared to film, video uses a wide depth of field. However, using a narrow depth of field can be a very useful technique to force the codec to spend bits on the elements you care about. If producing in an environment where the background can't be controlled, such as outdoors or on a convention floor, making sure your talent is in focus and irrelevant details aren't, will substantially improve the clarity of your communication.

    Controlling what's in focus and what's not may require some preplanning to get the camera far enough back from your foreground objects to achieve a narrow depth of field.

- **Reduce complex background detail.**

    One of the most difficult things to compress is foliage blowing in the wind. While this is pretty boring content, the continual motion of the highly detailed leaves is incredibly demanding. For optimal compression, compose your shots with boring backgrounds. If that isn't possible, use a narrow depth of field as described above. It will look a lot more natural than the blocky artifacts that will result from such demanding content.

*See also*

Refer to Figure M on page IX of the color section.

- **Use a slow shutter speed.**

    If you're shooting on video, stick to a shutter speed of 1/60 (NTSC) or 1/50 (PAL) of a second. These are the defaults for most cameras. Show shutter speed produces very natural looking motion blur. Film cameras have a natural 1/48 second shutter speed, which is one of the reasons film source content looks so good when delivered on the Web. A slow shutter speed can produce video that's very soft when freeze-framed, but the reduced compression artifacts are worth it.

- **Use tight shots.**

    A long panoramic shot may look great in the viewfinder, but the talent in the middle of the screen may vanish into insignificance when taken down to Web resolutions. An initial establishing shot may be effective, but the bulk of your shots should be close up on what you want the viewer to focus their attention on.

- **Use a tripod or dolly.**

    Handheld camera work can be extremely difficult to compress. For Web delivery, shoot everything with a locked-down tripod, so the overall frame moves as little as possible. When that's not an option, pan with a good fluid-head tripod, or use a smooth dolly. If handheld camera work is absolutely required, try to use a Steadicam or similar stabilization system. However, handheld shots with lots of changing angles require a lot of bits to encode. Can you rely on an optical stabilization system? As a last resort. A skilled, steady operator might be able to produce usable results.

# Postproduction

Postproduction is where a bunch of raw footage gets turned into finished video, through editing, the addition of effects and motion graphics, voice-overs, audio sweetening, and so on. Compared to production, postproduction is marvelously flexible—with enough time and big enough budget, a project can be tweaked endlessly. Such situations include the luxury of being able to make several passes of test compression to locate trouble spots. Again, video that will compress well is best achieved by a very sedate, classic editing style.

## Postproduction tips

- **Use antialiasing.**

    Antialiasing is the process of softening the edges of rendered elements, like text, so they don't appear jagged. By eliminating those jagged edges, antialiasing looks better when compressed. Most apps either have antialiasing turned on by default or make it an option. The highest quality render mode in After Effects uses antialiasing, as does the Hi Rez mode in Final Cut Pro. If importing graphics made in another app, such as Photoshop, make sure those files are antialiased as well.

    > This text is not anti-aliased
    >
    > This text is anti-aliased

    **5.1** Text with and without antialiasing. Note how much smoother the edges are when antialiased.

- **Use motion blur.**

    Earlier we learned codecs spend bandwidth on complex images and motion, and hence the most difficult content to compress is rapidly moving complex images. Motion blur is a natural way to reduce this problem. With motion blur, static elements are sharp and moving elements become soft or blurred. Thus, there are never sharp *and* moving images at the same time. If you're importing graphics or animation from a tool such as After Effects, make sure those images use motion blur. If you're compositing graphics over video, the graphics' motion blur value should match the shutter speed of the source. The default with film and video cameras is one-half the frame rate, which translates into a 180° shutter.

    Some motion blur effects spread one frame's motion out over several frames or simply blend several frames together. Avoid these—you want natural, subtle motion blur. Extreme blur doesn't help compression significantly, and looks very strange.

    Motion blur is usually found as a rendering option, as a plug-in, or as a filter in 2D and 3D graphics packages as well as video editing software. It can take longer to render motion blur.

**5.2** The same image with motion blur *(left)* and without *(right)*. From After Effects, with the curves showing the direction of motion.

- **Render in the highest-quality mode.**

  By default, you should always render at highest quality, but it's doubly important to render any effects and motion graphics in the highest quality mode when you're going to deliver compressed footage. Depending on the application, the highest-quality rendering mode activates a number of important features, including motion blur and antialiasing.

  If you're creating comps in a faster mode, don't make any decisions based on test compressions made with those comps—the quality can be substantially worse than you'll get from the final, full-quality product.

- **No rapid cutting.**

  At each video cut, the change in the video content requires a keyframe to account for the big change in the content. A cut every 10 seconds isn't a big deal, but MTV-style editing, with a cut every second or less, can cause that section of video to be reduced to incomprehensible mush after encoding.

- **No complex motion graphics.**

  Complex motion graphics, where multiple on-screen elements change simultaneously, are extremely difficult to encode. Try to make each element as visually simple as possible, and only move things that you have to. Flat colors and smooth gradients compress more easily than textures.

- **Don't rotate or zoom.**

  Modern codecs support *motion estimation*. This is the ability of the codec to find a portion of one frame that has moved from another frame, enabling elements that are merely shifting around the screen to be encoded much more efficiently. However, this functionality only works for things that are moving around the frame two-dimensionally, not those rotating or changing size.

- **Use simple backgrounds.**

    Keeping the background elements simple helps with compression. The two things to avoid in backgrounds are detail and motion. A static, smooth background won't be a problem. A cool particle effects background will be.

- **Do not use fine print.**

    The sharp edges and details of small text are difficult to compress, especially if the video will be scaled down to smaller resolutions. Antialiasing can help preserve the legibility of text but it can't work miracles if the text is scaled down too much.

- **Avoid cross-fades.**

    A cross-fade is one of the most difficult kinds of content to compress. Most of the tricks codecs use don't work with blended frames of two different images. In these situations, each frame winds up as a keyframe. It's much more efficient to use hard cuts. Fades to/from black are only slightly easier to compress than a cross-fade, and should be avoided if possible. Wipes and pushes work quite well technically, but aren't appropriate stylistically for many projects. A straight cut is often your best bet.

- **Use animation source.**

    3D animation software makes a great source for compression. Animation typically already uses the 0–255 dynamic range of computer RGB displays, and isn't necessarily interlaced. Animation can also be composited with video, in which case it's treated as any other source exported from a finishing app.

- **Match motion blur.**

    For all the reasons motion blur matters previously, it matters again with animated source. Motion blur in composited 3D elements that doesn't match that of the video is one of my pet peeves. It destroys the illusion that the 3D elements and the video elements coexist within the same reality. And it looks stupid. Make sure the motion blur settings of rendered animations match the source footage.

- **Use mipmapping.**

    Mipmapping is a type of scaling used by 3D programs to dynamically scale textures relative to an object's distance from the camera. The farther away an object, the less detail you'll see in its texture. If mipmapping isn't on, textures may develop sharp edged artifacts and otherwise look bad.

- **Be wary of frame rate.**

    Animators frequently work at 12fps, and 3D programs frequently do not support drop-frame rates of 29.976 and 23.976. This isn't so much a compression issue as it is a sync issue, but it's worth noting should you encounter lipsyncing that seems off in animated source.

# Chapter 6

# Capture and Digitization

*Capture* is the process of getting your source audio or video files off their original media (tape, film, whatever) and onto media your computer can access (usually a hard drive). Of course, the goal is for your source to look and sound *exactly* the same after capture and digitization as it did in its original form. Capture is probably the most conceptually simple aspect of the entire compression process, but it can be the most costly to get right, especially when you're starting with analog sources.

*Digitization*, as you've probably guessed, is the process of converting analog source to digital form. That conversion may happen during capture or it may occur earlier in the production process, say during a video shoot with DV format cameras or during a film-to-video transfer. When you capture video off a DV tape with a 1394 connector, you aren't digitizing because you aren't changing the underlying data.

Let's look at some of the technologies and concepts that come into play with capture and digitization.

## Broadcast Standards

Globally, there are two major broadcast standards that encompass every aspect of video production, distribution, and delivery: NTSC (National Television Systems Committee) and PAL (Phase Alternating Line). Signals produced from one will be incompatible for delivery on the other, because NTSC and PAL have different native frame rates and numbers of lines. Luckily, a lot of video tools—especially postproduction software—can handle both formats. Alas, it's not as common for tape decks and monitors to be format agnostic. Most video production facilities are either PAL- or NTSC-centric. When called upon to deliver content in a non-native format, they do a format conversion. This conversion is never perfect, as going from PAL to NTSC results in a loss of resolution. Going from NTSC to PAL loses frames. In both cases, motion is made less smooth.

## NTSC

The first National Television Systems Committee consisted of 168 committee and panel members, and was responsible for setting standards that made black and white analog television transmission a practical reality in the United States. The standards the NTSC approved in March, 1941 are still in use today throughout North America and in about 10 other countries. Those standards serve as the framework on which other broadcast television formats are based.

Some aspects of the NTSC standard—the 4:3 aspect ratio and frame rate—had been recommended by the Radio Manufacturers of America Television Standards Committee back in 1936. One of the most important things the NTSC did in 1941 was ignore the arguments of manufacturers who wanted 441 lines of horizontal resolution and mandate 525-line resolution.

The original frame rate of NTSC—30 interlaced frames per second—yielded 60 fields per second, and was chosen because U.S. electrical power runs at 60Hz. Electrical systems could interfere quite severely with the cathode ray tubes of the day. By synchronizing the display with the most likely source of interference, a simple, nonmoving band of interference would appear on the screen. This was preferable to the rapidly moving interference lines that would be produced by interference that was not in-sync with the picture tube's scan rate.

When color was later added to NTSC, it was critical that the new color broadcasts be backwards compatible with existing B&W sets. The ensuing political battle took more than a decade to resolve in the marketplace (from 1948, when incompatible color TV was developed, through 1953 when backward compatible techniques came to be, until 1964 when color TV finally caught on with the public), but that's a tale for another book. Politics aside, backward compatibility was a very challenging design goal. The NTSC worked around it by leaving the luminance portion of the signal the same, but adding color information in the gaps between adjoining channels. Since there was limited room for that information, much less bandwidth was allocated to chroma than luma signals. This is the source of the poor color resolution of NTSC, which is often derided as standing for Never Twice the Same Color. The frame rate was also lowered to 30000/1001, truncated to 29.97 (59.94 fields per second).

The canonical number of lines in NTSC is 525. However, many of those lines are part of the *vertical blanking interval* (VBI), which is information between frames of video that is never seen. Most NTSC capture systems grab either 480 or 486 lines of video. It's important to understand that systems capturing 486 lines are actually grabbing six more lines out of the source than 480 line systems.

## PAL

PAL is named for the color-coding system used in Europe, Phase Alternating Line. By reversing the phase of the reference color burst on alternate scan lines, PAL corrects for shifts in color caused by phase errors in transmission signals. For folks who care about such things, PAL is often referred to as Perfection At Last. Unfortunately, that perfection comes at a price in circuit complexity, so critics call it Pay A Lot.

PAL was developed by the United Kingdom and Germany and came into use in 1967. European electrical power runs at 50Hz, so PAL video runs at 50 fields per second (25 frames per second). Roughly the same bandwidth is allocated to each channel in PAL as in NTSC, and each line takes almost the same amount of time to draw. This means more lines are drawn per frame in PAL than in NTSC. PAL boasts 625 lines to NTSC's 525. Of those 625 lines of horizontal resolution, 576 lines are captured. Thus, both NTSC and PAL have the same number of lines per second, and hence equal amounts of information (576×25 and 480×30 both equal exactly 14,440).

The largest practical difference between NTSC and PAL is probably how they deal with film content. NTSC uses convoluted telecine and inverse telecine techniques. To broadcast film source in PAL, they simply speed up the 24fps film an additional 4 percent so it runs at 25fps, and capture it as progressive scan single frames. This makes it *much* easier to deal with film source in PAL. This is also why durations for European videos are shorter than their U.S. equivalents.

## SECAM

SECAM (Systeme Electronique pour Couleur Avec Memoire) is the French broadcast standard. It is very similar to PAL, except it offers better color fidelity, but poorer color resolution. SECAM today is only a broadcast standard—prior to broadcast, all SECAM is made with PAL equipment. Thus, you'll never see a SECAM tape. By the way, SECAM is sometimes referred to as Something Essentially Contrary to the American Method.

## ATSC

The Advanced Television Standards Committee created the U.S. standard for digital broadcasting. That Digital Television (DTV) standard comes in a somewhat overwhelming number of resolutions and frame rates (see Tables 18.2 and 18.3, page 294). As a classic committee-driven format, it can provide incredible results, but its complexity can make it extremely expensive to implement. And even though Moore's Law is bringing chip prices down, the expense necessary to build display devices (be they projection, LCD, or CRT) that can take full advantage of high definition quality remains stubbornly high.

Resolutions at 720×480 and below are called Standard Definition (SDTV). Resolutions above 720×480 are called High Definition (HDTV).

When and how DTV will wind up being implemented in the U.S. or other countries remains to be determined. Mindful of the coming transition, an increasing number of production outfits are shooting HD as a way to future-proof their content. Perhaps the biggest push to HD has come from the U.S. Public Broadcasting System, which has mandated 16:9 HD production for the last few years (which is why so much PBS programming is letterboxed when viewed on a conventional TV).

# 80   Chapter 6: Capture and Digitization

Capturing DTV is becoming possible, with products such as Pinnacle's CineWave board. While SDTV-quality Web video still sounds like science fiction, it could become a reality in the next few years as bandwidth and codec efficiency both improve. And, as happened with CD-ROM video, broadcast quality is only a stop along the quality growth curve. We'll start chafing at Standard Definition source before long, and we will be able to take advantage of DTV resolutions for capture.

 *See also*

Refer to Table 18.2, "Compression format constraints." on page 294.

## Tape Formats

**VHS**   VHS (Video Home System) is a horrible, horrible video standard. While some folks talk about Web video approaching "VHS quality," they're just revealing how low they've set their sights. VHS has extremely poor color resolution, and is extremely noisy. People just watching a VHS movie for fun don't notice how nasty it looks. Personally, I find the poor quality too distracting to be able to enjoy anything on VHS (which my wife gently mocks me for). But when the content is captured and displayed on a computer screen, the results are wretched.

Sometimes content is only available on VHS, and if that's all you got, that's all you got. But any kind of professionally produced content was originally mastered on another format. It's almost always worth it to track down a copy of the original master tape, be it on Beta SP or another format.

VHS supports several speed modes, the standard SP, the rare LP, and the horrible (and horribly common) EP. Because EP runs at one-third the speed of SP, you can get three times as much video on a single tape, but triple the amount of noise. Tapes recorded in EP mode are more likely to be unplayable on VCRs other than the machine they were recorded on due to tracking errors. Most professional decks can only read SP tapes, so if you have a tape that won't work in your good deck, try putting it into a consumer VHS deck and see if it's in EP mode.

VHS can have linear and HiFi audio tracks. The HiFi is generally much, much better, approaching CD quality. Linear tracks may be encoded with Dolby Digital—if so recorded, playing back in Dolby mode can help quality and fidelity.

**S-VHS**   S-VHS (the 'S' stands for Super) handles saturated colors better than VHS through improved signal processing. S-VHS was never a popular consumer standard, but was used quite often in prosumer/industrial production environments by wedding and corporate event videographers. Before the advent of miniDV, it was used as sub-masters from which viewing copies of master tapes were duplicated. S-VHS's quality is often compared to Hi8, but S-VHS is a *much* more durable tape format that lives on 1/2-inch instead of 8mm tape, so it was used for editing as well as acquisition.

If you're thinking of buying a VHS deck for your facility, I suggest spending a little more to get an S-VHS deck. Beware of consumer "Pseudo-S" decks, which can play back S-VHS tapes, but with VHS quality. Note VHS decks cannot play back S-VHS tapes, although S-VHS tapes can play back and record normal VHS.

Good professional/industrial S-VHS decks support device control, have an internal Time Base Corrector, and sport balanced outputs. Consumer decks aren't suitable for anything beyond light, non-critical capture.

*8mm*  8mm is better than VHS, although not by a lot. Again, 8mm should never be used for production, but if a client hands you a project on 8mm, you'll need to be able to play them back. Note Digital-8 players can play both 8mm and Hi8 tapes.

*Hi8*  Hi8 is an enhanced version of 8mm and uses the same type of signal processing as S-VHS to achieve its better color resolution. However, the physical tape format itself wasn't originally designed for carrying information as dense as the format calls for. This means the video is very sensitive to damage, often referred to as *hits*—long horizontal lines of noise lasting for a single frame. Every time the tape is played, the wear can cause additional damage. If you're working with a Hi8 tape, it's often a good idea to first dub the source to another tape format, and then capture from the fresh copy, so you don't have to worry about progressive damage. *Do not* spend a lot of time shuttling back and forth to set in- and out-points on a Hi8 tape.

The Hi8 market was quickly replaced by DV, but there's a lot of legacy Hi8 content still out there. This is mostly home movies and the like, but in the early days of the Web, Hi8 was the Web monkey's format of choice.

*3/4 Umatic*  3/4 (named after the width of its tape) was the Betacam of its day in that it was the default production standard. Although not particularly high quality, it didn't suffer much from generation loss, and the tapes themselves were very durable. There's lots of archive content in this format.

*3/4 SP*  3/4 SP was an enhancement of 3/4 offering much improved color. There's a lot of footage in this format floating around from before Betacam was the standard.

*Betacam*  Betacam, based around the Betamax tape format, provided much higher quality by running several times faster, increasing the amount of tape assigned to each frame of video. The small tape size of Betacam made it a hit with ENG (Electronic News Gathering) crews. Betacam has been largely replaced by Betacam SP.

There are two sizes of Betacam tapes, one a little smaller (though thicker) than a standard VHS tape, the other about twice as large. Generally, cameras only use the smaller size, but decks can use either size.

*Betacam SP*   Betacam SP, as you probably guessed, is an enhanced version of Betacam. It quickly eclipsed the original Betacam. While S-VHS and Hi8 arguably have higher color resolution, SP's low noise, tape durability, and native component color space make Beta SP the professional analog tape format of choice for the past decade. For producing content that will be compressed, Betacam SP is the only "good enough" analog standard.

You need to capture in component analog to take real advantage of Beta SP.

*Digital Betacam*   Digital Betacam (often referred to as D-Beta or Digi-Beta) is Sony's current high-end tape format. It offers effectively uncompressed digital quality. As an added bonus, D-Beta stores 10-bit per channel color, a step up from the 8-bit component video standard on miniDV, DVCAM, DVCPRO, and other digital formats. Ten-bit per channel offers a quality improvement that's most visible in subtle gradients and the like. Alas, most NLE and compression systems only process 8-bit per channel.

D-Beta decks can read analog Beta SP tapes, although they can't write to them.

*Betacam SX*   This weird hybrid format sits somewhere between Betacam SP and Digital Betacam. It uses interframe MPEG-2 compression at a high bitrate. SX offers digital connectivity via SDI, but with lower quality (and prices) compared to D-Beta. The DV format explosion occurred at nearly the same time as the introduction of SX, and kept SX from becoming particularly popular.

*DV25*   The result of collaboration among Hitachi, JVC, Sony, Matsushita, Philips, Sanyo, Sharp, Thompson, and Toshiba, DV25 is the generic name of DV (digital video) running at 25Mbps. DV25 comes in a variety of flavors (described as follows) that share the same underlying bitstream. DV compression is roughly 6.5:1 and DCT-based. The differences among the DV variants are mainly in tape durability and length.

DV is an 8-bit component video format. The color sampling for NTSC DV25 is 4:1:1; the PAL version is 4:2:0.

DV cameras offer good quality at affordable prices. Combined with computers and built-in IEEE 1394 (FireWire) support, DV25 led the digital revolution in video production. Before DV, "good enough" digital video required a minimum $80K investment. Now, "good enough" can be had for barely $5K.

*MiniDV*  Officially, miniDV is the consumer format of DV25 tapes. MiniDV is sometimes referred to simply as DV. It's used on all consumer DV cameras and on some professional models, including the Canon XL1S and GL1.

MiniDV tape runs at two speeds, SP and LP. *Do not* record with DV in LP mode. In theory, LP mode should produce satisfactory results, as it slows the tape down but retains the same bandwidth. In practice, tapes recorded in LP mode are an interoperability nightmare—they don't play in equipment from other vendors and sometimes don't play on other machines from the same vendor. Why? One possible cause could be that in SP mode the video track size is 10 microns. Running in LP mode both increases record times from, say, 80 minutes to 120 minutes, and reduces track size to 7.9 microns. Cramming more bits into a smaller space dramatically increases the frequency of analog errors, which turn into digital errors that the error correction circuits can't compensate for.

If you have a miniDV tape that suffers from horrible, blocky quality, odds are it was recorded in LP mode. If you simply must use it, try to track down a compatible machine and dub it to a tape running in SP mode.

Audio on miniDV is switchable between two 16-bit, 48kHz or four 12-bit, 32kHz channels. Always produce in the 16-bit mode.

*DVCAM*  DVCAM is Sony's professional DV format with native DV 5:1 compression at 25Mbps. Tapes come in mini (the same as miniDV) and the slightly larger standard size. Sony's DVCAM runs 50 percent faster than miniDV, so the longest miniDV tape will only record a maximum of 40 minutes of DVCAM footage, where the same tape in DV format would hold 80 minutes. Compared to miniDV, DVCAM offers a wider, more durable tape formulation (ME, metal evaporated) and longer maximum record times (there is a DVCAM tape that can hold 184 minutes of material). Note DVCAM records with a 15-micron track size, so while it's possible to put a miniDV tape recorded in DVCAM format in a miniDV camera or deck, playback of DVCAM format material won't work due to the difference in track size.

Some DVCAM tapes have a microchip on them. Some cameras use these chips to store settings such as aperture, f-stop, and other data. While they were an interesting idea, not many cameras make use of them. Save some money and buy chip-free tapes.

DVCAM audio, like miniDV audio, is switchable between two 16-bit, 48kHz or four 12-bit, 32kHz tracks.

*DVCPRO*  DVCPRO is Panasonic's professional DV format. At 18 microns, its track size is slightly larger than that of DVCAM. DVCPRO runs faster than DVCAM, and requires a different tape formulation (MP, metal particle). Also, the physical size of a medium-length tape is between that of miniDV- and standard-sized DVCAM tapes. Note DVCPRO decks can play back (but not record) miniDV tapes via an adapter.

DVCPRO tapes don't sport chips like some DVCAM tapes do. Panasonic jumped the gun a bit in their PAL version, and it initially came out in 4:1:1, like the NTSC version. Once the PAL DV standard was defined as 4:2:0, Panasonic added a mode that supports 4:2:0 to their decks.

DVCPRO audio is two 16-bit, 48kHz tracks and an analog cue track.

***DV50*** DV50 is the generic name for the 50Mbps, 3.3:1 compression, 4:2:2 bitstream, introduced with JVC's Digital-S format (now called D9) and used in DVCPRO50.

***DVCPRO50*** DVCPRO50 is Panasonic's trade name for its DV50 solution. It uses standard DVCPRO tapes, although they store only half the duration as they would in DV25 because the data rate is twice as high.

***DVCPRO HD (DVCPRO100)*** The 100Mbps, 1280×720 HD variation of DVCPRO50. Its color sampling is 4:2:0, and video compression is 6.7:1. It features eight 16-bit, 48kHz audio tracks.

***D1*** D1 is an uncompressed 4:2:2, 8-bit per channel tape format of ITU-R BT.601. D1 tapes hold up to 94 minutes of content. D1 is extremely high-end and expensive, although its quality cannot be faulted. For most uses, D-Beta is just as good, and quite a lot cheaper.

***D2*** D2 is an anomaly—it's a digital *composite* format. D2 just digital sampled the waveform of the composite signal itself. While this was a lot cheaper to implement than D1, it also meant that the signal had to be converted to composite first, throwing a lot of quality out. D2 VTRs haven't been produced in years, so you'll rarely encounter it in production, only as a legacy format.

***D3*** A version of D2 that runs on half-inch tape cassettes. Tapes can be up to 245 minutes in length; even rarer than D2.

***D5*** D5 was Panasonic's attempt at reaching the D-Beta market. It's great, but never really caught on. It uses the same half-inch cassettes as D3 and D5 decks can play back D3 tapes and provide component output. D5 is a 10-bit, 4:2:2 format. The tape format is now being used in the increasingly popular D5 HD.

***D5 HD*** The popular high-definition version of D5 is a 10-bit, 4:2:2, component video format using roughly 4:1 compression.

***D9 (Digital-S)*** Originally named Digital-S, D9 from JVC was originally touted as an upgrade path for S-VHS users and was the first DV50 format. D9 runs on half-inch metal particle tape that uses the same housing as VHS tapes. Two 16-bit, 48kHz digital audio channels are supported, along with two cue tracks and four audio tracks.

## Connections

While NTSC video is intrinsically Y'CbCr, the manner in which the video signal is encoded, both on a tape and for transmission, can have dramatic effects on the quality and range of colors being reproduced. Both the type of connection and the quality of the cabling are important.

*Coaxial*   Anyone with cable television in their home has dealt with coaxial connections, which consist of BNC (for Bayonet-Neill-Conselman, Bayonet for the business end of the male connector, Neill and Conselman being the guys who invented it) connectors with a wire protruding from the center of the male end and a twist top that locks. Coax carries the full bandwidth and signal of broadcast television on its single copper wire surrounded by shielding, so it can carry multiple channels simultaneously. Coax carries not just composite video signals—audio signals are mixed with the video signal for broadcast television. This can cause a lot of interference—intense video can leak into the audio channel.

*Composite*   The composite (video) format has historically been the method of choice for transmitting and storing video in most consumer video devices. In composite video, luminance and chrominance information are combined on a single signal, in exactly the same way that they are combined for broadcast. Alas, the way they are combined makes it impossible to separate the signals perfectly, so information can erroneously leak from one domain to the other. This is the source of the colored shimmer on Paul's coat in *Perry Mason* reruns. You may also see strange flickering, or a checkerboard pattern, in areas of highly saturated color.

How well equipment separates the signals is one of the big factors in determining how good consumer video equipment is. Cheaper stuff uses a notch filter, and higher end gear uses the often mentioned, rarely explained, and dramatically superior comb filter. There are a lot of different types of comb filters, and just as many arguments between engineers on the price/performance of different approaches. Fortunately, there's a simple answer. No video encoding or compression system should use composite signals.

In consumer equipment, composite is transmitted through an RCA jack, which is usually color-coded yellow. Professional systems use a BNC connector without color-coding.

*S-Video (Y/C)*   S-Video (also called Y/C) is a leap in quality over composite. With S-Video, luma and chroma information are carried on separate connections. This eliminates the *Perry Mason* effect when dealing with a clean source. However, since the chroma signals can interfere with each other, it still isn't quite optimum.

Virtually all consumer and prosumer analog video tapes (VHS, S-VHS, Hi8) record luma and chroma at separate frequencies. When going to a composite output, luma and chroma are mixed back together. So, even when dealing with a low-quality standard like VHS, you can see substantial improvements by working with signals sent via S-Video, because you can avoid continually mixing and unmixing the luma and chroma info.

S-Video connectors are also called mini-DIN connectors. Tip: in a pinch, old Macintosh ADB keyboard cables work as S-Video connectors, and vice versa.

*Component Y'CbCr* Component (video) is the only true professional analog video format. In component, Y, Cb, and Cr are stored and transmitted independently of each other, preserving very high quality. Component looks great, but you pay for it in substantially higher equipment and media costs. Betacam SP is the only modern tape format that uses component analog.

When color-coding cables for Y'CbCr, Y=Green, Cb=Blue, and Cr=Red. This is based on the major color contributor to each signal, as described in Chapter 3, "Fundamentals of Compression."

*Component RGB* As its name implies, Component RGB is the color space of component video. It isn't used as a storage format. It is a transmission format for all computer monitors and some high-end consumer video gear. Unless you're trying to capture from one of those sources, there is generally no need to go through Component RGB.

*IEEE 1394* IEEE 1394 (1394 for short) is a high-speed serial bus protocol developed by Apple Computer. IEEE 1394 is also known by two trademarked names, Apple's FireWire and Sony's i.Link. The initial versions of 1394 ran at 100 or 200Mbps. Current implementations run at 400Mbps, more than enough bandwidth in theory for even uncompressed video. Upcoming versions will support 800 and 1600 Mbps. 1394 is best known as the standard cabling for DV25 systems. Strikingly elegant compared to the tangles of cable other formats require, 1394 supports bidirectional digital transmission of audio, video, and device control information over a single cable. The plug-and-play simplicity, combined with the good enough quality of the DV format itself, has led to a revolution in prosumer, industrial, and event videography industries. Panasonic and Apple are working on adding support for DV50 and DV100 over FireWire.

All Macintosh computers have had built-in FireWire since 1999. It is also available either built-in or as an option on many Windows systems. 1394 is certainly the value leader for digitization systems, in that it provides low-end professional quality at a wonderfully low price.

1394 connectors come in two flavors, 4-pin and 6-pin. Most DV cameras require 4-pin connectors, while computers and other 1394-equipped devices use the 6-pin variety. Camera-to-computer connections require a 4-pin-to-8-pin cable. Note the extra two pins on 6-pin cables are used to carry power to devices chained together via 1394.

*SDI*   Serial Digital Interface connections transmit uncompressed 720×486 (NTSC) video at either 8-bit or 10-bit per channel resolution at up to 270Mbps. SDI can handle both component and composite video, as well as four groups of four channels of embedded digital audio. Thus, SDI transfers give you perfect reproduction of the content on the tape. Due to the wonders of Moore's Law, you can now buy an SDI capture board for less than a good component analog card. And SDI prices are going to continue to drop rapidly.

All modern digital formats are intrinsically component (some older formats like D2 simply digitized the composite signal stream). So even if you wind up in the analog domain after doing digital acquisition, you'll see better results than with component analog.

SDI uses BNC connectors and coax cable.

## Audio Connections

Like most things in the video world, audio often gets the short shrift. However, it is critically important to pay attention to audio when targeting compressed video. Audio is often the only element you have a hope of reproducing accurately. Plus, extraneous noise can dramatically hurt compression, so it's even more important to keep audio captures noise-free.

*Unbalanced*   Unbalanced analog audio is the format of home audio equipment, where RCA plugs and jacks are the connectors of choice. In audio gear designed for home and project studios, unbalanced signals are often transmitted over cables fitted with 1/4-inch phone plugs. Those audio inputs on computer sound cards are unbalanced mini jacks, like those used for portable headphones. Note mini connectors come in two sizes. It's nearly impossible to see the slight different in size, but they aren't compatible.

Unbalanced audio cables come in both mono and stereo flavors. Mono mini and 1/4-inch plugs use tip (hot), ring (ground) wiring. Stereo mini and 1/4-inch plugs use the same configuration, but sport two rings.

You've no doubt noticed stereo RCA cables use separate wires for left and right channels. Their connectors are generally color-coded red for right and black or white for left.

So what's an unbalanced audio signal? Unbalanced cables run a hot (+) signal and a ground down a single, shielded wire. Such cables have the unfortunate ability to act as antennas, picking up all sorts of noise including 60Hz hum (50 if you're in a country running 50Hz power) and RF (radio frequency interference). The shielding is supposed to cut down on the noise. The longer the unbalanced cord, the more it's likely to behave like an antenna. Keep your unbalanced cables as short as possible. And when confronted with the choice, buy shielded cables for audio connections between devices. Some say you can skip the shielding for speaker wires.

**Note**

**Jacks and Plugs.** It's easy to confuse these two terms, and they're often incorrectly used interchangeably. So, for the record: *Plugs* are male connectors. *Jacks* are female connectors that plugs are "plugged" into.

*Balanced*   Balanced, like unbalanced, sends a single signal over a single cable, however that cable houses three wires, not two. The signal flows positive on one wire, negative on another, and to ground on the third. Note the signals aren't strictly positive or negative, but rather alternating current, but they are 180 degrees out of phase. I won't burden you with the details, but this design enables balanced wiring to reject noise and interference much, much better than unbalanced.

Balanced connections are highly desirable in electromagnetically noisy environments, such as live production environments or video editing studios. They're also helpful in situations where long cable runs are required.

Balanced audio connections are most often thought of as using XLR connectors, but quarter-inch phone connectors can also be wired to carry balanced signals. XLRs are also called Cannon connectors after the company that invented them—ITT-Cannon. The company made a small X series connector to which they added a latch. Then a resilient rubber compound was added to the female end, and the XLR connector was born.

The difference in quality between balanced and unbalanced audio isn't as pronounced as the difference between component and composite video, but balanced audio can be a lifesaver in noisy environments.

All professional video cameras feature balanced audio connectors.

 **Note**

**Levels.** It's worth noting that matching levels plays an important role in getting the best audio quality. Professional audio equipment typically runs at +4dBm (sometimes referred to as +4dBu) and consumer/prosumer gear runs at –10dBV. Without getting immersed in the voodoo of decibels, suffice to say that these two levels (though measured against different reference points) represent a difference in voltage that can lead to signals that are either too quiet (and thus noisy) or too loud (and very distorted).

*S/PDIF*   S/PDIF (Sony/Philips Digital InterFace) is a consumer digital audio format. It's used for transmitting compressed or uncompressed audio. Because it's digital, there's no quality reason to avoid it in favor of the professional standard AES/EBU, although that format offers other useful features for audio professionals that don't relate to capture.

S/PDIF cables are coax terminated in either phone or BNC connectors. For most purposes S/PDIF and AES/EBU cables are interchangeable. If, however, you're connecting 20- or 24-bit devices, you'll likely want to differentiate the cables by matching their impedance rating to that of the format. 75-ohm for S/PDIF; 110-ohm for AES/EBU.

*AES/EBU*  AES/EBU, from the Audio Engineering Society and the European Broadcast Union, is the professional digital audio format. Of course, its actual content is no different than that of S/PDIF. The advantage of AES/EBU comes from being able to carry metadata information, such as timecode, along with the signal. If you're using SDI for video, you're probably using AES/EBU for audio.

AES/EBU 110-ohm impedance is usually terminated with XLR connectors.

*Optical*  Optical audio connections, under the name TOSLink are proof P. T. Barnum's theory that a sucker is born every minute is true. The marketing behind optical proclaimed because it transmitted the actual optical information off a CD, it would avoid the distortion caused by the D/A conversions. Whatever slight advantage this may have once offered, audio formats like S/PDIF offer perfect quality from the digital source at a much lower cost.

That said, some consumer equipment (like my DishNetwork PVR) offers an optical TOSLink connection as its only digital audio format.

# Capture Resolution

The three standard NTSC full-screen capture resolutions are 640×480, 720×480 (DV), and 720×486. In PAL, the resolutions are 720×576 (DV) and 768×576.

Defining resolution when going from analog to digital is tricky. In essence, the resolution of the source is explicit vertically, in terms of numbers of lines, since each line is a discrete unit in the original. But resolution is undefined horizontally, since each line itself is an analog waveform. So, 640×480 and 720×480 capture the same number of lines. And 720×486 captures six more lines from the source than 720×480—each line of those 480 is included in the 486, and six must be added to convert from 480 to 486. However, capturing 720 pixels per line doesn't mean you're capturing more of each line than in 640 pixels per line—you are just capturing more samples per line, and so have tall, skinny pixels.

This is arguably a good thing. Because there is no way to increase the resolution vertically—you can't take more line samples—the only way to improve resolution is to anamorphically increase the resolution horizontally.

For 4:3 source, 640×480 is the simplest resolution to work with, as its pixels are square. Some, but not all, authoring tools can automatically correct for non-square pixels, so they display the video at its actual aspect ratio. Unless you're going to particularly high resolutions, having more samples on each line doesn't have a significant effect on quality. This is especially true with video source, where the constraints of having 240 or so lines of source per field are a much larger limitation.

## Device Control and Timecode

### Device control

The ability to control a deck (or camera in deck mode) directly from a computer, without having to actually press any buttons on the devices itself, is called *device control*. Professional and an increasing number of consumer decks support device control. Device control and timecode go hand-in-hand. With both of these, you can do precise captures solely from a computer. With software able to batch capture, you don't even need to be there—the computer can automatically capture all the clips you need from a particular tape.

There are several different types of device control. 1394 has one built in. Most professional systems use RS-232, which is a good basic protocol. Higher end systems include RS-422 and RS-485, both of which support much longer cable lengths and multiple devices on the same signal chain. There are also consumer formats like Control-S, but these generally aren't frame-accurate. And frame accuracy is critical for professional work.

### Timecode

Timecode is a method of assigning each frame of video a unique address. The taxonomy of that address is: hour:minute:second:frame. Timecode used to keep audio and video in sync, as well as to provide a means of location events. Timecode is also called SMPTE timecode. There are two types of SMPTE timecode—LTC and VITC. LTC, for Longitudinal TimeCode, is a digital signal that's recorded either on a video deck's audio tracks or on a special track set aside for timecode. It's called Longitudinal because the tracks it can be recorded on run parallel to the video tracks.

LTC is pretty durable, and can be read while you're fast-forwarding or rewinding through a tape. But it disappears when the tape is paused or shuttling slowly.

VITC, for Vertical Interval TimeCode (pronounced "vit-see"), is a series of white dots (representing the same timecode as LTC) written between video frames. VITC has the benefit of coexisting with the video, so it can be read whenever the video is viewable, such as when it's paused or you're stepping through it one frame at a time. Most professional decks read both types of timecode.

If it weren't for timecode and device control, automated capture wouldn't be possible.

One problem you're likely to encounter frequently is tapes with non-continuous timecode. This occurs when, during a shoot, a camera is turned off between scenes, resetting the timecode back to 00:00:00:00. Most capture systems won't capture across a timecode break, so you'll either

## Data Rates

One critical issue with all capture is the data rate it's captured at. With DV, it's easy: DV is a locked 25MBps (3.125Megabytes per second). With uncompressed capture, it's also limited, generally, to 168MBps. Losslessly compressed systems like Media 100s average around 80MBps. Good enough quality with Motion-JPEG systems starts around 48MBps with most systems, and usable stuff can often be captured at 25MBps. Generally, the less noise in the source, the higher the quality. A 20MBps component capture easily beats a 30MBps composite capture of the same content.

 **Note**

***Dropframe or Non-dropframe.*** NTSC video if often described as running at 30fps. In the days prior to color television, NTSC *did* run at 30fps. These days, it runs at 29.97fps. When color was added to the signal, it was critical that color broadcasts be backward compatible with black and white sets. To accomplish this, they stretched the frame rate out to 29.97 and put the color signal in the extra space, which worked great until they invented timecode.

In the new, stretched-out 29.97fps rate, when the time clock is ready to count a new second, the last .97 of the 29$^{th}$ frame is still playing. If the timecode is allowed to roll right through this creep, to borrow an example from fellow *DV Magazine* Contributor, Jay Rose, a program set to start at 1:00:00 timecode and end at 1:59:59 would actually end a few seconds (3.6 seconds for you trivia buffs) after 2:00.

The solution to this conundrum was to periodically leave out two numbers from the timecode per minute. Counting reverts to normal six times an hour, on every minute ending in a zero to make up for the previous operation resulting in being 12 frames too short every hour. And that's what "dropframe" timecode is. No frames are actually dropped.

Complicated? Yep. And that's why nondropframe timecode is often preferred when run length isn't critical and for short form projects.

Both dropframe and nondrop run at 29.97fps, so their tapes are interchangeable. Better NLEs allow you to mix footage stamped with different timecode formats within the same project.

With compressed systems, the actual output data rate may be less than the requested data rate if some frames of video are very easy to compress, like black frames. This is a good thing, and no cause for concern. The higher the target data rate, the bigger the percentage gap between the requested and actual data rate. This is essentially like a free 1-pass VBR.

The three limiting factors in data rate are drive space, drive speed, and capture codec.

## Capture codec

Some capture codecs, especially in lower-end systems, have a maximum data rate they can pump out. There's no real way around that except to purchase a different board. In these cases, it's better to capture at 640×480 instead of 720×480 or 486, to increase the number of bits per pixel.

## Drive speed

If access to the hard drive isn't fast enough to capture all the video, frames may be dropped. This can be a serious, jarring quality problem. Always make sure your capture program is set to alert you when frames have been dropped.

Back in the day, this was The Big Issue, and the most expensive to address. High-end systems typically use RAID systems running off a high-speed SCSI drive. These may not be needed for straight captures—a lot of systems need to be able to read two simultaneous streams of video off the disk at the same time in order to perform real-time effects, which requires a lot more throughput than just recording a single stream.

Many people have reported good success capturing to inexpensive IDE RAID systems, and even capturing to a single very fast IDE drive. A single fast drive is clearly fast enough for DV today, and for some Motion-JPEG applications. If you're choosing a hard drive to capture to, be safe and get a drive that at the least meets the minimum speed defined in your capture card's system requirements.

## Drive space

The good news is that drives are cheaper than ever. The bad news is that they still aren't free. For long-form projects, like movies, captures can easily take up 80GB of space in a high quality mode. And typical high-end capture drives are SCSI, not IDE, so they're a *lot* more expensive. This is where IDE RAIDs start looking all the better.

## Optimizing capture

If drive speed or space is a significant issue, there are some things you can do to reduce the data rate. For one, use 640×480 instead of 720×486. This will let you drop data rate about 10 percent and preserve the same overall quality. If you are delivering to output resolutions below 320×240, a quarter-screen capture at 320×240 single field will cut the data rate by 75 percent and save on some preprocessing time. Capturing at a lower frame rate can also help. However, if you want to do high-end preprocessing, like inverse telecine, you'll need both fields of every frame for that to work.

## *The 2GB Limit*

One classic problem in digital video is that computer file systems have been unable to handle files larger than 2GB. When these systems, which included Windows and Mac OS, were originally designed, this kind of file size was considered absurdly large. My first digital video system had a 4GB RAID formatted into two partitions of 2GB, so it this problem didn't even come up.

With drive space so limited, most editors (or their assistants) would log the tape first, only capturing the actual scenes they wanted to use, keeping the individual files down to a reasonable size. And they probably wouldn't ever render to a file, choosing instead to output directly to tape.

But as folks started to look at capturing entire tapes and do the logging, the 2GB limit became a serious issue. With DV, 2GB lasts little over nine minutes. For uncompressed 601, it's less than two minutes. This obviously won't work!

Fortunately, modern software and operating systems are able to handle larger than 2GB files. Here's a shopping list of minimum requirements; just add compatible software:

### Macintosh:

Mac OS 9.04 or later or OS X

HFS+ (a.k.a. Apple Extended) formatted hard drive

QuickTime 4.1.2 or later

### Windows:

Windows NT, 2000, or XP

NTFS 64-bit file system

# Chapter 7

# Preprocessing

## Introduction

Preprocessing encompasses everything done to the original source between the time it's handed to you and the time you hand it off to a codec. Preprocessing operations include cropping, scaling, deinterlacing, aspect ratio correction, noise reduction, image adjustment, audio normalization, and so on. It's during preprocessing that the source is optimized for delivery as compressed video. While there's no absolute right or wrong way to preprocess source, and exact settings are determined on a case-by-case basis, there are a number of important rules of thumb that will help you achieve optimal results when compressing video.

## Cropping

Cropping is basically the act of specifying a rectangular portion of the source video that will be used in final compression. It usually includes trimming out unwanted portions of the frame. Eliminating unwanted portions of the frame occurs much more frequently than you might think. This is because content originally produced on or for television typically doesn't go all the way to the frame's edge, while video produced on or for a computer monitor does. Thus when you're working with content originally designed for playback on a TV screen and are retargeted to a computer monitor, you'll be cropping. However, you don't typically crop when going from one video format to another, for example when putting DV format footage on a DVD, unless there are more lines in the source than are available on your delivery medium. More on all this follows.

### Edge blanking

Broadcast video signals are overscanned—that is, they're larger than the viewable area on a TV screen. This is done to hide irregularities in the edges of aging or poorly adjusted picture tubes on old analog TVs. The area of the video frame that carries the picture is called the *active* picture. The areas outside the active picture area are called *blanking* areas.

When describing video frame sizes, for example 720×480 or 720×468, vertical height (a.k.a. the 480 and 486 in the above sizes) is expressed as *active* lines of resolution. The active line length in our two examples is 720 luminance samples. Active lines and line length equate one-for-one to pixels in the digital domain.

So what's all that mean in practical terms? Examine a full broadcast video frame and you'll see the blanking quite clearly. It appears as black lines bordering both the left and right edges, as well as the top and bottom of the frame. Sometimes you'll also see black and white dashes at the top of the screen. These borders exist to accommodate the foibles of television picture tubes. Note motion pictures are also cropped during projection, to present a clean edge on the silver screen.

Computer monitors and similar digital devices show every pixel. LCD displays physically have one pixel per pixel of the display. CRT displays are naturally analog, but instead of cutting off the edges of the video, the whole image is shown with a black border around the edge of the monitor. So, when displayed on a computer monitor or in a media player's window on a wireless device, blanking from the analog source is extraneous, and you should crop it out. The active width of 720-pixel wide source is 704 pixels, so crop eight pixels off the left and right sides of the video.

Another problem that occurs with VHS source is *tearing*. This is a distortion in the lower portion of the frame. Such distortion usually occurs outside the viewable area of a television screen, but is ugly and quite visible when played back on a computer monitor. Crop it out as well.

Note there are no rules that say you have to crop the same amount from every side of the video—if there's a big black bar on the left and none on the right, and there's VHS tearing at the bottom but the top looks good, crop only those problem areas. However, you should make sure your aspect ratio is accurate. More on that to come.

## Safe areas

While blanking is used to accommodate the foibles of television picture tubes, video *safe areas* were designed to further protect important information from the ravages of aging analog televisions. Typical analog TVs vary in the amount of active picture they show, both between models and over the life of each set. Generally, older televisions show significantly less of the image leaving out the border. Also, portions of the image that appear near the edge of the screen become distorted, making it difficult to read text or otherwise see fine detail. Because of this, video engineers defined Action Safe and Title Safe areas to ensure that important content would be preserved within those boundaries. The action safe area is inset five percent from the left/right/top/bottom edges of the frame. Assume anything falling within this zone will not be visible on all TV sets.

Cropping **97**

**7.1** A clip with both tearing and edge blanking. The minimal cropping you'd want to pick is shown in Cleaner v5.

 ***See also***

A color version of Figure 7.1 is on page X of the color section.

Title safe is an additional five percent area inset within the action safe area. Important text is not placed outside the title safe area, as it may not be readable by all viewers.

Any television should be able to accurately display motion in the action safe area, and display text clearly in the smaller title safe area. As you can see, these areas limit the overall video safe area by a significant amount.

Whether or not you crop out any of the safe area depends on your targeted output resolution. As discussed in brain-numbing depth in the later "Scaling" section (page 105), one of our goals is not to scale the video up, but only down on both axes. Because of this, when you're targeting higher resolution video, you want to grab as much of the source video as possible. However, if you're targeting lower resolutions (generally below 320×240 for NTSC or 384×288 for PAL), you're always going to be scaling down. In those cases, use a more aggressive crop to make the elements in the middle of the screen larger. This will make them easier to see at low resolutions. If your footage was originally shot for video, the camera operators made sure any important information was within the safe areas, so this aggressive cropping can usually be done without losing anything important. If necessary, action safe almost always can be cropped. Title safe works most of the time, although you should check your video thoroughly to make sure nothing important is lost.

# Chapter 7: Preprocessing

Table 7.1   Typical crop amounts.[1]

| Resolution | Clean Aperture | | Action safe | | Title Safe | |
|---|---|---|---|---|---|---|
| | Left/Right | Top/Bottom | Left/Right | Top/Bottom | Left/Right | Top/Bottom |
| 640×480 | 0 | 0 | 32 | 24 | 64 | 48 |
| 720×480 | 0 | 0 | 36 | 24 | 72 | 24 |
| 720×486 | 0 | 4/2 | 36 | 26 | 72 | 26 |
| 720×576 | 0 | 0 | 36 | 28 | 72 | 58 |
| 768×576 | 0 | 0 | 38 | 28 | 76 | 58 |

1. Note: To keep the field order correct, you need to crop top and bottom by even numbers, which is why the 720×486 crop uses different values on the top and bottom.

## Letterboxing

Originally, standard film production was done in the 4:3 aspect ratio that television uses. But once TV caught on, film companies decided they needed a technology to differentiate themselves, and so implemented a number of widescreen formats (including Cinemascope, Vista-Vision, and Panavision) for film production. And now television productions are being shot in 16:9 in anticipation of the HDTV future.

Most content ends up getting displayed on traditional 4:3 televisions. So, how do you display a widescreen image on a smaller display? There are two basic strategies: pan and scan and letterboxing. Pan and scan entails chopping enough off the edges of the film frame for it to fit the 4:3

7.2   The safe area boundaries for a typical 4:3 source, as drawn within After Effects.

**7.3** Our sample picture is processed into 192×144 outputs with none *(upper left)*, action *(upper right)*, and title safe source cropping *(lower left)*.

window. On the high end, operators spend a lot of time trying to make this process look as good as possible, panning across the frame to capture motion, and scanning back and forth to catch the action. If you receive pan and scan source, you can treat it as any other 4:3 content in terms of cropping, except that you'll want to crop out the edge blanking. Be sure you're not removing more than the edge blanking, because some of the image is still left outside the safe area during the pan and scan process.

But, especially when dealing with source beyond 1.85:1, film purists do not find pan and scan acceptable. The initial solution was letterboxing, which displays the full width of the film frame on a 4:3 display and fills the areas left blank above and below the 16:9 image with black bars. This preserves the full image at the cost of making the overall image smaller while not filling the available on-screen real estate. Starting with Woody Allen's *Manhattan*, some directors began insisting their films be made available letterboxed. Letterboxing has since been embraced by much of the public, especially when DVD's higher resolution made the loss of image area less of a problem. DVDs can also store anamorphic content at 16:9, which reduces the size of the black bars needed for letterboxing, and hence increases the number of lines of source that can be used.

**7.4** Some standard production ratios, overlaid on a 4:3 source.

When compressing letterboxed source for computer delivery, you'll generally want to crop out the black bars and deliver the content at its native aspect ratio. One of the wonderful things about delivering on a computer is that you can use virtually any aspect ratio, so a 2.35:1 film can be delivered in its full widescreen majesty. Not only is this the more correct solution, it impresses viewers.

If a client requests you deliver the video at 4:3 or some other specific frame aspect ratio, you may have no choice but to either pan and scan the video yourself, or deliver the video with the black bars intact. Apps such as Discreet CineStream and Adobe After Effects provide good pan and scan tools, although the pan and scan process is inevitably labor intensive, if you want to do it well. I've found that clients who insist all content be 4:3 tend to quickly change their minds when they see a realistic pan and scan bid.

If you need to deliver in a set window size, a better option can be to encode the video at the correct frame width and let the height be less. By scripting the Web plug-in, you can usually specify that the video always appears vertically centered in the frame. This will keep the frame size constant, while still preserving the full aspect ratio.

If you need to leave the bars in, use a black restore filter to make the video signal that contains the bars a single, mathematically uniform color for more efficient encoding. It also enables the black bars to match other on-screen black elements. However, leaving the black bars visible is not the ideal solution. Because of the sharpness of the edge, codecs often have difficulty correctly encoding the edge area. The result is some video leaks into the letterboxing area and some of the black leaks below it.

If you have to letterbox, try to get the boundary between the black and the image to fall along the 8×8 grid most DCT codecs use for luma blocks. And for a slight improvement beyond that, align along the 16×16 grid of macroblocks. This will allow complete blocks of 0s, which are much easier to compress. Even with this, some bandwidth and CPU power is wasted on encoding and decoding the section.

## Deinterlacing and Inverse Telecine

As we learned in Chapter 5, most video is interlaced at its inception. This means that the even and the odd lines of video were captured 1/59.94th of a second (or 1/50th in PAL) apart. Because this is the norm for video viewed on a television screen, leave the video interlaced if your target playback device is DVD. However, all the other delivery formats discussed here require delivery in progressive scan mode.

Depending on the nature of your source, you will need to do one of three things to preprocess your footage for compression—deinterlace, inverse telecine, or nothing. Failure to deinterlace an interlaced frame results in severe artifacts for all moving objects, where a doubled image is seen.

**7.5** With and without aligning letterboxing edges along 16×16 blocks. Note the errors above and below the black edge with the unaligned image *(bottom)*.

At higher resolutions, the actual interlacing pattern can be seen as a series of alternating lines showing the video in both source fields. This looks bad in and of itself. It also hurts compression in two ways. First, the high detail in the interlaced area requires a lot of bandwidth. Second, moving objects split into two and merge back into one as they start and stop moving, wreaking havoc with motion estimation.

### See also

The same frame of video, as interlaced source, deinterlaced, and then both compressed to JPEG at the same file size is shown in Figure O on page XI of the color section.

## Interlaced Video—Deinterlacing

You can tell if content was originally shot on video because every frame with motion in it will show interlacing. Prior to compression, content produced and edited as interlaced should be deinterlaced. The most basic form of deinterlacing is to simply throw out one of the two fields. This is very fast, and so is used by most hardware and low-end software solutions. The drawback is that half of the resolution is simply thrown away. For NTSC DV source, this means you're really starting with a 720×240 bitmap of each frame. Throw in a little action safe cropping, and you're actually having to scale *up* vertically from 648×216 to make a 320×240 image, even though you're scaling down horizontally. This produces a stair-stepping effect on sharp edges, especially diagonals, when going to higher resolutions. Such aliasing, however, looks better than deinterlacing artifacts.

Some advanced deinterlacing algorithms, like Cleaner's Adaptive Deinterlace and After Effects' Motion Detect mode in Interpret Footage, are able to selectively deinterlace only those parts of the image that are moving, which provides a good option when outputting 240 lines or more. Note that systems like this are never 100 percent accurate. If you're having strange quality problems in areas of intense motion, you might try rendering out that section to an uncompressed codec to see if the issues are with adaptive deinterlacing or with the codec.

Hardware-based transcoding systems use motion-tracking in adaptive deinterlacing, which means the two fields of moving objects can actually be reassembled. Hopefully we'll see this incorporated in software encoding tools of the future.

One solution I *don't* like for deinterlacing is "blend" algorithms. These simply blur two fields together. This blending eliminates the fine horizontal lines of deinterlacing, but leaves different parts of the image still overlapping. While better than no deinterlacing at all, this blending makes motion estimation very difficult for the codec, substantially reducing compression efficiency. Plus it looks dumb.

# Deinterlacing and Inverse Telecine 103

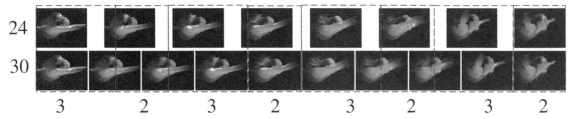

**7.6** The relationship between frames of film source *(top)* and output video *(bottom)*. Note the 3:2 pattern of which video fields are based on which film frames, and the repeating pattern of three frames of progressive and two of interlaced video output.

## See also
The horror of a blended deinterlace is shown in Figure P in the color section (page XII).

## Telecined Video—Inverse Telecine

Content originally produced on film and transferred to NTSC video via the telecine process should be run through an inverse telecine. As we learned in Chapter 6, telecine takes the progressive frames of film and converts them into fields, with the first frame becoming three fields, the next frame two fields, then three, then two, and repeating. This is called 3:2 pull-down. The great thing about telecined content is that reversing it restores a progressive scan source to its native resolution and speed, meaning the output will actually have more accurate motion than it did on video. You can recognize telecined content by stepping through the source one frame at a time. You'll see a repeating pattern of three progressive frames followed by two interlaced frames. This 3:2 pattern is a good mnemonic to remember with 3:2 pulldown.

Bear in mind the film and its corresponding audio are slowed down by 1001/1000, the same amount 29.97 is slowed down from 30fps. This means the 3:2 pull-down pattern can be continued indefinitely, without any adjustments for the 29.97. However, it also means the inverse telecine process leaves the frame rate at 23.976, not 24. That 23.976 value is the frame rate you should use for any further compressions of inverse telecined content.

One gotcha to watch out for is with video that was originally telecined, but was then edited and composited in a postproduction environment set up for interlaced video. The two big problems are cadence breaks and graphics mismatches. In a cadence break, a video edit doesn't respect the 3:2 pull-down order, and might wind up with only a single field remaining from a particular film frame. In a graphics mismatch, a frame of graphic doesn't match the movement of the film source underneath. This could be due to the video having been field-rendered, so you could have a title that moves every field, while the underlying film source only changes every two or three fields. The result of this is an odd mish-mash of both types. There isn't really any straightforward way to deal with this kind of source beyond having an inverse telecine algorithm smart enough to reassemble the source based on the underlying video.

## Mixed source

What to do with mixed film/video source? If the video content is predominant, your most likely choice is to treat all of your source as interlaced video. If the film content is predominant, you could try to use an inverse telecine filter. The only one I've seen that can deal with substantially messed-up source is Sorenson Video's Squeeze. The inverse telecine filter in After Effects is infamous for not being able to recover from *any* cadence breaks.

Ideally, a compression tool would be able to vary the output frame rate between sections when going to a format such as QuickTime that supports arbitrary frame timing.

Delightfully, film transferred to PAL doesn't have to run through any of that 3:2 rigmarole. Instead, the 24fps source is sped up 4 percent, and turned into 25fps progressive. Yes, it's progressive! And sports no decimals. The only warning is that if someone does compositing or other processing to that source with fields turned on, it'll ruin the progressive scan effect. Generally, adaptive deinterlace filters à la Cleaner do an excellent job with this kind of footage, restoring almost all the original 25p goodness. This is only one of the many delights of PAL that leave me fantasizing about moving to Europe.

# Progressive Scan—Perfection Incarnate

The best option, of course, is to get your video in progressive scan in the first place. This is generally done by shooting and editing the video in progressive scan mode (which is increasingly easy with modern cameras and NLE systems). Note capturing in progressive is unnecessary—as long as the frames on the tape are progressive, you can treat the footage as progressive. Except for DV, most formats don't support an explicit progressive mode. Instead, progressive mode just ensures that the content going into each field comes from the same capture.

While it is safe to shoot projects intended for compressed delivery in progressive, programs meant for both interlaced and progressive playback should be shot as interlaced, because progressive video on an interlaced display looks quite odd. It's not that it looks bad, but it looks different from what viewers are used to seeing, and they'll usually sense something is off about the video even if they don't understand what.

When working with progressive source, make sure any compositing and rendering is done in progressive scan as well. All it takes is one field-rendered moving title to throw the whole thing off.

# Scaling

Cropping defines the source rectangle. Scaling defines the size and the shape of the output rectangle. Scaling can be a little trickier than it sounds because source is often not square-pixel or may have an unusual aspect ratio. The basic rule, when targeting square-pixel formats (as is the norm), is to make the aspect ratio of the output resolution match that of the input resolution.

Anyone who has tried to take a low resolution still photograph (like a DV still) and blow it up larger knows scaling can hurt image quality. Even a 2X zoom can look painfully bad with detailed content. The good news is that scaling down, when done correctly, looks great (see Figure 7.8).

Different scaling algorithms provide different scaling quality. Most tools including the Adobe Premiere and After Effects use bicubic scaling. Bicubic is fine when scaling down to at least 50 percent or up to at most 200 percent. However, bicubic scaling starts having difficulties when getting down to much bigger or smaller resolutions. The problems become most evident as shallow diagonal lines get very blocky. The sine algorithm does much better outside bicubic's useful range. Cleaner users sine scaling, and it's an option in Equilibrium DeBabelizer.

There are a few older algorithms you want to stay away from completely. Nearest neighbor is the simplest, fastest, and ugliest scaling algorithm. It simply figures out which source pixel is closest to the output pixel, and just uses that value. It doesn't sample any other pixels, and yields an extremely blocky image when going up, and can lose a lot of detail when going down. Bilinear is better, but still less than ideal. In bilinear scaling, the output pixel can be a mixture of pixels on the left and right of the source, but not up and down (see Figure 7.7).

Upscaling isn't a big issue in most compressed video today, but will matter once we're asked to start delivering HD content from SD sources.

The biggest problem we're likely to face with scaling is when taking interlaced source to progressive output. As discussed earlier, when deinterlacing with a non-adaptive system, or with a clip with high motion where adaptive deinterlacing isn't able to help, in NTSC you're really starting with 240 lines of source at best. This is the big reason progressive scan and film sources are so well suited to computer delivery. 16:9 is also good for this reason, as you spend more of your pixels horizontally than vertically. If you've only got 240 real lines, 432×240 from 16×9 can be a lot more impressive than the 320×240 you get from 4:3.

# Aspect Ratio

In recent years, many standard definition video cameras have added support to shoot in the 16:9 widescreen format of HD. This enables content to be somewhat more future proof, making it

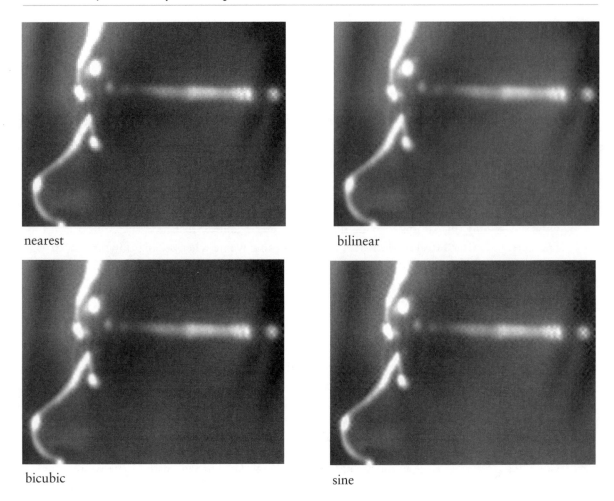

**7.7** Different scaling algorithms, showing scaling up 400%. My algorithm of choice is sine *(lower right)*, although everything is a lot better than nearest neighbor.

easier to scale up for HD in the future. Also, 16:9 is becoming increasingly common in European broadcasts.

If you are producing content for compressed delivery, and your subject suits a wider panorama (great for beach volleyball, bad for rock climbing), producing in 16:9 can give your final output a look and feel different from that of 4:3 television. And remember, everything you can do to distract the user from comparing Web video to television, the more impressive you'll look to your clients!

In figuring output resolution for computer playback, don't pay attention to the source resolution, which may or may not be square-pixel. Instead, make sure your output resolution matches your target aspect ratio. A major indicator you're doing something wrong: Compressing 720×480 source to 360×240—that's a 3:2 aspect ratio, not supported in any camera. If you have 4:3

## Aspect Ratio 107

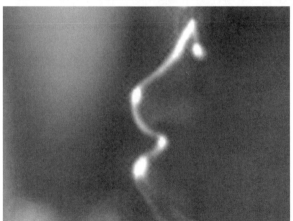

**7.8** The sample frame scaled down *(lower right)* and up *(lower left)* (well, just a section of it scaled up). Scaled down, the video looks good, but smaller. Scaled up, the limitations in the source video become painfully obvious.

source, both 640×480 and 720×480 scale down to 320×240. Lots of folks tie themselves up into algebraic contortions trying to figure out the right math for output ratio. The good news is the easiest way to do this right is to completely ignore the source resolution. Just make sure your output aspect ratio matches your source aspect ratio. If you're delivering in a square pixel format, which is typical unless you're doing MPEG-1 for VideoCD or MPEG-2 for DVD, it's easy to check. Divide the output resolution of each axis by the ratio for that axis, and make sure the numbers are identical or nearly so. So, with 4:3 going to 320×240, divide 320 by 4 and 240 by 3. In this case, both values are exactly 80.

## Pixel Shape

For strange historical reasons, the default pixel shape used in digital video is *not* square, although the pixels in all modern (post 1986) computer systems are. If you're coming from a non-square pixel format, you'll need to correct for the aspect ratio. You'll be glad to hear the best way to correct for it is to ignore the source resolution, and just pay attention to the source aspect ratio.. Define the proper output resolution for your aspect ratio, and then use it. Don't get confused by the fact the source pixels aren't square. If you're delivering to 4:3, 320×240 is a great resolution, no matter if the source video was 640×480, 720×480, 720×486, or 768×576. It's pretty common to see content from 720×480 delivered at 360×240 or 352×240, but this is almost always a mistake. If the video is 360×240 when it should be 320×240, everything is going to be 10 percent too wide—and the husky among us get rather ticked at that. The one exception is when delivering to formats that are also use ITU-R.BT601 pixels, like MPEG-1 in VideoCD mode, and MPEG-2 in DVD mode. MPEG-1 is typically 4:3, 352×240 in NTSC and 352×288 in PAL. MPEG-2 on DVD can be both 4:3 and 16:9 where appropriate. I see folks getting this wrong a lot, so I'm going to repeat it below. (See Figure 7.9.)

 *Note*

If you are scaling 720×480 down to 360×240, you're doing it wrong.

## Resolution Constraints

Once you figure out the math for the perfect size window, realize that there are some restraints in the resolution you pick. While most modern codecs let you pick any arbitrary number, it is advisable to pick a resolution divisible by four, and it's ideal to use one divisible by 16. These work better with the hardware acceleration that video cards use, and some codecs require them outright. Even if codecs can handle resolutions divisible by four, they generally calculate internally in 16×16 blocks, and store blocks that size even if the resolution isn't divisible by that. This generally means that going to the next divisible-by-16 jump in resolution is "free" quality. For example, many codecs store 240×180 internally at 240×192, with the bottom 12 pixels padded with dummy data and then not drawn on output. Going to 240×176 instead only loses four pixels on the screen, but saves 16 in compression and decompression.

**7.9** A 4:3 *(upper left)* and a 16:9 *(upper right)* source, shown initially as 720×480 square pixel, and corrected to both 4:3 and 16:9. Note both the raw pixels and using the wrong correction give bad results.

Targeting resolutions divisible by 16 is enough of a jump that it can be tricky to get the aspect ratio accurate enough for the video to look good at lower resolutions. Sometimes the best solution is to pick the closest aspect ratio, and then adjust your cropping slightly to match. So, if you're coming from 640×480 and targeting 240×176, you'd crop an extra six pixels each off the top and bottom from the source to maintain the aspect ratio. Generally, errors in aspect ratio of less than 5 percent are difficult to detect, so if you're using an encoder that can't crop, leaving those six lines in probably would look okay. If you're cropping with a tool that lets you graphically set cropping with a constrained aspect ratio, use your output resolution for the ratio.

## Noise Reduction

Noise reduction refers to a broad set of filters and techniques that are used to remove errors from video frames. These errors encompass all the ways video goes wrong—composite noise artifacts, film grain and video grain from shooting in low light, film scratches, artifacts from DV compression, and so on. While noise reduction can help some, it never results in content as good as if it were produced cleanly in the first place. But when all the video editor hands you is lemons, you've got to make lemonade, even if there is never enough sugar.

The noise reduction features in most compression-oriented tools tend to be quite basic—they're generally simple low-pass filters and blurs. Cleaner has a filter called Dynamic Noise Reduction that tries to only blur noisy parts of the image while leaving the cleaner parts alone. Cleaner's Dynamic Noise Reduction includes a Temporal Processing option, which compares the pixel not only to surrounding pixels, but to pixels in the previous and following frames. This can be very effective with the kind of fine random noise from shooting in low light.

The drawback in all noise reduction filters is that they make the video softer. Of course, softer video is easier to compress, but loses detail, quite a lot with a lot of noise reduction worse.

What the world needs is a Y'CbCr-native noise reduction filter. Most video has more noise in its chroma tracks than in luma. However, because filters tend to operate in RGB, there is no good way of applying noise reduction to select luma. Oddly enough, the only Y'CbCr-native noise reduction filter I know of is free—the fxVHS filter from VirtualDub. Like most VirtualDub software, the fxVHS filter is powerful, but unpolished. If you don't have this filter, one workaround is to convert the whole file to numbered still image frames, and run each frame through Photoshop in Lab color mode, applying blur on the A and B channels, but leaving L alone. But it would be a rare project worth that much hands-on effort.

## Lowpass Filtering

A lowpass filter lets low frequency information pass through (hence lowpass) and cuts out high frequencies. For video, frequency is a measure of how fast the video changes from one pixel to the next. Thus, a lowpass filter tends to soften sharp edges, but leaves other stuff alone.

## Sharpening

I generally don't recommend applying sharpening to video. Sharpening filters do make the image sharper, but sharper images are more difficult to compress. And sharpening filters aren't selective in what they sharpen—noise gets sharpened as much as any other element.

For the usual strange historic reasons, Unsharp Mask is another name for sharpening, typically seen in applications whose roots are in the world of graphic arts.

## Composite Noise

Composite noise is caused by high-frequency luma information leaking into the chroma channels when luma and chroma channels are combined into a single signal. This is one of the reasons you should avoid capturing video through a composite connector. If source already has bad composite errors, capturing it through an excellent comb filter, which separates the luma and chroma channels, can help. You'll often see high-end consumer equipment marketing boasting of "3D adaptive comb filters" and that sort of thing. Professional gear has better-yet circuits that can help significantly.

 **See also**

A close-up showing typical composite noise artifacts is shown in Figure Q on page XII of the color section.

Once the file is captured, there really aren't any good ways of removing composite noise that don't cause some blurring. However, because the error is only in the chroma channels, you can use tools that allow filtering on only single channels to blur them, while leaving the luma channel intact. This is one of the big reasons I want the Y'CbCr native noise reduction filter described previously.

## Motion Tracking

Sometimes slight motion occurs in the video image, from the camera having been handheld, from tripod vibration, or from the use of a long telephoto lens. After Effects, Pinnacle Commotion, and similar tools allow you to define anchor points and correct for slight motion from frame to frame. This can make the video easier to compress and watch. The biggest drawback is that this approach inevitably requires additional cropping around the edge of the frame. Running the calculations can also be fairly CPU intensive, which translates to long render times.

The other tool I'm fond of for very detailed color work is the Color Finesse After Effects plug-in from Synthetic Aperture (see Figure 7.11). It provides a huge number of different modes and tools for adjusting color. It's overkill for most compression projects, but it's unrivaled as a software product for working with very troubled source, or doing very fine color work. It really provides the flexibility of a telecine suite on a desktop computer (though not in real time).

## Luma Adjustment

As its name implies, luma adjustment is the processing of the luma channel signal. While many applications actually process everything in RGB, they still provide tools that describe things in luma, (the Y in Y'CbCr) terms.

One important luma adjustment consideration: be sure to view the video being worked on in the gamma mode of your target platform. So, for example, if you're targeting Windows from a Mac, put your display in Windows gamma mode with a tool like GammaToggle.

## Remapping from ITU-R BT.601

The normal dynamic range of ITU-R BT.601 video is 16 to 235, and in NTSC the darkest black is at 7.5 IRE. Conversely, video on a computer has a range of 0 to 255, where black is 0 IRE. Changing or remapping ITU-R BT.601 optimized content to computer delivery optimized content usually involves expanding the range of levels available. This enables blacks in the image (such as those on a title card) to match other black elements on the screen. It also adds a pleasing

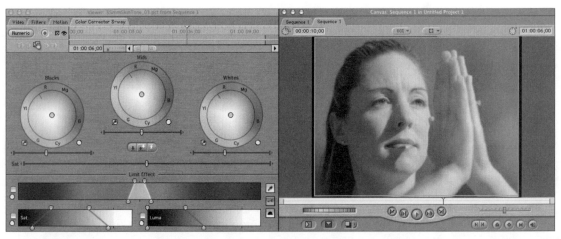

**7.10** The Color Correction Three Way interface of Final Cut Pro. It is fully Y'CbCr native, offering great performance and quality.

vivid contrast to the image. Some applications such as Apple's DV codec, do this automatically. In others, for example Motion-JPEG codecs, a contrast control is required to adjust contrast.

If no other luma filters are run, you need only to adjust contrast to expand from the 16 to 235 to the 0 to 255 range. Use a value of +27 (Cleaner) or 1.27 (Adobe apps). However, tools may often apply brightness *before* contrast, which means doing a brightness adjustment makes it more difficult to do a clean remapping. In such cases, it's important to monitor the overall levels.

## Black levels at zero

An important aspect of the remapping is getting parts of the screen that are black down to mathematical Y=0. Because computer screens are so much better than television screens, noise that wouldn't be noticed on a TV screen can be painfully obvious on a computer screen. This is especially problematic with Mac gamma (See "Gamma Adjustment" on page 115) because the difference between Y=0 and Y=1 can be apparent. Also, a large field of 0s is extremely easy to compress mathematically, saving a lot of bits. Because DCT-based codecs have a lot of trouble with ringing on text with sharp edges, getting blacks down to zero lets the codec spend the bandwidth on giving clean, crisp edges to the letters.

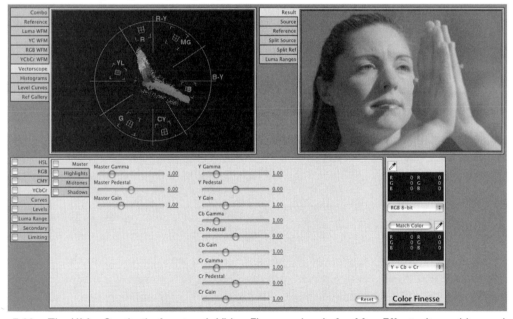

**7.11** The UI for Synthetic Aperture's Video Finesse plug-in for After Effects. It provides unrivaled control of the color correction process for desktop software. Note how the flesh tones are clustered along the flesh tone axis in the vector scope.

### See also

Turn to page XIII of the color section to see color versions of Figure 7.10 and Figure 7.11.

Getting black levels to 0 correctly can really pay off in final compressed quality. Lastly, having the black levels at 0 means the video's black will match other black elements on the computer screen, instead of appearing an icky gray.

Getting black levels down to 0 is much more difficult with noisy source, as many pixels that should be black have values well above 16. The solution is to get those brighter pixels down to 0 as well. This is done with a combination of filters (described in the following). Noise reduction can also help, because it will bring the higher values closer to 0. However, care must be taken that parts of the image that should truly be dark gray, not black, don't get sucked down as well.

## Brightness

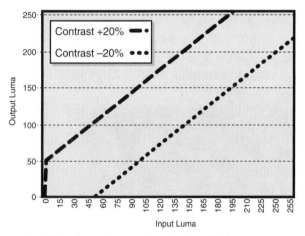

**7.12** This chart shows the relationship between before and after luma values with different brightness values.

Brightness filters adjust the overall intensity of the light by raising or lowering the value of each pixel by a fixed amount. If brightness is used in compression, it is normally used to make the video darker overall, instead of lighter. Unless there is a really huge contrast adjustment, raising the brightness will leave black levels above 0. (Refer to Figure 7.12.)

## Contrast

Contrast increases or decreases the values of each pixel by an amount proportional to how far away it is from the middle of the range. So, for an 8-bit system, 127 (the middle of the range) stays the same, but the farther away from the center value a pixel is, the larger its change.

The most basic use of contrast is to increase the range from video's 16 to 235 to the computer display's 0 to 255. In Cleaner, a value of +27 will do a precise remapping. With Adobe applications, a value of 1.27 achieves the same result.

With noisy source, it can often help to use a somewhat higher contrast value to lower black levels. This technique is often mixed with a corresponding decrease in brightness to keep the whites from getting blown out, and to get a combined effect in getting the dark grays down to black without hurting the rest of the image too much.

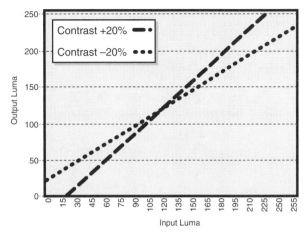

**7.13** This chart shows the relationship between before and after luma values with different contrast values.

## Gamma Adjustment

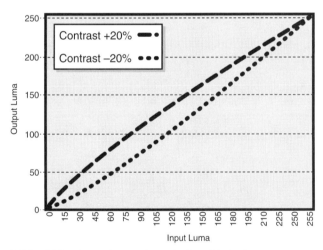

**7.14** This chart shows the relationship between before and after luma values with different gamma values.

Gamma adjustment has been a pain in my butt since I tackled my very first CD-ROM compression job. Remember: the important thing is that video and Mac OS systems both have defined gammas, but they're different (2.2 and 1.8 respectively). The various Microsoft Windows operating systems don't have a clearly defined standard gamma, although its values tend to range around that of video, from 2.2 to 2.5. If you encode video so it looks good on a Mac, the midtones will look too dark on Windows, and vice versa. To add to the agony, some source codecs, like Apple's DV codec, apply their own gamma curve to the video so the video on the Mac's monitor will match that on the video monitor, meaning that any gamma adjustment must both account for the platform conversion you are trying to do, and compensate for whatever gamma adjustment the source codec is making. More aggravation: the source codec may behave differently in different circumstances, so it does gamma correction when shown on the screen, but not when processed by an encoding tool or when viewed on one platform but not another. For example, within QuickTime, the DV and MPEG-4 codecs are automatically corrected to 1.8 gamma on Mac and 2.5 gamma on Windows.

Assuming there is no autocorrection going on, gamma correction is pretty simple math when you know the source and target gamma. Mac has a gamma of 1.8. Video's gamma is 2.2. But Windows' gamma is undefined, and can vary quite a bit between systems, typically in the range of 2.2 to 2.5. I generally assume the value of 2.2, which matches video. However, if you're targeting a corporate environment where a standard video card and monitor is used, you can calibrate the video so it looks great on that standard platform. Table 7.2 shows the correct values to use for Cleaner and After Effects.

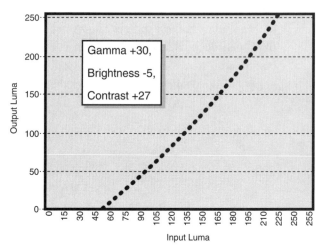

7.15 This chart shows the cumulative effects of combined gamma, brightness, and contrast filters at some typical settings.

Table 7.2     Gamma correction values.

|  | Mac gamma output | Video gamma output | Windows gamma output |
| --- | --- | --- | --- |
| Mac gamma input | +0 or 1.0 | +30 or 1.3 | +30 or 1.3 |
| Video gamma input | −30 or 0.8 | +0 or 1.0 | +0 or 1.0 |
| Windows gamma input | −30 or 0.8 | +0 or 1.0 | +0 or 1.0 |

Beyond getting cross-platform issues squared away, gamma can be a useful tool in its own right. If the blacks and whites look just right, but the midtones seem a little washed out or dark, adjusting gamma can often improve quality significantly.

## Chroma Adjustment

Chroma adjustment, as you've already guessed, refers to tweaks made to color. The good news is that with most projects, you won't need to make chroma adjustments. And when you are required to do so, it likely won't be with the complexity that luma adjustments often require.

## Saturation

Saturation controls affect changes in the intensity of color, analogous to luma. Saturation changes affect both color channels equally. Most tools only offer a simple, linear saturation. Alas, a simple saturation slider is often a poor tool for image optimization for compression—it would be much better to have tools analogous to brightness/gamma/contrast controls for adjusting saturation. This allows specifying the middle of the saturation range becomes more intense, while leaving the extremes where they are, avoiding color errors in unsaturated portions of the screen, and peaks in the most saturated portions. This filter would be analogous to a gamma curve.

There are several ways and reasons to adjust saturation. As you know, computer monitors support much better saturation than television screens. The 601 digital video range for saturation is also narrower than the –127 to 128 range. Expanding the saturation range slightly to fill this full range can bring some vibrancy to the video. Still, the quality improvement this can provide is much less than what results from expanding the luma range.

## Hue

Incorrect hue is when the entire color range of the video is shifted in the wrong direction, either due to white balance not being set correctly when the video was shot, or from capturing through badly calibrated composite video systems.

Typically the problem shows up when flesh tones and pure white don't look correct. Adjust the hue until whites look correct. Ideally, use a full-featured color-correction tool when such problems abound in your source. The hue of flesh tones falls along a very narrow range (different ethnicities vary mainly in how saturated or dark the skin color is, not the hue). A vectorscope will show you the distribution of colors in a frame and a line indicating ideal skin color tone.

## Color Correction

Standalone tools for making bad video good refer to the process of color correction. This, despite the name, includes a lot of luma control as well.

Lots of applications simply provide RGB sliders for color correction. These aren't ideal, as they change luminance as well as color, and such tools are pretty hard to control.

My favorite color correction tools are—no surprise—Y'CbCr-savvy. Apple Final Cut Pro 3 introduced excellent color-correction features within a single affordable desktop application, previously only available to owners of high-end da Vinci dedicated color-correction systems. If you need even more control beyond what Final Cut provides, check out Synthetic Aperture's "Color Finesse" plug in for After Effects. It offers a full range of controls in many different modes and

**118** Chapter 7: **Preprocessing**

**7.16** Doing hue adjustment to fix skin tone in the Final Cut Pro 3 Basic Color Corrector.

**See also**
The color version of Figure 7.16 is on page XIV.

many different scopes. Both products also include excellent manuals that discuss the process of color correction clearly and concisely.

# Frame Rate

While frame rate is typically specified as part of the encoding process, conceptually it really belongs with preprocessing. The critical frame rate rule is the output frame rate must be an integer division of the source frame rate. Thus:

- NTSC: 59.94, 29.97, 15, 10, 7.5, 6, 5
- PAL: 25, 12.5, 8.67, 6.25, 5
- FILM: 24, 12, 8, 6

And that's it! After so many subjective concepts, I'm happy to finally present a hard and fast rule.

We need to use these frame rates to preserve smooth motion. Imagine a source that has an object moving an equal amount on each frame. At the source frame rate, everything looks fine. If you

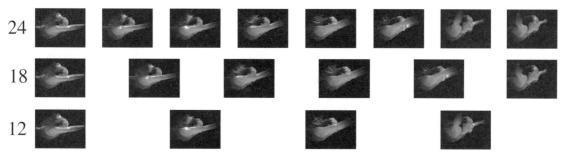

**7.17** Patterns of dropped frames. Note 12 aligns with 24 smoothly, but 18 does not.

drop every other frame, the amount of movement in each frame doubles, so it is consistent. Same with dropping two frames out of every three, or three out of every four. But if you use any other frame rate, you'll be dropping an uneven number of frames. Take for example, going from 24 to 18fps. That will drop one frame out of every three, so you'll have two frames that show equal motion, then one that shows double motion. (See Figure 7.17.)

Of course, there are some complexities. For one, how do you treat source that is mixed, like a promotional piece for a movie in which video interviews are intermixed with telecined film scenes? If you're outputting QuickTime, which allows variable frame rates, you could set the frame rate up or down as appropriate. But Windows Media requires a fixed frame rate (although frames can be dropped), meaning a given frame can have its duration doubled, tripled, and so on, filling the space where several frames would have normally gone.

## Audio Preprocessing

There's a lot less to do in audio preprocessing than video because there isn't as huge a difference between a stereo system and a computer with speakers as there is between a TV and a computer with monitor.

Well-produced audio aimed for home stereo system or television delivery will already be normalized and clean, and it might not need anything further for successful compression. Raw captured audio might need quite a bit of sweetening. How to produce professional quality audio is well beyond the scope of this book. Fortunately, Jay Rose has written an excellent book on the subject, *Producing Great Sound for Digital Video*, (also from the DV Experts Series, CMP Books, 2001).

### Normalization

Normalization adjusts the level of an audio clip by finding the single loudest sound in the entire file, and then raising or lowing the volume of the whole clip so the loudest sound matches whatever level has been specified. This doesn't change the relative volume at all, just the absolute overall volume.

Normalization is of vital importance to the overall listening experience. It's all too typical to listen to some poor, overly quiet Web audio, that requires you to keep cranking up the volume to hear it clearly. And then, when the clip is over, you forget to turn the volume back down, so the next system beep is a thunderous, ear-splitting BOOM! Not nice. Normalizing your audio tracks eliminates this problem.

## Compression

In a source of perennial confusion, there is an audio filter called a compressor. In audio, extreme transients (very loud, very short noises), such as explosions or cymbal crashes, can cause distortion in audio playback and recording devices. Likewise, extremely quiet passages can get lost in the noise floor (the always present hums and hisses). Audio compressors act on dynamic range, smoothing out peaks and valleys of a signal making it seem larger and louder—they're the reason television commercials seem so much louder than regular programs. Audio compression, when used properly, can tame wild level changes.

Audio compression used to be critically important back in the 8-bit audio era, when already lousy CD-ROM video was playing through tiny internal speakers. Today, it's much less of an issue. If you're dealing with source mixed for television delivery, it will likely have audio compression on it already.

## Noise reduction

Noise is no better for audio than video. And audio can have quite a lot of it: tape hiss, air conditioners in the background, hum from power lines, wind in the mic, and so on. A number of professional audio tools are available to reduce this noise (although the results are never as good as if the audio had been noise-free in the first place).

The scope and use of such audio noise reduction systems is a broad topic, well beyond our scope here. However, if you have bad sounding audio, there is often (but not always) something a good audio engineer can do to help the quality. The results are always be better when the source is clean from the get-go.

# Chapter 8

# Video Codecs

## Introduction

You've probably noticed there are roughly six bazillion video codecs out there. While the rest of this book offers specifics on how to use each of those individually, there are many aspects of video codecs and compression we can investigate in general terms. Be aware that many of these technologies use different terms to describe the same feature, and sometimes call different features by the same name.

## Codec Settings

The most basic codec setting is the setting that determines which codec is being used. Some formats, such as MPEG-1, offer but one codec. QuickTime, at the other extreme, is a codec container that holds literally dozens of codecs, most of which are wildly inappropriate for any given compression project.

Picking the right codec is of critical importance. You'll find details on each codec's strengths and weaknesses in the chapters devoted to each format. When evaluating new codecs, compare their features to those described in the following to get a sense of where they fit into the codec ecology.

### Resolution

While not typically listed as a codec setting, resolution is one of the most important parameters to consider when choosing a codec. With a given source and frame rate, a given codec needs a certain number of bits per pixel to provide adequate quality, so there is an important balancing act between resolution and bitrate.

Some codecs can use any resolution. Others may require each axis be divisible by two, by four, or by 16. Others may have maximum resolutions, or support only one or a few resolutions.

## Aspect ratio/pixel shape

Most Web codecs only support square pixel playback, hence the aspect ratio is determined by the resolution. MPEG-1 and MPEG-2 are the major exceptions. They can be flagged with different pixel shapes, e.g., NTSC DVDs are always 720×480 video, but can be in either 4:3 or 16:9. QuickTime supports having the encoded and playback resolution be arbitrarily different, which effectively changes pixel shape, but this isn't commonly used.

Note square pixel compression and display offers the best compression efficiency, so anamorphic compression should only be used when required for a particular delivery platform. MPEG files meant for computer playback, not VideoCD or DVD playback, are best encoded as square pixel. Windows Media Player ignores the aspect ratio flag.

## Bit depth

Modern codecs for compressed video content are all 8-bit per pixel. However, there are exceptions in special-use and older codecs.

The ancient Apple Video and Microsoft Video 1 codecs are both 16-bit RGB (five bits each for Apple Video, but six for green in Microsoft Video).

Some codecs support indexed modes like 8-bit. Indexed modes shouldn't really be considered a color space because the 256 available colors are all described as 8-bit RGB values.

## Color space

Most codecs support only a single color space. For modern Web codecs, that color space is generally 4:2:0, although a number are YUV-9, including the Indeo codecs and older versions of Sorenson. Others, such as VP3 and H.263, are 4:2:2. For almost all real-world applications, 4:2:0 looks a lot better than YUV-9. The theoretical advantages YUV-9 offers in compression aren't enough to make up for how things look in real-world conditions. 4:2:2 looks good, but is slightly less efficient for progressive content than 4:2:0.

Authoring codecs often use 4:2:2, and NTSC DV's 4:1:1 is incredibly popular.

A few (but not enough) codecs support an 8-bit grayscale mode for encoding black and white. This doesn't really help compression efficiency much (with black and white source, almost no bits are spent on the Cb and Cr channels), but being able to tell the codec to just ignore Cb and Cr channels on encode and decode would provide meaningful speed improvements.

A few very old video codecs like Video 1 and Cinepak support 8-bit indexed modes. These almost always offer horrible quality. They don't do a good job of 8-bit palette generation, and the lossy algorithms used in video compression are designed for continuous tone images (a.k.a.

photographs), not indexed color images. Using indexed colors works much better in lossless RGB codecs like PNG, MNG, and Apple's Animation and Graphics.

Only a few codecs—MPEG-2 and the ancient TrueMotion v2X come to mind—allow a choice between 4:x:x color spaces.

## Frame rate

Modern formats let you specify frame rate. Some older formats like VideoCD only offer a specific frame rate, and MPEG-1 and MPEG-2 have a limited number of options (native frame rates for PAL, NTSC, and film).

Most formats and tools allow only one frame rate per file (see the Note on "Mixing Frame Rates" on page 124 for more details). If you have source that mixes frame rate (like sequences of both 23.976 film and 29.97 video) you'll need to decide which frame rate to favor, depending on which is most important to your communication goal.

Some tools (RealProducer and Windows Media Encoder, but not Cleaner) support delivering all fields from interlaced source, e.g. 59.94 fields per second for NTSC and 50 fields per second for PAL. These can be extremely vivid with sufficient bandwidth and playback is on fast computers.

As previously discussed, the output frame rate must be an integer division of the source, meaning there are only a handful of valid output frame rates for any given source frame rate. Try to retain the full frame rate of the source. This is generally possible for broadband Web and CD-ROM applications. Increasing the frame rate beyond that of the source doesn't do any good. Lowering it provides more bits per pixel, increasing spatial quality. However, dropping the frame rate in half doubles the difference between frames because twice has much time has passed between them. These greater changes means more bits are required for a delta frame, which reduces the compression efficiency gains. Also, at the lower frame rate, each compression artifact is on-screen longer, meaning frames of the same quality will appear more objectionable at lower frame rates. Lastly, the lower the frame rate, the greater the number of frames that should be keyframes (more to come on that).

Altogether, you can't drop data rate by the same amount you drop frame rate if you want to keep the same apparent quality. Roughly speaking, you can cut data rate by 30–50 percent if you cut frame rate by half. This means frame rate is generally not the first place to go looking to cut complexity from the file, except at the very lowest data rates.

We are much more attuned to changes in frame rates when viewing a large video rectangle, and less apt to notice changes when the rectangle is a smaller portion of the field of view, so sitting closer to the monitor makes it easier to notice a lower frame rate. Thus, using lower frame rates is less objectionable at lower resolutions.

**Note**

***Mixing Frame Rates in a File.*** Most formats only allow a single frame rate per file. In most formats, frames can be dropped, which causes the previous frame to double, triple, quadruple, and so on in length. The frame can only get longer to replace a particular number of missing frames. This means you can't arbitrarily mix 24fps and 29.97fps segments in the same file. In theory, one could mix 30 and 24 by running at a frame rate of 120, and dropping four out of five frames for 30fps source and five out of six frames for 24fps source. However, neither Windows Media or Real are tested against such high frame rates and real-world results are poor.

QuickTime pioneered the approach of letting each frame have its own unique duration, which MPEG-4 has inherited from the QuickTime file format. This means QuickTime and MPEG-4 can vary the frame rate arbitrarily within a file, so you can mix 29.97fps and 24fps in the same file, if you have an authoring tool that supports doing that. Today, the only reliable way to author mixed frame rate content is to encode each segment as its own file, and then paste them together via QuickTime Player Pro or LiveStage Pro. It's a little labor intensive, but works great.

## Keyframe rate

Virtually all codecs let you specify the frequency of keyframes. In most formats this is better defined as "keyframe at least every" instead of "keyframe every." Thus the natural keyframes automatically inserted at cuts and such will change the pattern of keyframes. For example, having a keyframe rate of "every 100" doesn't mean you'll have a keyframe at frames 1, 101, 201, 301, etc. In this case, if you had automatically inserted keyframes at 30 and 140, you'd get keyframes at 1, 30, 130, 140, 240, and so on.

This is the optimal behavior. The main use of keyframes is to ensure the file has sufficient keyframes to support random access, and to keep playback from being disrupted for too long after a dropped frame or other hiccup. The drawback of keyframes is that they don't compress as efficiently as delta frames. Some codecs may also be prone to "keyframe flashing," where the visual quality of a keyframe is different from that of the surrounding delta frames.

Keyframe rate is generally specified as "keyframe every X frames" or "keyframe every X seconds." Converting from one to the other is straightforward arithmetic.

Typical keyframe values vary wildly between formats and uses. In interframe-only authoring codecs, every frame is a keyframe. MPEG-1 and MPEG-2 typically use a keyframe every half second or second. This is because those formats are typically used in environments very sensitive to latency (set-top boxes and disk-based media), and formats have a much smaller difference in size between keyframes and delta frames than modern codecs.

Web formats use from 1 to 10 seconds by default. Beyond that, the cost to random access is high, but the compression efficiency gains aren't significant. However, if keyframe flashing is a significant problem, reducing the number of keyframes will make the flashes less common.

## Natural keyframe sensitivity

Except for MPEG-1 and MPEG-2, all modern codecs will automatically insert keyframes when the size of an delta frame is nearly as big as a keyframe. In MPEG-1 and -2, the function is optional. This is to help random access (it puts keyframes at places you'd be likely to want to jump to—video cuts) and quality (keyframes can look better than delta frames with some codecs). The default threshold for many codecs is 85 percent, although this can vary (and often isn't documented). Some codecs let you change the sensitivity. With more sensitivity, keyframes are inserted after smaller changes. This can help random access and quality if the video includes a lot of dissolves instead of hard cuts. With less sensitivity, fewer keyframes are inserted. This can help if there is very fast motion as with explosions.

In most cases with most video, there won't be any need to change the value—the default will work just fine.

For MPEG-1 and MPEG-2, this feature is typically an option that needs to be activated, and is called something like "automatic I-frame injection" or "insert I-frame on scene change." Unless you're targeting an unusual application (like a multi-angle DVD), turning this option on will improve compression efficiency without any drawbacks.

## Inserted keyframes

A few tools let you manually specify particular frames to be keyframes. Like natural keyframes, inserted keyframes reset the counter for Keyframe Every settings.

Back when Cinepak was the dominant codec, being able to insert keyframes was critical to achieving maximum video quality, because the automatic keyframe insertion of Cinepak was so lousy, and the quality of Cinepak keyframes was so much better than delta frames. Modern codecs are a lot better at inserting keyframes automatically, so you should only do manual keyframing if there's a particular problem you're trying to address. For example, a slow dissolve might not have any single frame that's different enough than the previous to trigger a keyframe, so you can make the first full frame after the dissolve a keyframe. Or, if you need to be able to randomly access a particular frame, you can force it to be a keyframe.

Manual keyframing is supported by Cleaner (via EventStream) and more elegantly through Adobe Premiere with its "Keyframes on Markers" feature, which has been in Premiere since at least version 4.0. Many MPEG encoders also support keyframe insertion; it's usually called "manual I-frame insertion" or something similar.

### Bi-directional frames

Most codecs only support keyframes and forward-predicted frames (frames based on the previous frame). Some support bidirectional frames (B-frames). Depending on the format, bidirectional frames may be based on the frame after instead of before, or might combine elements of both the previous and last frame. Either way, B-frames improve compression efficiency by giving the codec more options for data on which to base encoding. While B-frames can be based on the frame before and after, no frame is based on a B-frame, because that would mean frame X would be based on frame Y, which would be based on frame X; an obvious impossibility.

Because no frames are based on them, B-frames can be dropped on playback if playback speed or streaming bandwidth is a problem. This can be very useful for playback and data rate scalability.

B-frames are on by default in MPEG-1 and MPEG-2, and are optional in Sorenson Video 3 Pro and in some MPEG-4 profiles. Windows Media Video and RealVideo do not use B-frames. For most B-frame formats, two B-frames per P-frame (delta frame) are the (appropriate) default. Sorenson can only do one B-frame per P-frame.

The QuickTime file format can't deal with video frames that refer to later frames, which means it can't deal natively with B-frames. This is why MPEG-1 behaves so strangely in QuickTime, and why Sorenson Video 3 with B-frames gives sync errors (specifically, using SV3 B-frames delays video by two frames relative to audio—a full explanation and a simple workaround are in Chapter 14. QuickTime simply doesn't play any B-frames in MPEG-4 files.

 **See also**

Refer again to Figure 3.19 on page 55 showing the interrelationship between frames and B-frames.

## Data Rate

Data rate is the single most critical parameter when compressing files for delivery on the Web, CD-ROM, or any place where there are limits on bandwidth or file size. Different architectures measure data rate in different ways. QuickTime's native export dialog measures it in kilobytes per second (KBps), while most Web codecs measure things in kilobits per second (Kbps), and MPEG-2 is typically measured in Mbps. Converting between KBps and Kbps is easy—there are eight bits in a byte, so 8Kbps = 1KBps. Note compression tools normally say that a Kbit is 1,000 and an Mbit is 1,000,000 bits, not the 1,024 ($2^{10}$) and 1,048,576 ($2^{20}$) values typically used in computer science. In some tools you specify a total data rate from which the audio data rate is subtracted to determine video data rate. In others, the two are determined independently.

Within a codec's sweet spot, quality improves directly with increased data rate. Beyond a certain point (at the same resolution, frame rate, and so on), increasing the data rate doesn't appreciably improve quality. The codec itself may have also have a fixed maximum bitrate. Below a certain data rate (again, with all parameters fixed), the codec won't be able to produce files as requested—either the data rate comes out higher than requested, or the frame rate lower. Some codecs also reach a point where slight drops in data rate lead to catastrophic drops in quality. MPEG-1 is especially prone to this kind of quality floor; it lacks a deblocking filter, so things can get very ugly very quickly.

Compression efficiency is the killer app for new codecs. When people talk about a codec being "best" they usually mean "most efficient." When a codec is described as "20 percent better than X" this means it can deliver equivalent quality at a 20 percent lower data rate. Compression efficiency measures how few bits are needed to achieve good enough quality in different circumstances. Codecs vary wildly in typical compression efficiency, with modern codecs being able to achieve the same quality at a fraction of the data rates required by older codecs. Different codecs have different advantages and disadvantages in compression efficiency with different kinds of sources and at different data rates. The biggest differences are in the range of "good enough" quality—the higher the target quality, the smaller the differences in compression efficiency.

Some authoring and special-use codecs don't offer a data rate control at all, either because the data rate is fixed (as in DV) or because it is solely determined by image complexity (like PNG). Apple Video also didn't have a data rate control, but hey, what do you expect in a product that predates Windows 3.0?

 **Note**

Files encoded for the Web often include multiple versions of the same content at different data rates, so a given encode would use multiple data rates. More on this in Chapter 11 "Web Video."

## Data rate modes

All that 1-pass, 2-pass, CBR, VBR stuff is about terminology to describe data rate modes. These are one of the key areas of terminology confusion, with the same feature appearing under different names, and different features being given identical names. We can classify data rate control modes in a few general categories.

## VBR versus CBR

The VBR (Variable Bit Rate) versus CBR (Constant Bit Rate) distinction is especially confusing. All modern video codecs are variable in the sense that not every frame uses the same number of bits as every other frame. This is a good thing—if every frame were the same size, keyframes would be terrible compared to delta frames. Even codecs labeled "CBR" can vary data rate quite a bit throughout the file.

## Chapter 8: Video Codecs

> **Tip**
> The only way I know of to get a truly constant bitrate out of a video codec is with Cleaner's "Flat" encoding mode.

### 1-pass versus 2-pass

By default most, codecs function in 1-pass mode, which means compression occurs as the file is being read. As each frame is input, a compressed version of it is output. Live broadcasting/Webcasting requires 1-pass codecs.

The limitation of 1-pass codecs is they have no foreknowledge of how complex the content after the frame they're currently processing is. Thus, they don't know enough about the overall complexity of the entire file they're being asked to compress to hit a specific data rate and still deliver an image quality that's consistent across the file. Although a lot of work has gone into finding optimal data rate allocation algorithms for 1-pass codecs, there will always be files that different implementations of these algorithms will guess wrong on, either over-allocating or under-allocating bits. An over-allocation can be especially bad, because when the data rates go up above the average they will have to eventually go below the average by the same amount. If this happens when the video gets more complex, quality falls dramatically.

2-pass codecs first do an analysis pass, where they measure the relative complexity of each frame. Once the entire file has been analyzed, the codec figures out the optimal way to spend the available bits over the entire file to achieve the highest possible average quality. In essence, this lets the codec see into the future, so it always knows when and how much to vary the data rate. 2-pass compression can yield substantial improvements in compression efficiency, typically from 10 to 50 percent.

### Buffered versus whole-file data rate control

The two different ways that the bitrate can be allocated over the file are buffered and unbuffered (whole file). In buffered, the average data rate must be maintained over a certain period of time, although the bitrate can vary within that time period as much as is appropriate. For example, for a five-second buffer, the data rate for each five seconds must not be higher than the requested average. This isn't a question of multiple, discrete five-second blocks, but that any arbitrary five seconds plucked from the file must be at or under the target data rate. This is also called *sliding window* and Constant Bit Rate (CBR). Buffered control is critical for real-time streaming because those formats all have a buffer time in the server. If the target data rate is exceeded for longer than the duration of the buffer, the client can run out of data received from the server and have to pause until more data arrives. The trade-off is that the longer the buffer time, the higher the average quality, but the longer the latency can be in starting the clip and doing random access within it.

For Real and Windows Media, the server is able to read what buffer size the file was encoded at and respond accordingly. QuickTime and MPEG-4 don't support this yet.

With modern servers (QTSS4 and WMV9 and later), using larger buffer sizes is less of a problem because the servers will aggressively expand the buffer when there is excess bandwidth. Thus, if a file is encoded at 300Kbps with a 30-second buffer, a 900Kbps connection might only actually take 10 seconds for buffering.

Default buffer sizes range from 3 to 25 seconds in different platforms. Due to ongoing improvements in server buffer management, I tend to use longer buffers for quality reasons. Generally speaking, the longer the file and the less random access you expect, the larger the buffer you should use. Real maxes out at 25 seconds, which I often use.

There aren't any true 1-pass unbuffered modes that control data rate, because an implicit buffer is always used to control the data rate. Otherwise, you'd get arbitrarily long data rate spikes with difficult to compress content, potentially exceeding well beyond the target data rate. The ability of 2-pass encoding to catalog an entire file's complexity is required to redistribute bits from the end credits to the opening titles.

### Limited versus quality limited data rate control

For almost all compressed video delivery, you want to control data rate, and have the codec provide the best quality it can within that constraint. However, many formats also support quality limited encoding. This is called 1-pass VBR in Windows Media, and in QuickTime is achieved by just turning off the data rate control. In both cases, the Quality control is used to define a quality target for each frame, and hence the data rate.

Quality limited is mainly useful for archiving content for local use, especially as intermediate files for later compression. If making an intermediate, the highest quality value should be used. For most codecs, just using insanely high data rates gives the same effect, except for formats like MPEG-1 and MPEG-2, which will pad the file with zeroes to give you the requested data rate.

Quality limited's advantage is you get the smallest possible file that meets your quality target because easy frames are very aggressively compressed. The disadvantage is that data rate is largely unpredictable—at the same settings different content will have radically different output data rates.

A 2-pass unbuffered compression is mathematically identical to a 1-pass quality limited file where you figured out in advance what quality value would have yielded your target file size. Back in the day, I commonly used Indeo in 1-pass quality limited mode as a pseudo-VBR, successively reencoding the file until I hit a data rate I liked (typically with a quality value around 75 to 85).

Table 8.1    Names for different data rate modes.

| Encode mode | QuickTime | Windows Media | RealVideo |
|---|---|---|---|
| 1-pass buffered | basic mode or 1-pass VBR | 1-pass CBR | 1-pass VBR (default) |
| 2-pass buffered | None | 2-pass CBR | 2-pass VBR |
| 2-pass unbuffered | 2-pass VBR | 2-pass VBR | None |
| 1-pass quality limited | Data rate control unchecked | 1-pass VBR | None |

## Spatial quality

Lots of codecs have a control labeled "Quality," but what it actually does varies from format to format. The traditional use, as established in the first version of QuickTime, is to specify spatial quality. In ancient codecs like Apple Video, the spatial quality slider would specify a quality target for each frame. Interframe authoring codecs like the various JPEG derivatives also use the term Quality in this way. For these JPEG codecs, using the spatial quality slider performs a 1-pass quality limited compression. In other codecs, notably Cinepak, a Spatial Quality control specifies the quality of keyframes, while the Temporal Quality slider specifies that of delta frames.

## Temporal quality

A few older QuickTime codecs also provided a Temporal Quality slider (you can access this in the standard QuickTime dialogs by clicking on the Quality slider with the Option/Alt key held down). In those codecs that support it, temporal quality controls the quality of the delta frames, where spatial quality controls the quality of the keyframes.

## Spatial quality threshold

Many codecs and formats have an option called Quality that is actually a spatial quality threshold. This threshold sets a minimum value below which the quality of any given frame can't drop, and so the data rate of that frame is raised. Most codecs that implement this also automatically reduce the frame rate to compensate for the data rate increase, while a few codecs make this frame rate reduction an option. When frame dropping is turned off, the file just gets larger instead. The net effect of spatial quality threshold functions is that when the video gets more complex, the quality of each frame remains high, but the frame rate drops, sometimes precipitously.

Most formats offer a threshold range of 0 to 100. RealVideo gives you five levels of quality/frame rate.

Most codecs have an implicit minimum quality, and so they might engage in frame dropping at lower data rates, even with the threshold set to 0.

Many codecs only make spatial threshold quality an option in 1-pass modes, or in buffered modes. 2-pass algorithms do much better in avoiding frame dropping, especially as the buffers get larger.

Most users find irregular frame rates more irritating than a lower, but steady frame rate. For that reason, I normally set spatial quality thresholds to their absolute minimum. If I find the spatial quality isn't high enough, I reduce the frame rate a notch.

## Frame dropping

A few codecs, notably those produced by Sorenson Media, let you specify whether or not frames should be dropped to maintain data rate. If these features are turned off, the data rate target can be missed, especially if a spatial quality threshold is specified. When using such codecs, I typically turn this parameter on, but leave the spatial quality threshold at the minimum so it'll only kick in when absolutely needed to maintain the target data rate. In many cases with higher data rates or unchallenging content, it may not kick in at all over a whole file.

## Compression speed

Some codecs offer a choice between a high quality, slower encoding mode, and a faster encoding mode with somewhat lesser quality. The primary technique to speed up encoding is to reduce the size and precision of the motion search. Depending on the content, the quality difference may be imperceptible or substantial; content with more motion yields more noticeable differences. Speed improvements can vary from slight to many times faster (especially with MPEG-1 and MPEG-2). However, compression speed only applies to the codec's encoding portion of the compression process, not to source decode and preprocessing. So, if the bulk of the compression time is spent on these other tasks, huge codec speed gains would only give a moderate improvement in overall speed.

My philosophy is I can always buy more computers, but I can't buy my customers faster connections. Thus, I always do final encodes with codecs set in their slowest, highest quality modes. I occasionally do comps in faster modes, just to verify that the preprocessing is correct, the overall compression seems correct, and to test integration. But I only use comps for internal testing.

When used for real-time encoding, codecs may also make trade-offs to internally optimize themselves for speed. Speed is much more critical in real-time compression, because if the codec isn't able to keep up, frames will be dropped on capture.

## Motion search range

A few MPEG encoders allow you to specify the motion search range. Many other codecs implicitly specify this via fast/slow encoding modes.

While your gut reaction may be to jack up the number of pixels to the resolution of the file, ensuring a full motion search, this generally isn't the right solution. A full motion search is very, very slow, and with MPEG-1 and MPEG-2 can give erroneous false positive matches and actually hurt quality. Also, with larger motion search areas, additional information may be needed to describe motion, which can also hurt compression efficiency. Keep in mind the motion search value applies both vertically and horizontally, so a search of 64 pixels will cover four times the area of 32 pixels.

I normally stick to the defaults for high quality encoding, unless I know something about the particular codec or source that would lead me to change the default values. If I'm going to tweak, I'll use larger values for high resolution, because with higher resolution, the same motion will take more pixels. So, if I used a value of 32 for 320×240, I'd use 64 for 640×480.

## Hinting

QuickTime and MPEG-4 need additional data, called *hint tracks*, if the file is to be RTSP streamed. A hint track is essentially an index to the compressed video and audio data in the file that tells the server the optimal way to serve the file. The hint tracks themselves aren't transmitted, and so don't affect bandwidth, but they do bulk up the file on the server quite a bit. This means you can't extrapolate the data rate of a hinted file by dividing the file size by its duration.

There are two basic hinting modes: Normal and Optimized for Server. In Normal mode, hints are an index of the data contained in the file's tracks, be they video, audio, or other type of track. In Optimized for Server mode all the data referenced by the hint tracks is copied into the hint tracks themselves. This saves the server the trouble of parsing non-hint track data, but makes a file almost twice as large as a file with normal hinting.

In the real world of modern servers, Optimized for Server doesn't offer much of a performance advantage. Typically, bandwidth becomes a limiting factor well before parsing files stresses the CPU.

Hinting is an either/or thing, depending on whether the file is intended for RTSP streaming or file-based playback. Hint tracks are absolutely required for RTSP streaming. If the file isn't going to be streamed, don't use hint tracks; they just make the file larger without doing anything useful.

RealVideo and Windows Media files stream natively, and don't require additional information for streaming. Nor is there data to remove, should you want to not stream the files and reduce file size.

## Error resilience

Error resilience is anything added to the file to make it more able to recover should packets be dropped. Different codecs have different error resilience modes and methods. One such method is called *block refresh*. It's described in the next section.

In RealProducer, error resilience is an option during encode. Windows Media doesn't allow you to control error resilience. MPEG-4 has a variety of solutions among its different Profiles.

## Block refresh

Block refreshing is a particular form of error resilience popular in QuickTime codecs, particularly those from Sorenson Media and H.263 (where it's called Cycle Intra Macroblocks).

You can think of block refreshing as a Keyframe Every value for each (typically 16×16) block in the video frame. If a given block isn't refreshed naturally (due to a keyframe or motion) within a certain period of time, a new copy of that block is transmitted. This means if you tune into a stream when it's between keyframes, the frame will be completely restored by the end of the block restore interval. Block refreshing paces out the blocks being refreshed, so you don't get a whole screen worth hitting at once.

The drawback of block refreshing is that for fairly static shots it wastes bandwidth, as bits are spent refreshing unchanging parts of the scene. However, very low motion is easy to compress in the first place. With difficult, fast motion, block refreshing doesn't kick in, because all the blocks are naturally refreshed anyway. Thus block refreshing only hurts compression efficiency when compression efficiency is least required.

Block refresh is normally only used with RTSP streaming, where a user can tune into the middle of a stream and packets can be lost. Typically, block refreshing wastes bandwidth with file-based viewing where data integrity is assured. The one exception to that is if the video is suffering from "trails," that is, moving objects leaving visible distortion in their wake. Block refresh would erase those trails. Fortunately, trails aren't much of a problem in modern codecs.

## Postprocessing filters

Many advanced codecs are able to apply a variety of postprocessing techniques to improve quality on playback. The three main classes of these techniques are deblocking, deringing, and frame rate interpolation.

- *Deblocking* is the most common, and offers the biggest quality boost—it smooths out blockiness in video caused by encoding artifacts. The source stream is tagged to indicate which blocks should be smoothed and which should not, so actual flat blocks in the video should look fine after deblocking.

- *Deringing* reduces the errors around sharp edges, particularly common in DCT codecs and with fine-lined text.

- *Frame rate interpolation* is today only a feature of RealVideo. It actually interpolates additional frames to make up for frames dropped during encoding and to increase output frame rate so it's higher than the input frame rate. In RealVideo, this is done by interpolating the motion vectors, which indicate where things move between frames, and by calculating intermediate values. This works well, but sometimes suffers errors.

In most codecs, postprocessing is automatically applied if there is sufficient CPU power left after the base decode of the video. If the codec supports multiple postprocessing filters, deblocking is usually applied first, followed by deringing, and lastly frame rate interpolation if there is sufficient CPU power.

### See also
See Figure U on page XIV of the color section to see an MPEG-4 file with and without deblocking and deringing.

The Sorenson Video codecs allow deblocking to be specified in the file. In the case of Sorenson Video 3, the codec may override the file setting and turn off deblocking if the computer isn't able to play the file with deblocking on. DivX 4 and later does something similar—a file can be tagged with a target postprocessing level that can be further reduced on playback.

Overall, postprocessing is a critical feature of modern codecs, which offer much more graceful quality degradation at lower data rates. MPEG-1 and MPEG-2 lack postprocessing, which is one of the reasons their quality drops off so quickly below a certain data rate.

## Achieving Balanced Mediocrity with Video Compression

We introduced the concept of "balanced mediocrity" in Chapter 4, "The Digital Video Workflow." Balanced mediocrity is supposed to be an amusing name for a very serious concept: getting all the compression settings in the right balance. You'll know when you've achieved the right balance when changing any parameter, in any direction, reduces the quality of the user's experience. "Quality" in this case equates with "fitness for use." A quality video compression is one that's most fit for its intended purpose—"quality" only exists in context.

For example, take a file compressed at 15fps. If 15fps is the optimal value, raising it to 29.97 would mean the increased artifacts or increased CPU requirements hurt the file more than having smoother motion would help it. Conversely, reducing the frame rate to 10 would hurt the perception of motion more than any improvement in image quality would.

In talking about each setting, I'm going to break down the trade-offs of different values of each setting. If you have a project that's giving you trouble, and you aren't sure what to tweak, run through this list, and make sure each setting is at its optimal value for your specific needs. If you're not sure, encode it a couple different ways, and see what happens!

## Choosing a codec

Picking a codec is so basic and so important, it can be difficult to conceptualize the trade-offs. There are a number of things you're looking for in a codec, but they all revolve around solving the problem poised by our three questions:

1. What is your content?
2. What are your communication goals?
3. Who's your audience?

You want the codec that can do the best job at delivering your content to your audience while meeting your communication goals.

The three big issues you face in picking a codec are compression efficiency, playback performance, and availability.

*Compression efficiency*   Codecs vary radically in compression efficiency. New ones are much, much better than old ones. For example, Windows Media Video v9 can achieve quality at 100Kbps that Cinepak requires more than 1,000Kbps to reach. Some codecs work well across a wide range of data rates, others have only a narrow range between "good enough" and "more bits don't help."

However, modern codecs can also have higher playback performance requirements, due to their use of more complex compression techniques. When targeting high resolution or slower machines, a less efficient but faster codec may be appropriate.

*Playback performance*   Playback performance determines how fast the playback computer needs to be to decode the video. The slower the codec, the faster a computer you need for playback, or the lower resolution and frame rate you need. Playback performance is typically proportional to total image area and fps (which can be described as pixels per second; determined by height × width × fps). Data rate can also have a significant effect. So, the higher the target resolution, frame rate, and data rate, the faster the playback computer will need to be. For example, Sorenson Video 3 at 640×480, 29.97fps can't play back reliably on a 500MHz G4, while a file of the same specs encoded in Cinepak can be played back on a much slower computer. Note slower CPUs may allow less postprocessing, hence a file may display at lower quality, but at full frame rate B-frames may also be dropped on playback on slower machines. The worst case would be that only keyframes would be played.

*Availability* A great looking file that can't be seen doesn't count. When you think of availability, the questions you should ask yourself are: What percentage of the audience already has the codec installed? How many of the rest are willing to download it? Does the download happen automatically?

There is a gap in time between when a great new codec is released and when its installation is ubiquitous. The more compelling your content, the less of a problem that gap will be. The *Star Wars Episode 1: The Phantom Menace* trailer proved millions of people were willing to download and install a beta of QuickTime v4 to see a single video clip.

*Resolution* Resolution primarily affects data rate and playback performance, both of which are proportional to area. Going from 320×240 to 640×480 gives you four times the pixels, so you'd need about four times the CPU speed to play the video back to achieve the same level of quality.

You'll also need to increase data rate by four times to maintain the same quality per pixel. However, as you raise resolution, you can afford less quality per pixel—there's more to look at, so the artifacts become less visible. How much varies by codec and other factors. Our 320×240 to 640×480 example generally requires around three times more bandwidth to achieve the same perceived quality.

Remember, you normally don't need to play back the video at the resolution at which it was encoded. Each of the Big Three formats lets you specify a different playback resolution, through dialog boxes in QuickTime and via scripting the plug-in with RealVideo and Windows Media. However, when video is played back at higher resolutions than those at which it was encoded, artifacts become much more visible, so bits per pixel must increase somewhat.

*Aspect ratio/pixel shape* There's no real trade-off in anamorphic compression (with non-square pixels)—only use it when you're going to a format that requires non-square pixels. Square pixels give you the best compression efficiency, and they're easiest for computers to play back.

*Bit depth* You'll only specify bit depth in a few unusual codecs, mainly when compressing screen captures. While 24-bit is the default screen depth for modern computers, most on-screen software works great in 16-bit, which gives a substantial data rate reduction. 8-bit can sometimes be more difficult to compress than 16-bit, because 8-bit displays often use dithering.

*Color space* Very few codecs offer a choice of color spaces. The only common one is MPEG-2, which uses 4:2:2 for authoring but 4:2:0 for delivery. Other codecs might give you the option of encoding black and white content in 8-bit grayscale.

*Frame rate*   Frame rate is directly proportional to decode performance, but not bitrate. Cutting the frame rate in half doesn't let you drop the data rate by anything near 50 percent for any interframe codecs. It's often better to reduce resolution to at least 320×240 than to reduce frame rate. Below 320×240, it's a balancing act. Remember that lower frame rates are less objectionable at lower resolutions.

*Keyframe rate*   The Keyframe Every rate is about balancing compression efficiency and the elimination of keyframe flashing against the need for random access. For lower bitrates, setting a keyframe every 10 seconds is good. At higher data rates, one every one to 10 seconds is a good range. Less than one every 30 frames or so starts causing substantial degradation in compression efficiency, so the lower the frame rate, the more seconds you'll want between keyframes.

MPEG-1 and MPEG-2 expect one or two keyframes per second, and may not perform well with the lower keyframe frequencies typical of Web video.

*Natural keyframe sensitivity*   For most files, you can leave Natural Keyframe Sensitivity at its default setting. Increasing sensitivity means there will be more keyframes in the file, increasing random access, while potentially reducing compression efficiency. Increasing keyframe sensitivity will mean there are fewer natural keyframes. Natural keyframes rarely cause keyframe flashing because there would be enough change in those frames to obscure that effect. Thus, raising natural keyframe sensitivity may reduce keyframe flashing in all but the most static content.

*Inserted keyframes*   Manually inserted keyframes act just like any other keyframes. They're rarely used unless you're trying to address a particular problem, such as when the codec doesn't produce a keyframe during or after a dissolve, or when you need to facilitate random access. Because manual keyframes aren't often inserted in great numbers, they don't have much effect on compression efficiency.

*Bi-directional frames*   Bidirectional frames improve compression efficiency and playback scalability in the formats that support them (such as MPEG-1 and MPEG-2), and so should generally be on in formats that can use them. There are some specific complications with B-frames in Sorenson Video v3.1 Pro. These complications are discussed in Chapter 14.

*Data rate*   For real-time streaming applications, where you have hard data rate ceilings, you almost always have to make trade-offs to hit your target data rate. The closer the actual data rate is to the target data rate, the higher the quality, but the more fragile the stream will be if there's any disruption to or slow down of the connection. Offering your end users multiple bitrate versions of a given file can be very helpful in such situations.

For progressive download, the trade-off is between download time and file quality. The larger the file, the longer the wait for the file to download, but the higher the quality. There are many more details on this in Chapter 11.

*Data rate modes*   There is usually one ideal data rate mode to match your content. For normal compressed video delivery, your first choices should be buffered for streaming and unbuffered for file-based playback. If a 2-pass encoding mode is available, use it. Two-pass compression takes twice as long to encode, but the quality boost is usually worth the wait, unless you're extremely strapped for time.

You'd only use quality limited for special cases. Remember, 2-pass unbuffered is the same as using quality limited with a target data rate.

*Spatial quality*   Most modern video codecs don't use spatial quality without any data rate encoding. For intraframe authoring codecs, higher values provide higher quality but have higher data rates; the converse is true for lower quality.

For Cinepak and Apple Animation, spatial quality gives relatively more quality to keyframes, which can either help or hurt keyframe flashing.

*Temporal quality*   Temporal quality is supported by very few codecs. In Cinepak and Apple Animation, it controls the relative quality of delta frames compared to keyframes. Higher values have higher quality and data rates, and vice versa.

*Spatial quality threshold*   Spatial quality threshold increases the spatial quality of difficult to compress frames, at the expense of either raising the data rate or lowering the frame rate. The stuttery look this can give the video means it's generally best to set spatial quality to its minimum and reduce frame rate to the point where the output frame rate is consistent.

*Frame dropping*   If you need to hit a target data rate, enable frame dropping if it's available in your codec of choice. Without frame dropping, data rate might be substantially overshot in the most difficult sections. However, it's preferable to find a resolution, frame rate, data rate, and so on where significant frame dropping isn't required.

*Compression speed*   Unless your needs require real-time (or just plain fast) encoding, codecs should always be run at the highest quality, slowest encoding mode available.

*Motion search range*   Larger motion search ranges slow compression, and depending on the codec, may cause trouble with slow-motion sources. However, they generally improve quality with a lot of motion. The motion search range should be large enough to capture the full distance of motion of the fastest moving objects between frames. So if the fastest moving objects move 50 pixels between frames, a motion search range of more than 50 should be used. Most tools indirectly control motion search range with high/low speed encoding.

*Hinting*   If you're making QuickTime and MPEG-4 files for RTSP streaming, hinting needs to be on. For QuickTime and MPEG-4 files for progressive download (or other file-based uses), hinting should be off, as hint tracks aren't used and add to the data rate.

*Error resilience*   Error resilience slightly increases data rate (or reduces compression efficiency) to make the file more capable of handling dropped packets. If it's available in the codec, error resilience should be on for any files that are going to be real-time streamed. Error resilience is not necessary for file-based playback because file-based systems don't drop packets.

*Block refreshing*   Block refreshing can significantly reduce compression efficiency, but makes the file better capable of handling lossy connections. As a form of error resilience, block refreshing should be on for real-time streaming files, and off otherwise.

*Postprocessing filters*   Postprocessing improves quality on playback, but increases CPU load. Most modern formats apply it automatically. For those codecs in which postprocessing needs to be manually activated, it should be used when there is sufficient power in the target platform to decode the file with the option on. For codecs where preprocessing can be automatically turned off by the codec when it's running on slow machines, preprocessing can be maxed out in most cases.

# Chapter 9

# Audio Codecs

## Introduction

This chapter is about the common attributes of audio codecs, and the decisions to be made with them. It follows the structure of the previous chapter on video codecs.

Remember, though video may take 85 percent of the compression time and bandwidth, audio still makes up half the experience. Getting the audio right is just as important as getting the video to look good. It's also quite a bit easier.

## Audio Codec Settings

### General purpose codecs versus speech codecs

Audio codecs fall into two main camps: general purpose codecs and speech codecs. General purpose codecs are designed to do well with music, sound effects, speech, and everything else audio people listen to.

Speech codecs do speech well, and simply can't reproduce other kinds of content. However, speech codecs can go to lower data rates than general purpose codecs, and generally provide better quality with speech below 32Kbps. However, they don't go to higher bandwidths. Speech codecs generally only support monophonic, single channel sound. And, of course, any non-speech content in the audio will either be distorted or removed. Most speech codecs have a pretty low maximum bitrate and quality. The notable exception here is RealAudio Speech, which goes up to 64Kbps and sounds nearly perfect for audio book types of applications.

## Sample rate

As we discussed in Chapter 3, "Fundamentals of Compression," the audio sample rate controls the range of frequencies that can be compressed. 44.1kHz is "CD-quality" and anything much less than 32kHz starts having audible reductions in quality. 22.050kHz is about as low as you can go for "entertainment" quality music content. 8kHz is telephone quality, and about as low as you can go for speech.

Target at least 32kHz, and ideally 44.1kHz sampling rates. 48kHz is increasingly common as a source format, but isn't well supported by older audio cards, and doesn't offer any meaningful quality boost on the user side.

## Bit depth

All audio codecs worth using are at least 16-bit. While 8-bit compression was common in the paleolithic age of computer multimedia, The IMA codec, widely distributed starting in 1995, offered much better compression and higher quality than 8-bit, happily leading to 8-bit's swift demise.

While most codecs only support 16-bit input and output, a few support input and output at higher bit depths. Mathematical models the codecs use may not think in terms of bits, and so they may be able to output 20- or 24-bit, even when 16-bit content is input.

There is one 12-bit codec: DV's four-channel mode. 12-bit simply doesn't have enough bits for high-quality encoding, and shouldn't be used for professional work.

## Channels

Modern computer audio codecs support either mono or stereo audio. Home theater codecs like AC-3 and DTS support multichannel output from DVDs, but they don't really integrate with the multimedia frameworks. This will change with the introduction of Microsoft's Windows Media Audio (WMA) Professional, which will support six-channel audio. However, it will be some time before there is a substantial amount of multichannel audio to encode, and a substantial audience of end users with multichannel audio playback compatible with WMA Pro (which doesn't support existing Dolby Digital-based audio output) as part of their computers. The main initial use of WMA Pro is likely to be with Windows Media-enabled high-definition playback via forthcoming WM-enabled DVD players.

Older codecs allocated equal amounts of bandwidth per channel, which meant stereo required twice the bandwidth of mono. Modern codecs are able to use the redundancy between the channels to reduce the additional bandwidth required (more on that later). Still, going to stereo from mono requires around 20 to 50 percent more bits to achieve the same sound quality.

## Stereo encoding mode

MPEG-derived audio codecs often support a variety of stereo encoding modes. The most basic are joint stereo, normal, and split stereo. Most other codecs are either locked to a particular mode, or pick which mode to use based on data rate.

*Joint stereo*   In joint stereo, the L and R channels aren't encoded directly. Instead, R+L and R-L are encoded. Most stereo audio contains a lot of elements in common, so R+L will typically be a lot more complex than R-L. Thus, more bits can be spent on R+L than on R-L while still providing good quality. This is sometimes called M/S for Mid/Side encoding. Joint stereo provides optimal compression efficiency, increasing when two channels become more alike. However, joint stereo can cause some shifts in the phase relationships between the two channels, which can hurt Dolby Pro Logic encoding, which uses phase relationships to encode surround sound information.

These phase errors keep the audio from becoming truly CD-quality, so if you're shooting for maximum quality without concern for data rate, use normal stereo.

Most Web codecs are only joint stereo, or they automatically turn it on for lower data rates (typically anything under 100Kbps). For Web use, you'll almost always use joint stereo.

*Normal stereo*   In normal stereo, each channel is encoded separately. Some codecs are able to balance bandwidth between the two, so that if one channel requires more bits than the other, it'll get more bandwidth.

*Split stereo*   Split stereo is like normal stereo, except that the bandwidth allocated to each channel is the same. Most older codecs like IMA are split stereo, not normal.

## Data rate

Older audio codecs don't specify a data rate. Instead, their size is determined by the number of channels and the sample rate. The math is simple, sample rate in K × channels × bits is kilobits per second for uncompressed. Divide that by the compression rate for compressed codecs. So IMA 44.1kHz stereo is 44.1 × 2 × 16, divided by 4 for IMA's 4:1 compression ratio. Thus, 352.8Kbps. Easy stuff.

## CBR versus VBR

Traditionally, all audio codecs have been CBR, in that each unit of audio (anywhere between 1/50th of a second to a second) were the exact same size. Even CBR codecs have a little flexibility—MP3 uses a "bit reservoir" where some unused bandwidth from previous blocks can be used by future, more difficult to encode blocks.

The main reason why audio codecs were originally CBR is that it made random access much easier—if you knew what time you wanted, you could figure out automatically where in the file it was. However, video codecs solved this problem a long time ago, since their frames varied a lot in size, and the same techniques were applied to audio.

There are many kinds of VBR in audio codecs. There are even different implementations within the same codec.

*Quality-limited*   Like a quality-limited video codec, a quality-limited audio codec provides consistent quality with highly variable data rates. This form of codec is mainly useful for archiving content.

*Average data rate*   An average data rate VBR raises and lowers the data rate with quality, but attempts to hit a target. This is analogous to 1-pass video data rate-limited video codecs. Options for maximum and minimum bitrates might also be provided. Some codecs may be dead-on accurate for the data rate; others can vary quite a bit (the data rate is more likely to go down than up). AAC is the first common codec with average data rate VBR.

There aren't any 2-pass VBR audio codecs yet, just 1-pass.

*Additive VBR*   An additive VBR, common to MP3, lets you set a target data rate, but it will raise the data rate above that point for difficult audio. Therefore you know the minimum data rate, not the maximum. Still, with an output file at a given final bitrate, the Additive VBR should sound better than a CBR of the same file.

Additive isn't as good as an Average Data Rate VBR for most uses, since neither the final data rate nor the quality can be controlled.

*Subtractive VBR*   In a subtractive VBR, the audio data rate can go down for easy sections, but never goes up higher for the difficult sections. Subtractive VBR can be a good mode for encoding Web video, since you can assure some minimum quality, but can save bits on easy to encode sections.

## Encoding speed

A few codecs have different speed modes. QDesign Music 2 Pro has a simple faster/higher quality option. The LAME and Fraunhaufer MP3 codecs offer 10 levels of quality versus data rate. In the case of MP3, the last few levels get quite a bit slower, but in most cases don't add much quality. Still, because audio typically takes up such a small portion of encoding time, just maxing out the quality options and forgetting about it is a fine approach. You probably won't notice the speed difference, but you may notice a quality difference.

# Trade-offs

Generally speaking, you'll have to choose between speech and general purpose codecs for low bitrates. At broadband and CD-ROM rates, general purpose codecs can be used for just about every kind of audio. For narrowband connections, with luck the content you're trying to compress will be dominated by either speech or general purpose sound. If not, you need to decide which is more important to emphasize. The classic example is narration over background music. In such cases, try encoding with a speech codec first, because the narration is the content that is most important to your communication. If the background music causes too many artifacts, try the general purpose codec.

- **Sample rate**

    For broadband and CD-ROM uses, set your sample rate to at least 32kHz. But keep in mind that higher frequencies may cause artifacts. Lower sample rates are also faster to encode and play back.

- **Bit depth**

    The rule of thumb with bit depth is simple—only use 16-bit codecs. If you have to use 8-bit for some esoteric reason, you'll have to use 8-bit. If you have a choice, avoid it like the plague.

- **Channels**

    For most listeners, stereo is less important than having a high sample rate and few artifacts. Only go to stereo if the stereo signal is critical to your communication goals, or if you have enough bandwidth to do 44.1kHz stereo without many compression artifacts.

- **Stereo encoding mode**

    Most codecs don't offer a choice of stereo encoding modes. Of those that do, joint stereo is the best option for compression efficiency, but normal stereo is best if you need to keep accurate phase relationships for Dolby Surround encoded content, or are doing a high bitrate archive.

- **Data rate**

    Higher data rates make audio sound better, but they require larger files. Typically, the data rate of a file allocated to audio is much less than for video; maybe 1:3 at modem rates, up to 1:10 and higher for CD-ROM rates. Getting the data rate balanced correctly between audio and video is all about balancing mediocrity. You've got it right when the quality gains of raising the audio don't match the quality loss of dropping the video data rate by the same amount.

- CBR versus VBR

    Most audio codecs are still CBR. If you have a codec that supports VBR, you probably want to use it to get maximum compression efficiency.

- Encoding speed

    Encode in the slowest, highest quality mode your codec provides when you're encoding to a file. For live broadcasting, you may need to use a faster mode.

# Chapter 10

# CD-ROM, DVD-ROM, and Kiosk Delivery

## Introduction

Compressed video delivery can be broken down into two broad categories—disk-based and Web-based. Video delivered on disk-based media of one sort or another has been around far longer than Web video. Disk-based media traditionally meant CD-ROMs, but has come to include all the different flavors of DVD as well as playback from hard drives. In this chapter, when talking about DVD delivery, I'll only be talking about DVD-ROM playback—DVD and VideoCD are covered in-depth in the MPEG-2 and MPEG-1 chapters, respectively.

## Characteristics of Disk Playback

The two biggest advantages of disk playback over Web playback are low latency and high bandwidth. Because the video files reside on local storage devices, random access is nearly instantaneous—there's no buffering, no waiting.

CD-ROM bandwidth is also huge. Back when the 9,600 baud modem was at the cutting edge of online access, even 1x CD-ROM drives were able to provide a full 1,000Kbps of bandwidth. Today's CD-ROMs provide more bandwidth than delivery codecs are able to use. This makes better-than-broadcast quality video delivery a perfectly reasonable goal.

Another advantage is the disk can be used when the user is offline or has a slow Web connection. You can generally include an installer with your CD-ROM-based video, reducing the risk of losing a user if they don't want to wait for a big download of either your media or specific media player.

On the downside, there are two big limitations to disk playback. First, the user must have access to the media. If the video is a cut scene in a video game, no problem—your audience already has to have the CD. But if your audience is looking for training materials, delivering those materials on disk means an end user will need to track down that disk before they can experience your content. Hence, the computer-based training industry is moving away from CD-ROM delivery in favor of Internet and intranet-based delivery. Second, no matter how large the hard drive, there's a limited amount of local storage available. The biggest CD-ROMs today can only hold 870MB, max. Compared to Web or intranet delivery, CD-ROMs offer much higher bandwidth, but servers and server arrays can usually hold as much video as you might want to stick on them. A single-layer DVD can hold up to 4.7GB. A double-sided, double-layered DVD-ROM can hold up to 17GB, enough to hold 25 CD-ROMs, but DVD still has a limit, as does a local hard drive when used for kiosk.

## Authoring for Disk Playback

Authoring for disk playback is usually a lot easier and more forgiving than authoring for delivery over the Web. For any computer built in the last five years, there will be ample bandwidth for transferring data between disc and computer.

### A few terminology notes

Traditionally, CD-ROM data rates were measured in kilobytes per second (KBps), not in kilobits per second (Kbps). However, for consistency's sake, I'm going to continue to discuss data rates in Kbits for this chapter. Older software or documentation (and, even today, QuickTime's default export dialog box) still measure data rates in KBps. So if a CD-ROM-oriented resource talks about a "200K/sec" video, it might mean 1,600Kbps (remember there are eight bits in a byte).

You might also see speeds measured in "X" terminology, which refers to multiples of the 150K/second (or 1,200kbps) speed of a first generation CD-ROM drive. So, 4x video targets a 4x CD-ROM drive, which has a theoretical rate of 600K/sec, or 4,800 kbps. However, 4x video will be encoded at a total data rate somewhat less than the theoretical maximum to provide sufficient headroom.

An "X" for a DVD-ROM is about 10Mbps, so even a 2X DVD-ROM drive provides huge amounts of sustained bandwidth.

Even the slowest modern internal hard drive provides at least 50Mbps sustained bandwidth.

### Peak data rate limits

Back in the paleolithic days of video compression, data rate was all about how fast the computer could get data off the CD-ROM. When I started doing CD-ROM-targeted video compression, 2x CD-ROM was only then becoming standard, and there were a lot of 1x drives out in the wild. Targeting older machines meant using a maximum data rate that would fit in the theoretical 1,200kbps. At the time, Mac OS systems all used SCSI CD-ROM drives, and were generally much better multimedia machines, so we would cut 15 percent off the maximum data rate for Mac OS projects and 30 percent off for Windows.

Look for the data rate the drive is capable of sustaining, which will be much lower than the "burst" rate often listed in marketing hype. The minimum sustained data rate the system is capable of producing must be higher than the highest data rate peaks in the streaming file. CD-ROM drives do have a buffer, so a single large keyframe won't kill performance, but the average should be over any period of a few seconds.

The "X" terminology has long since stopped being meaningful. These days, drives are advertised as "up to 56x" or whatever. You shouldn't care what the peak data rate is; pay attention to the lowest sustained data rate that is going to work.

The good news is that virtually any recently manufactured computer comes with a CD-ROM drive capable of at least 8x speeds, or 9,600Kbps. This is more than ample for almost any codec, and is comparable to MPEG-2 peak data rates. Modern codecs are typically limited to a maximum data rate that is much lower than this. The limiting factors are total disk space and how fast your target computer is.

## Disk space limit

Since there's so much bandwidth available for disk-based projects, the real limit is in how much space you have available. If you have a full disk on which to put video, and only have 10 minutes of content, you can use huge data rates.

First, some handy math:

```
Data rate=space/time
```

Or, in more conventional units

```
Data rate in Kbps=space in MB/minutes of video*133
```

Thus, if you have 600MB available and 60 minutes of video, your total data rate can't be higher than 1,330Kbps. Of course, in a given project, different clips may be of differing relative importance. Remember, some of the disk will likely be used for applications, installers, and other content beyond the video files themselves. Ideally, you'll know the size of all those assets in advance. I like to leave an extra 20MB of headroom—something always seems to come up at the last second of mastering that requires more space.

### 2-pass VBR

The constrained average data rate with high peak bandwidth model fits unbuffered 2-pass VBR perfectly. Good CD-ROM codecs like Sorenson Video 3 and DivX provide 2-pass encoding for optimal quality. Of course, 2-pass VBR only works across a single file. In an ideal world, performing 2-pass VBR encoding across multiple files would allow you to obtain the highest average quality.

>  **Note**
>
> **A cunning trick for multi-file 2-pass VBR.** Here's a clever trick that works with QuickTime to use 2-pass VBR over a whole disc's worth of files.
>
> - Edit together all your source files, either in an NLE or by pasting them together in QuickTime Player Pro.
> - Compress the file in a tool that supports both manual keyframes and 2-pass VBR. Today, the only such tool is Discreet Cleaner. Put a manual keyframe at the start of each clip. Then encode the whole file normally.
>
> Once compressed, you can extract each section into its own file via QuickTime Player Pro. Since there's a keyframe at the start of each file, they are fully self-contained. Alternatively, you can just put one big file on the disc, and use Director or similar authoring environment to control random access to the various sections. Either way, you gain the benefit of a 2-pass VBR over an entire disc's worth of files.

## Interactivity

Video on discs almost never just files. A UI is usually provided. Web interfaces are often used, but in such cases interactive features rely on the particular browser installed. You'll gain more predictable results and supply your audience with better control of your content if you build your own player application.

The leading tool for CD-ROM authoring is Macromedia Director. Director can provide rich, complex interfaces, works well with QuickTime, and provides basic DirectShow playback. With plug-ins (called Xtras), it can do much more complex DirectShow and MPEG-1 playback. I talk about using Director with the various formats in the sidebar beginning on page 153.

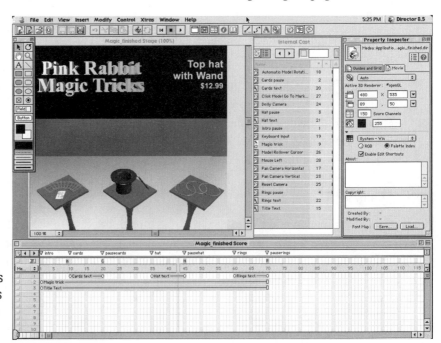

**10.1** The Director user interface. Director's central metaphor is a timeline, as seen at the bottom.

## Cross-platform discs

Many CD-ROMs are designed to play back on both Mac and Windows machines. This is easily accomplished with the right tools.

Windows CD-ROM discs use the ISO 9660 format, with recent discs using the Joliet extensions to allow file names of more than eight characters. Macs natively use either HFS or HFS+, and can also read ISO 9660 and Joliet discs (only the first 31 characters of the filenames work under Mac OS v9, although all appear under Mac OS X). DVD-ROM uses the UFS format, which can be read by any computer that supports a DVD-ROM drive.

For basic cross-platform discs, you can get away with just using ISO 9660+Joliet. However, Mac systems can use more complex icons and can have better control over icon placement than Joliet provides, so most authors make discs that are hybrids consisting of HFS and ISO 9660 portions.

The tool of choice for making hybrid discs is Roxio Toast Titanium for Mac (formerly from Adaptec and before that Astarte). This once-$500 tool is now only $99, and hides a lot of its awesome power under a bubbly consumer interface, but it remains an essential tool for anyone making discs to be played on Macs.

One great feature of Toast is the ability to share content between the HFS and 9660 sections of the disc. Thus, the platform-specific installers, applications, and notes are exclusive to each format, but the huge media files can be shared instead of having to be duplicated for each section.

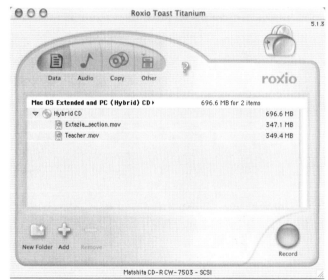

**10.2** Toast Titanium, set up for a simple hybrid CD burn.

Toast does a lot more than this. It's a tool that gives a critical advantage to the Mac platform.

## Installers

It's good practice to put installers for your playback software on your CD-ROM discs. This is easy to do with QuickTime, but somewhat more tricky with the DirectX libraries used to play back Windows Media, AVI, and other formats on Windows.

## Formats for Disc Playback

While every format is capable of doing disc-based playback, only a few are well suited for it. These are QuickTime, AVI, MPEG-1, and Windows Media. There are a lot more notes about the strengths and weaknesses of each format for this kind of work in the format-specific chapters, but we can briefly outline a few things here.

*QuickTime*   QuickTime is the most mature, and has the best functionality for more involved tasks. QuickTime provides a good range of audio and video codecs.

Apple provides a free, licensable OEM version of the QuickTime installer that supports scripted installs. This means you can create an application that will automatically install or update required QuickTime components.

*AVI*   As you'll read later, AVI is a file format that works in multiple media architectures including QuickTime (if the codecs are installed), DirectShow, and a smattering of UNIX apps.

AVI's big advantage over QuickTime is that Windows machines don't typically need additional software installed if you stick to standard codecs. However, if you can't install any software, you may be limited to ancient codecs such as Indeo and Cinepak.

*MPEG-1*   If an installer isn't an option, your best choice for a file format is MPEG-1. MPEG-1 offers decent compression efficiency and ubiquitous playback on almost all platforms. It plays back natively in QuickTime on Mac and DirectShow on Windows, and in a variety of players on other operating systems.

I strongly recommend using square-pixel MPEG-1 (e.g. 320×240 instead of 352×240) for disc playback—more on that in Chapter 17.

*Windows Media*   Because Windows Media can play via DirectShow, applications that support DirectShow can play .wmv files like any other media type. Windows Media has excellent codecs and a great 2-pass VBR mode. However, integrating Windows Media in this fashion limits playback to only recent versions of Windows.

In a staggering act of atavism, Director, at least up to v8.5, uses the ancient VfW (Video for Windows) API instead of DirectShow. For more on that, check out the following sidebar on Director.

*Flash MX*   Macromedia's new Flash MX platform with Spark Pro video is becoming very interesting for CD-ROM delivery. With Flash MX, you can convert a Flash movie into a stand-alone app for Windows or MacOS Classic and X. Video playback performance isn't as good at high resolutions as Director, but no installer is needed for video playback.

# Director

Macromedia Director ruled the world of interactive CD-ROM, and is still the tool to beat for interactive authoring tasks. With Director, you can build complex presentations using a visual authoring tool and a programming language called Lingo. Director can compile standard applications for both Mac (Classic only so far, not MacOS X) and Windows.

## Formats

Director embraced the Web from v4 on, introducing Shockwave as a way to deliver platform-neutral, compressed multimedia presentations. Books can (and have) been written about Director. Many folks love it for its power and extensibility (it supports a wealth of third party plug-ins called Xtras). Other folks hate it for its cell-animation–derived UI and indifferent adoption of object-orientation. I'm going skip the debates and focus on Director as a video playback platform.

On Mac, Director uses QuickTime for all CD-ROM video playback. On Windows it can use QuickTime and VfW, not DirectShow. However, DirectShow support is possible with plug-ins from Tabulerio Producoes Ltda. (they're Brazilian). Among others, they provide plug-ins for MPEG-1, DirectShow, and Windows Media streaming inside Director ($199.95 each, with a 25 percent discount if you already own one). Director can also embed RealVideo streams, but this feature is more important for Web use.

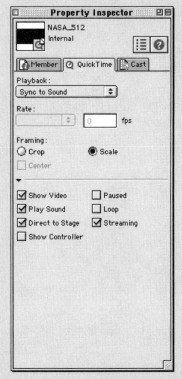

**10.3** The QuickTime Property Inspector in Director. Director offers a variety of useful controls for working with QuickTime projects.

Overall, I recommend using QuickTime inside Director unless you have a compelling need for another format. It works great cross-platform, and Lingo supports a number of features in QuickTime that don't work for AVI. Note this is a feature of the QuickTime interface, and you can tell Director to use QuickTime for AVI files as long as QuickTime also supports the codec in question.

### *Optimizing Director for video performance*

Director can do a lot of excellent tricks with video, like real-time alpha channeling, having multiple actions happening at the same time, and so on. However, all those activities can cause quite a performance hit, especially at higher resolutions. If you want maximum performance, eliminate any unnecessary Director activity while video is playing. You also want to use "Direct to Stage" mode. With Direct to Stage, Director lets QuickTime or VfW push pixels straight to the screen, without Director touching them. This enables YUV overlays to work, and vastly improves performance.

### *Casts*

Traditionally, video would be kept in files on the CD-ROM. This, however, allows users to directly access the video, which is not always welcome. To avoid this, incorporate the video into a cast, which bundles multiple media assets together into a single file (including inside the application itself).

Casts can also compress bitmap images used in the Director presentation, either losslessly or as JPEGs. Casts can also encode audio via Shockwave Audio (which is based on MP3). Compressing these assets (if you have a lot of them) can save some room for video.

# Chapter 11

# Web Video

## Introduction

For most people, video compression equates with putting video on the Web. Back when CD-ROMs ruled, Web video was "a year away" for a lot of years. But in 1998, my video compression business went from being 80 percent CD-ROM to 80 percent Web over a mere 12 months. This was frustrating in some ways—we'd finally gotten CD-ROM performance to approach broadcast quality, so optimizing video for tiny windows and narrow bandwidths was like starting over with QuickTime v1. Four years later, better codecs, faster computers, and faster connections have brought Web video to nearly the same point CD-ROM video was in 1998.

That people were willing to put up with postage stamp video meant Web video had to be doing something right. And it was. The promise of the Web was always to provide a searchable, immediately available network of content. No having to try to find the CD-ROM at a store. No limit to how much video could be put in a given project.

The Web is good stuff. I'm hoping in a few short years that missing an episode of *Buffy the Vampire Slayer* when I'm on the road just means I'll pay $1 for a Web-based video-on-demand repeat. The technology is here today, although the business models haven't been worked out yet.

When we finally get it to work, Web video should change how the entertainment and education industries work. Unfortunately, a lot of crappy content coupled to insane hype gave this vision a bad name. The Web video revolution isn't about making everybody Orson Welles; easy distribution does not talent make. But Web video will transform the way we experience content from something we synchronize our life to something that synchronizes to us.

## Connection Speeds on the Web

The critical factor in putting video on the Web is connection speed. Compared to CD-ROM, the available bandwidth can be minuscule, especially with modems and mobile devices. Fortunately, broadband has been becoming increasingly mainstream over the last few years. This is good, because modems really suck for video. A 56K modem will at best run at 53Kbps, and if it's on a noisy phone line, or calling through a hotel, even 24Kbps may be hard to sustain. In late 2001, broadband finally accounted for more total hours of Web access than modems (in part because people with broadband stay on the Web longer).

Of course, just because someone is accessing Web video through a T1 line at work doesn't mean they're able to access video at T1 speeds. Office bandwidth may be shared among dozens or hundreds of users, not to mention other functions like Web and email servers.

| Connection type | Rated bandwidth (Kbps) | Notes |
|---|---|---|
| PCS wireless modem | 9.6 | |
| 28K modem | 28 | |
| 56K modem | 53 | Bad connection can drop down to 28K speeds |
| Single ISDN | 64 | |
| Dual ISDN | 128 | Two single channels paired together |
| 3G wireless | 150-450 | |
| Satellite | 400 | Lower during peak times of day |
| XDSL | 256-4000 | Stable to rated speed |
| Cable modem | 200-5000 | Shared bandwidth through neighborhood, can vary with time of day |
| T1 | 1500 | Typically shared by many users |
| T3 | 45,000 | Typically shared by many users |
| Local 802.11b | 11,000 | Shared bandwidth, can drop rapidly with interference |
| Local 100Base-T | 100,000 | Switched network can get within 80 percent of rated performance |

# Four Kinds of Web Video

Web video is a single phrase that can mean a lot of different things. First, we'll break down Web video into its major categories, and then talk in more detail about how it works.

## Downloadable file

A downloadable file is the simplest form of video on the Web. The audience uses a file transfer mechanism such as FTP to download the file. No attempt is made to play it in real time.

This method is mainly used for pirating movies à la Napster and Gnutella, or for video game cut scenes. The advantage of downloadable files is that there is absolutely no expectation of real-time performance or interactivity. The limit to data rate is how big a file a user is willing to download.

## Progressive download

Progressive download sits between downloadable files and real-time streaming. Like file-based content, progressive download files are served from standard Web and FTP servers. The most important characteristic of progressive download is that it is not real time, and it uses lossless file transfer protocols like HTTP and FTP. These two attributes go hand in hand.

All content transferred over the Web is broken up into many small packets of data. Each individual packet can take a different path from the server to the client. HTTP and FTP always deliver every packet of the file. If a packet is dropped, it is automatically retransmitted until it

**11.1** QuickTime Player doing progressive download. The portion with the solid line has been transferred, and the user can do random access within this area. The portion with the dashed line has not yet been transferred.

arrives, or until the whole file transfer process is canceled. Because of this, it is impossible to know in advance when a given packet is going to arrive, though you know it will arrive. This means playing in real time can't be guaranteed, but video and audio quality can be.

Modern progressive download systems can start playing the file while it's partially transmitted. This means less waiting to see some video, and gives the user the ability to get a taste of the content, and the option to terminate transmission should they decide they don't want to see the whole thing. This ability to play the first part of the video while the rest is being transmitted is the biggest advantage progressive download has over downloadable files.

At the start of the progressive download, the user initially sees the first video frame, movie transport controls, and a progress bar indicating how much of the file has been downloaded. As the file is downloading the user can hit play at any time, but only that portion of the file that's been transmitted can be viewed.

Most progressive download files are set to start playing automatically when enough of the file has been received that downloading will be complete before the file finishes playing. In essence, the file starts playing, assuming the transmission will catch up. This works well in most cases, but when a file has a radically varying data rate, a data rate spike can cause the playhead to catch up with the download, and thus pause playback for a time. The major players indicate how much of the video has been downloaded by tinting the movie controller differently for the downloaded portion of the movie. As long as the playhead doesn't catch up with the end of the gray bar, you won't have playback problems. Typically you can do rapid random access within the downloaded section, but can't move the playhead beyond it.

The formula most systems use to determine when to start playing a file is when the remaining amount of data to be transmitted, at the current transmission rate, is less than the total duration of the file. From this, we can derive the following formula to determine the buffering time, that is, how long the user will have to stare at the first frame before the video starts playing. Note connection speed is the actual connection speed, not the rated one (so a 56K modem might have an actual value of 48Kbps or 20Kbps).

$$\text{BufferTime} = \left(\frac{\text{FileDataRate}}{\text{ConnectionSpeed}} - 1\right) \times \text{Duration}$$

This formula has some interesting implications. First, if the connection speed is greater than the file data rate, then the buffer time is nil—a progressive download clip can start playing almost immediately at low data rates or high connection speeds. Second, as the ratio between the clip's data rate and the connection speed changes, buffer time can change radically; a 5-minute 300Kbps clip will play in real time at 300Kbps, have a 2.5-minute buffer at 200Kbps, have a 5-minute buffer at 150Kbps, and have a 45-minute buffer at 30Kbps. Longer duration hurts buffer time—a 2-minute movie trailer with a 6-minute buffer is much more palatable than a 2-hour movie with a 6-hour buffer.

## Real-time streaming

The defining characteristic of real-time streaming is, of course, that it works in real time. No matter the duration of the file, it'll start playing in less than a minute, and assuming stable and sufficient bandwidth, it will play all the way through. Real-time streaming also provides random access, so the user can fast-forward and rewind. Each time the playhead moves, the video will have to rebuffer.

Real-time streaming requires specific streaming video server software, vendor-specific for Windows Media and RealVideo, and with a variety of options for QuickTime and MPEG-4. Such servers are required to support the protocols used, MMS for Windows Media, and RTSP for all other formats.

All mainstream real-time streaming solutions are able to fall back to TCP or HTTP transmission as well, to pass through firewalls that don't support their streaming protocols. The real difference between real-time systems with fallbacks and progressive download is that progressive download always transmits the entire file, from start to finish, while real time from a server offers random access and bandwidth negotiation, even when falling back to other protocols.

A critical real-time streaming feature of Windows Media and RealVideo (and tragically lacking in QuickTime) is multiple bitrate control (MBR). The concept of MBR is to create multiple versions of a given content stream, dynamically switching among the versions based on the user's connection speed. Thus, if a 56K modem is actually running at only 24Kbps, a lower data rate version of the content is delivered.

MPEG-4 uses a variant of MBR in its scalable profiles, where there is a base layer providing minimum quality, and additional enhancement layers can be added if there is sufficient bandwidth, improving resolution, frame rate, and image quality.

## Multicasting

Multicasting is real-time streaming where multiple recipients can all view the same content at the same time. This contrasts with the unicasting model of normal real-time streaming, where a single on-demand stream is sent to each user who requests it.

The original vision of multicasting was to do it via IP multicasting, often called mBone after the experimental network that IP multicast was developed around. With IP multicast only one copy of any given packet needs to be sent to any given router (a connection node on the Internet). That router then sends a single copy of each packet to each additional router that is subscribed to the IP multicast. The net effect of this should be to dramatically reduce the bandwidth requirements of multicasting as the number of users scales up.

However, for multicasting to really work as advertised, every router between server and client needs to be multicast-enabled. Even though modern routers support multicasting, many older

**11.2** An example of how a unicast and a multicast transmission loads. Note that even one non-multicast router can eat up the majority of the bandwidth.

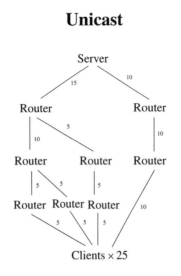
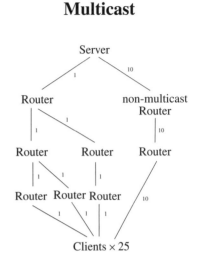

routers don't, or if they do, their operators don't have multicasting turned on. It only takes one non-multicast router to disable multicasting for every client connecting through that router.

Providing a sort of pseudo-multicasting is one of the features Content Distribution Networks (CDNs) like Akamai provide. CDNs handle the multicasting internally, so only one copy of the stream is sent to each edge server with one or more clients connected to it. More about this in the following section on hosting.

The biggest drawback of multicasting is that it doesn't support video-on-demand (VOD)—with television, multicasting requires users to tune into a broadcast. For time-critical information, this isn't a problem, but for entertainment content, people on the Web want to see content at a time that's convenient for them. After all, we already have a great broadcast quality, analog multicasting (it's called television).

Even simple things like pausing don't work in most multicast implementations. Other features might not be available in a given format when multicasting, e.g. retransmission of dropped packets, generally doesn't happen. Windows Media's Intelligent Streaming bandwidth negotiation feature isn't available for multicasting either.

## Live broadcasting

Live broadcasting is a special form of multicasting in which the video is captured and compressed as well as transmitted in real time. Webcasting used to be a difficult, finicky, and low-quality mechanism. However, improved hardware and software brought Webcasting to maturity in rapid fashion. Faster processors capable of faster, better real-time compression and capture cards that support higher quality connections and on-card preprocessing, leaving more CPU power for the codecs themselves, have made quite a difference in the state of the art of live Webcasting.

Remember, unless the content itself is live, you can make a TV-like experience with multicasting by capturing and compressing the content before broadcast. This is a lot easier and more reliable. You should only do live broadcasting when you absolutely have to be live.

Still, mission-critical live broadcasting is about the most frightening things you can do in streaming. One severed network connection, one person tripping over a power cord, and the whole production comes to a crashing halt—just like live television.

While some formats let the same box do both capture and serving, it's preferable to break up those tasks onto different machines. Should one module go down, it's easier to swap it for another than it is to swap out your whole production pipeline, even if it's a self-contained system.

 **Note**

Web formats—even when doing live broadcasting—introduce latency in the broadcast. This means that web broadcasters are simply not viable for videoconferencing because there will be a several second delay. This isn't a bug; the web formats use this latency to optimize quality, which is one reason why they look so much better than videoconferencing at the same data rate.

Videoconferencing requires videoconferencing products, generally based around the H.323 standard.

## Selecting the Right Transmission Mode

Choosing the right transmission mode is an important decision that will have substantial ramifications throughout the compression and delivery process. For most projects, the choice is between progressive download and real-time streaming. While multicasting and live broadcast have their place, you really only want to use those technologies when you have to for a specific project when nothing else will do.

The choice between progressive and real-time breaks down to being able to control how the content will look and sound, or when it will arrive. You can't have both. With progressive, you know precisely how each frame of video will look, and how it will sound. But you can't know how soon it'll play, unless the user has much higher bandwidth than the data rate. With real-time streaming, you can control time, but the quality of the file can vary enormously with the quality of the connection and its data rate.

Because progressive download is much more flexible to encode and host, I start with progressive download as the default, and look for reasons why real time might be a better choice. The biggest objection to progressive download is the long buffering times, so I look for things that could cause a long buffering—content with a long duration, and a potentially large mismatch between the file's data rate and the user's actual connection speed. If your audience has a significant number of both modem and broadband users, a good idea is to provide multiple versions of progressive download content so they can pick one appropriate to their connection (QuickTime can handle this automatically, with some caveats).

One old argument for real-time streaming is that it made copy protection a lot easier. This isn't as important as it once was—Windows Media's Digital Rights Management (DRM) solution works just fine for both progressive and real-time streaming files.

# Integrating Web Video

Web video is almost never seen by itself. It is launched from a Web browser or other type of application.

## Metafiles

One important bit of Web video technology is the metafile, sometimes called the *reference file*. A metafile is a short text file that lives on an HTTP server that provides a link to where the actual streaming file is on the streaming server. The metafile can be just a simple link, or can provide instructions (such as specifying playback resolution or defining what skin to use) for how the video should be played.

The Windows Media metafile format is ASX, with files with the .asx (old style), .wvx (video), and .wax (audio only) extensions. The RealVideo metafile format is .ram. In QuickTime, metafiles are just a different kind of .mov file (nice and simple). You can read about the details of each in their respective chapters in this book.

## In-browser

In-browser video is typically embedded into a frame on a Web page. This is the default mode of QuickTime, and is easily accomplished with Windows Media and RealVideo, which by default appear in their respective media players.

### File type issues

When a file type is only supported by one player technology (like RealVideo or Windows Media) things are easy—the file automatically plays in the proper player. Things get more complex with file formats supported by multiple players, like .mov and .mpg. This has historically been the biggest problem with QuickTime.

There have been two different ways to embed video into a Web page: EMBED and OBJECT tabs. EMBED was pioneered by Netscape. Later, OBJECT was introduced. It offers a better way to associate a particular player with an object, but OBJECT wasn't supported in older Netscape browsers. From IE v5.5 SP2 on, Microsoft's dominant browser no longer supports EMBED. This has forced content authors to use a turgid combined tag. You'll find more specific information on this is in the QuickTime chapter.

For files that are progressively downloaded, MIME types also need to be set up correctly on your server. MIME (Multipurpose Internet Mail Extensions) types are information on your server that tells the browser which application or plug-in is used to play that file type. This helps guide the file to the right playing environment, although the user's browser can still have the wrong local plug-in or application assigned to the file type.

For example, QuickTime's MIME type is

```
.mov   -   video/quicktime
```

And RealVideo's are

```
.ra    -   audio/x-realaudio
.ram   -   audio/x-pn-realaudio
```

Note that the RealVideo options MIME types are also used on Mac OS to set the proper application and file metadata.

### Frames

The normal reason to have video play in-browser is to have in a frame, with other content displayed onscreen at the time. While it's possible to have a movie fill the entire browser window, this generally looks silly—it'd be better to spawn to the player in such cases.

Playing video in a frame entails a few complexities. For one, any activity in other frames will suck CPU power available for decode, so machines that are only marginally able to play back the video will have an even more difficult time.

It's also important to make sure the resolution of the embedded object is accurate when using a frame. A classic problem is to use the resolution of the clip as the size of the object, forgetting to include room for the controller.

## Spawning to player

The default mode of RealVideo and Windows Media is for a link to launch and play the video in the respective media player. Launching the QT player is an option with QuickTime. This has the advantage that the architecture's creator can control and tweak the player experience, potentially improving performance, and adding additional features like being able to trigger full-screen playback.

One neat feature is skins (see Figure 11.3), where the player itself is given a custom appearance. This helps provide a complete branded experience. Each of the Big Three supports a skins file format to redefine the appearance of the player on the screen. QuickTime authoring tools like Tribeworks iShell and Totally Hip LiveStage Pro can go beyond this to provide amazingly rich, complete presentations delivered inside QuickTime Player.

## Custom applications

If a browser or skin doesn't provide deep enough support, a custom application can be built for playback. The Big Three architectures all provide SDKs for incorporating playback into third-party apps. Java also provides the Java Media Framework for pure Java video playback (QuickTime also has excellent Java support).

**11.3** A custom QuickTime Player skin created by Feelorium's Tattoo, over a flat gray background. Note the skin is actually casting shadows on the background.

Making a custom application is a lot of work, and as browser functionality and skinning improve, it's less and less necessary. Still, it may be the only way to provide a 100 percent predictable user experience.

The leading tool for making custom rich media applications with video is, again, Macromedia Director, either as an executable application, or as a Shockwave applet.

## Hosting

Web video, by definition, needs to live on a server somewhere. What that server is and where it should live vary a lot depending on the task at hand.

Server type is simple to define. Progressive download lives on a normal HTTP or FTP server, and real-time streaming servers use an appropriate server that supports the format. For RealVideo and Windows Media, this means a server from the respective vendor. For QuickTime and MPEG-4, there are a variety of servers to choose among, although in both cases many of them are based on Apple's open-source Darwin Streaming Server.

The biggest question is where to put your video: on an in-house server, a hosting service, or on your own server in a colocation service? In general you want your files to be as close to the viewer as possible.

## In-house hosting

In-house hosting makes most sense for content that is going to be mainly viewed on an internal network. In-house hosting only makes sense for providing content outside your network on extremely small or large scales. Because most businesses are provisioned with fixed bandwidth, your simultaneous number of users is limited to how much bandwidth you have. If you want to handle a high peak number of users, you'll need to buy far more bandwidth than you'd use at non-peak times.

This may not be a problem when the peak number of users is inconsequential (say a few per day), or you have such a tremendous and varied audience that usage averages out (say if you're Yahoo or AOL). For most folks in between, a hosting service or colocation makes a lot more sense.

## Hosting services

With a hosting service, you rent space on their servers instead of having to provide or configure servers yourself. Hosting services are the easiest and cheapest way to start providing files to the public Internet. Hosting services are also much easier to manage, and can provide scalability as your bandwidth usage goes up or down. You are typically billed on how much bandwidth you use in a month, so huge peaks and valleys are much less of a problem.

High-end hosting services like Akamai describe themselves as Content Distribution Networks (CDNs). A CDN isn't just a single server, but a network of servers. The connections between the servers are high-speed, high-quality, and multicast-enabled. The idea is to reduce the distance packets need to travel on the public Internet by putting servers as near to where customers are (a.k.a. all around the "edge" of the Internet) as possible. For multicasting, this means that multiple streams only need to be distributed from each local server. For static content, files can even be cached at the local server, so it would only need to be transmitted once from the center.

## Colocation

A colocation facility is a hybrid between in-house and hosting services. You provide your own server or servers, but put them in a dedicated facility with enormous bandwidth. The canonical company providing colocation was Exodus, now a part of Cable and Wireless.

Colocation makes sense when you have enough volume that it's worth the cost and complexity of managing your own servers. Because you provide more management, colocation is typically quite a bit cheaper per bit transmitted than a hosting service. The biggest drawback to colocation is that you need to physically add more machines to gain scalability. Another benefit/feature of colocation companies is that they can tap into a CDN's network of servers.

# Chapter 11: Web Video

# Chapter 12

# Rich and Interactive Media

## Introduction

Traditionally, compressed video has been treated as some kind of souped-up videotape or television broadcast-like experience, watched end-to-end in strictly linear fashion. However, such an experience doesn't take advantage of digital content's main advantage—no linear tape! Having video on a spinning disc in a computer means random access is possible, providing near-instantaneous searching. Coupled with the complex UI a computer-based system can provide, video can be made interactive. The most common example is found on modern DVDs, which offer animated menus, seamless branching among different versions of a movie, or different angles, access to still photos, and other neat features.

Such advanced features can be put in two classes: Interactive Media and Rich Media. Interactive media is any medium where you can do random access of the content in a more structured fashion than just fast-forward and rewind. Rich media is when other data types, like still images or text, are synchronized with video and audio playback.

Most formats that support one mode also support the other, although the rich and interactive features vary wildly between formats. The ways they're implemented are also quite different.

## Interactive Video

From one perspective, all video on demand is interactive, in that the user can pick whether and when to watch the file. Being able to pause, stop, rewind, and fast-forward is also useful. However, none of those really offer anything beyond what a VCR can do.

A basic level of interactive video breaks the video into chapters. This allows the user intelligent random access to particular parts of the video. This is a wonderful technique when dealing with long-form and instructional content.

## Chapter 12: Rich and Interactive Media

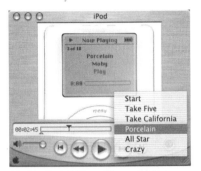

12.1 QuickTime Player's chapter track navigation system, in a demo for Feelorium's Textation.

More complex interactive video can offer branching between clips. In the early days of compressed video, the "Interactive Movie" was a holy grail for many. However, it quickly became clear users preferred either linear content (a movie) or content with immediate interactivity (a computer game). Branching can make sense for training or diagnostic projects; e.g., "If the head gasket looks like this, click on A, otherwise click on B."

## Rich Media

Non-video and audio content synchronized with video and audio presentation, a.k.a. rich media, can be a very effective way to deliver high quality content over slow connections, as each object only needs to be sent once. A classic example of this is synchronizing PowerPoint slides with video. By not sending difficult-to-compress PowerPoint through the video stream, the video data rate can be much lower. Instead, PPTs can be sent as much smaller still image files that take up the same file size no matter if they're on the screen for five seconds or five minutes.

The best known tool for doing these synchronized PowerPoint presentations is Microsoft Producer. It works well. Its big limitation is that IE v5.5 SP2 or later is required for playback, which means it only works under Windows, and not with other OSs or other browsers.

Another handy application of rich media is for doing subtitles. Incorporating legible subtitles as part of a video stream requires a lot of bandwidth, and the quality is often distractingly poor. But sending just the text data requires almost no bandwidth, and it can be rendered in very high quality on playback.

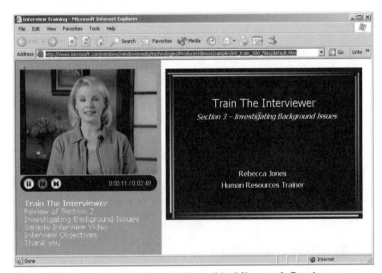

12.2 This is a presentation authored in Microsoft Producer, integrating PowerPoint with Windows Media.

# Authoring Rich and Interactive Media

The big drawback to these forms of media is that they require a lot more labor at the compression stage. Relative to the time spent producing and editing the content, this is generally trivial, of course. Each format has a very different way that the rich content is authored and deployed.

## SMIL

The Structured Multimedia Integration Language is a W3C standard for rich media. SMIL is structurally similar to HTML in that it provides a layout system, where different frames of the document store different kinds of data. Version 1 of SMIL was adopted by RealNetworks as the basis for interactivity in RealPlayer, and was then adopted by a number of other player technologies (many of which don't handle standard audio and video types).

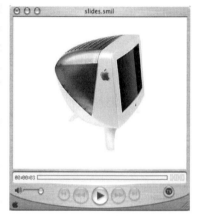

**12.3** The presentation SMIL made, in QuickTime Player. If it weren't a picture, you'd hear an audio narration play as the slide image changed.

SMIL can do a great job with basic layout needs, setting up the window as a series of rectangles, each containing some kind of media, be it video or a still. It also includes some very useful features, such as being able to switch among different settings and embedded files based on things like language, connection speed, screen size, and so on.

SMIL v1 is supported in RealPlayer from G2 on and in QuickTime from v4.1 on. However, most SMIL files with video in them only target RealPlayer. For a SMIL file to run in another player, some modifications are almost always required.

SMIL v2 added a number of very welcome features, like transitions (including most SMPTE wipes), animated position and elements attributes, enhanced interactivity, and advanced layout (including hierarchical combinations of objects), and control of objects in multiple windows.

SMIL v2 is supported in RealOne and later, and Internet Explorer v5.5 and later (providing SMIL functionality to Web pages that can also contain Windows Media, although not internal to the player). On the following page is a sample SMIL v1 presentation code (used by permission from Steven Guile's *QuickTime for the Web Second Edition*).

One of the nice things about SMIL is that v1 is a structured, text-based format, similar to XML. Version v2 is XML-based. This means it's relatively easy to dynamically generate SMIL presentations, changing values and files on the fly. A simple Perl script can radically modify or customize a presentation.

```
<smil>
<head>

<layout>
<root-layout id="rl" width="380" height="320" background-color="white" />
<region id="main" width="100%" height="100%" fit="fill" />
</layout>

</head>

<body>
<par>
  <audio src="narration.aif" />
  <seq>
        <img src="slide1.jpg" region="main" dur="5" />
        <img src="slide2.jpg" region="main" dur="5" />
        <img src="slide3.jpg" region="main" dur="5.1" />
  </seq>
</par>

</body>
</smil>
```

## HTML+Time

Microsoft has a very different take on rich media integration. Instead of adding such features to the player, most rich media features are added to the browser, which then can embed the various media types inside the browser, with video, still images, text, and so on as elements. There is nothing fundamentally wrong with this approach in which the browser is the player.

Instead of SMIL, Microsoft developed and adopted HTML+Time as their standard. It is similar to SMIL on a very basic level, but quite different in implementation. Its biggest limit is that support for Windows Media rich media is limited to Internet Explorer 4+ on Windows. Thus, there is no Mac, no Linux, and no Netscape playback.

Because HTML+Time is a markup language like SMIL, it shares SMIL's ability to be dynamically generated.

## Wired Sprites

Beyond SMIL, rich media in QuickTime is implemented by tracks, where the still image, subtitles, and other media types are stored in the same (or in another referenced) file. QuickTime provides a simple, powerful way to deliver this kind of content. There is a wide variety of QuickTime authoring tools, from simple things like QuickTime Player Pro and MovieWorks, up to full-scale professional authoring environments like LiveStage Pro and iShell.

One of the track types, called a sprite, is a small bitmap. QuickTime v3 introduced the Wired Sprite, a sprite that can interact with other sprites. Initially, this was used for simple "click here to play movie" kinds of stuff, but smart people quickly expanded Wired Sprites to handle a staggering array of functionality. One of the more extreme examples I've seen was a custom movie player that supported a text chat mode, rendered in real time on top of the video window. Complete interactive CD-ROM projects have been authored with Wired Sprites as well. I've even seem some pretty complex 2D computer games that rival any Shockwave movies built in Director.

**12.4** This sample movie from LiveStage Pro shows typical Wired Sprite movie controls.

The QuickTime file format is fully documented, so it's theoretically possible to do dynamic generation of wired sprite movies. However, the file format is much more complex than SMIL, so QuickTime projects that need to be generated on the fly tend to either use SMIL or have movies smart enough to fetch the information they need. There is much more on this in Chapter 14, "QuickTime."

# BIFS

The MPEG-4 model for rich media and interactivity is BIFS (BInary Format for Streams). While the basic MPEG-4 file format was derived from QuickTime, interactivity in MPEG-4 isn't based on Wired Sprites, but on VRML (Virtual Reality Markup Language).

BIFS seems to be a good format, although we're in the very early days of their use. Lots of very impressive e-commerce demos of BIFS have been shown, but there has been little in the way of real-world, delivered implementations. ISMA Profile 0 and 1 MPEG-4 don't mandate support for BIFS presentations. Instead, BIFS requires Main or Core, supported in the free Envivio plug-in available for QuickTime and RealPlayer for Windows—with Mac and Windows Media versions forthcoming. Envivio's Envivo Authoring Studio is the early leader in BIFS authoring tools, although many others are in development.

## Flash

Flash has long been a popular format for doing rich and interactive media, although it historically hasn't been well suited to video. There have been a number of Video-in-Flash technologies, but most offered poor quality and were difficult to use. Instead, Flash would either be embedded as another file type inside a SMIL presentation or a QuickTime movie.

This is all changing with Flash MX, which includes a video codec called Spark, created by Sorenson Media. This makes Flash a good container for video as well as the other rich media elements. It doesn't offer the compression efficiency of the leading codecs (the codec was designed to play back on the slow processors in mobile devices), and doesn't support real-time streaming, but it will offer playback on an extremely wide variety of platforms.

As an alternative, Flash can be embedded in a QuickTime movie (Flash v4 in QuickTime v5, and Flash v5 in QuickTime v6). Flash embedded in a QuickTime movie can interact with the other tracks and wired sprites, enabling some very cool seamless integration.

## EventStream

EventStream was Media 100's attempt to define a common subset of the rich media features of the different platforms, and to create a vendor-neutral format called TIES (Terran Interactive Event Stream) to store them. Alas, at the time only a few interactive features were supported cross-platform, and the technology hasn't been updated since 2000. Media 100i and iFinish as well as Discreet Cinestream editing software can create a TIES file, and Cleaner can render files using it.

Overall, the market for multiplatform rich media authoring hasn't developed as much as was anticipated. EventStream is generally only used for QuickTime authoring, but other, much more capable tools like LiveStage Pro and iShell are preferred.

| EventStream Tag | Platforms | Use |
| --- | --- | --- |
| Chapter | QT, WM | Sets a chapter marker—will show up in player |
| Display Text | QT, RM, WM | Shows text in a white bar at bottom of video frame |
| Go To Time | QT | Goes to that time in the file |
| Hotspot | QT, RM | Makes a clickable area that triggers another function |
| Keyframe | QT, RM | Makes that frame a keyframe |
| Open URL | QT, RM, WM | Opens the URL, in any target frame |
| Pause | QT | Pauses the video at that frame |

| EventStream Tag | Platforms | Use |
| --- | --- | --- |
| Play | QT | Unpauses the video |
| Web Poster | QT | Sets this frame as poster frame |
| Replace Movie | QT, WM | Replaces current movie with new movie |

# Skins

A skin is a custom controller that can run within the standard media player (as opposed to a custom player). Originally, skins were something that were mainly under user control, in that a user could pick their favorite look. Increasingly, custom skins are being delivered with particular media. Thus, when the user clicks on a link, the player that comes up can be precisely branded for the project in question.

Windows Media, QuickTime, and RealVideo all support complex skins, with QuickTime having a slight lead in some of the more advanced skinning features like transparent shadows. Windows Media skins only work on Windows, not with Windows Media Player on other platforms, while QuickTime and RealPlayer skins work well on both Mac and Windows.

A skin is built on a combination of bitmap graphics and some kind of layout document. None of the vendors provide a visual authoring tool for skins. Tattoo from Feelorium provides a good visual tool for authoring QuickTime skins. For RealVideo and Windows Media, you'll have to manually create graphics in a tool like Photoshop, and code up XML files that define the skin behavior.

**12.5** Sample QuickTime *(below)* and Windows Media *(right)* skins with branding.

# Chapter 13

# Choosing Platforms and File Formats

## Introduction

One of the most fundamental questions you face in implementing a video compression project is determining the proper file formats and platforms. It's important to distinguish between file format and platform. A file format is simply that—a combination of codecs and file structure. Some formats are platform-specific, like Windows Media or RealVideo. Others, such as MPEG-1 and MP3, can be played in multiple players and platforms.

While in some cases file format and playback platform are synonymous—Windows Media files only play in Windows Media Player or other tools built with the Microsoft library—many formats such as AVI and MPEG-1 can work in multiple players. Also, MPEG-4 is more format than a platform because many disparate platforms are able or will be able to incorporate MPEG-4 within them.

## OS Compatibility

One basic, unavoidable question is "can my audience run the software?" You may have good enough content that it can convince people to install additional software, but it's rare content indeed that would cause someone to buy a new computer.

OS compatibility isn't always an either–or thing. There are often subtle differences between support on different OSs, even within the same platform. Also, just because a player exists doesn't mean it's good. Just try to watch some broadband RealVideo on a first-generation iMac running a pre-MacOS X version of the operating system.

Historically, cross-platform meant Mac and Windows. However, we've seen an explosion of powerful PDAs based on PocketPC and other operating systems, and a growing interest in Linux and other UNIX desktops. Which platforms matter for your project can only be determined by you.

It's worth noting that there are several Linux compatibility layers, such as WINE or Codeweaver CrossOver, that allow Windows software to run on Linux. These make it possible for Linux viewers to use software not natively available for their platform (although these require Linux to be running on x86 CPUs).

*QuickTime*   QuickTime is the leader for two-platform compatibility, being virtually identical on Mac OS (both Classic and OS X) and Windows. QuickTime also runs on much older machines than some other formats—any Win32 (95, NT, or later) or any PowerPC Mac running at least MacOS 8.6. Between the platforms, the differences are few and subtle. The one I hear about most is that font sizes in text tracks are larger on Windows than on Mac OS. Some older codecs like Indeo are also missing from the Mac OS X version of QuickTime.

The QuickTime file format is fully documented, so there are a number of third-party applications that can play QuickTime movies. However, many of the codecs aren't documented, like the Sorensons, and so many of those files won't work on other platforms. The most common third-party QuickTime player is DirectShow, which only handles ancient QuickTime v2.1-compatible movies. More recent UNIX software like GStreamer and Mplayer does a lot better with compressed movie headers and better codecs like H.263.

*Windows Media*   A version of Windows Media Player is available on more platforms than any other such as Win32, Mac OS Classic, Mac OS X, PocketPC, and Solaris. However, differences among platforms are substantial. For example, the popular ACELP.net audio codec isn't available under Mac OS X and PocketPC. Also, VBR-encoded files don't work on the older players required by Windows 95/NT and Solaris. The latest video codecs are also not always available on the different platforms.

Performance across platforms is very impressive—Windows Media Video and Audio were designed to be easy to port to a variety of devices.

*RealVideo*   RealPlayer had support for different platforms from very early on. Currently, the RealOne (Real v9) player is only available for Windows (and in beta for MacOS X), with RealPlayer v8 for Mac OS 9, and a variety of UNIXs (Solaris, IRIX, AIX, Solaris, HP-UX, Unixware, and several Linux variants). A mobile version of RealOne is also available for PocketPC, Symbian-based devices, and some Nokia cell phones—although nothing is shipping yet. The Windows version works well and is well supported. The Mac OS 9.x version has suffered from very poor performance since RealPlayer G2 (a.k.a. v6); playing back even a 300Kbps stream well requires at least a 600MHz G4. However, the new MacOS X beta of RealOne has very good performance on G4 machines. The UNIX versions are also slow, and don't have any official support, just "community support." Still, RealVideo is the only one of the Big Three with support in UNIX for their current codecs.

RealOne and RealPlayer 8 use the same codecs, and other differences don't really affect content delivery.

*MPEG-1*   MPEG-1 has plenty of limitations, but OS availability isn't one of them. MPEG-1 is easily the most compatible file format available. It's very well documented, has only one video and two audio codecs, and has no license fees. Hence any computer shipped in the last half-decade came with MPEG-1 playback installed. QuickTime, Windows Media Player, and RealPlayer can all play back MPEG-1 files, as can most of the panoply of UNIX players. MPEG-1 playback isn't particularly common on handhelds yet.

*MPEG-4*   In the long term, MPEG-4 is poised to replace MPEG-1 as the default "play anywhere" format. In the short term, RealOne and Windows Media (both via a plug-in from Envivio), as well as QuickTime, support MPEG-4 playback, as will a variety of other tools on many platforms.

MPEG-4 does have a license fee of 25¢ per player after the first 50,000 distributed per year, so we may not see as many free MPEG-4 players, but there should be out-of-the-box MPEG-4 playback support in all significant OSs by 2003. There will also be profiles targeting low-power devices like cell phones.

The biggest limitation with MPEG-4 as a delivery format is that there are many different possible implementations of it. The ISMA Profile 0 and Profile 1 are likely to be the first mainstream formats widely used, but they lack support for the rich media functions that are critical to MPEG-4's appeal in many sectors. It remains to be seen what the MPEG-4 common denominator will be. We may wind up with a number of industry-specific MPEG-4 Profile@Level combinations.

*Flash MX*   Previous versions of Flash achieved remarkable penetration in all kinds of environments, with an oft-cited "98 percent of Internet connected devices." Macromedia has their marketing act well honed, and I expect Flash MX to achieve equivalent ubiquity over time.

At its release, Flash MX supported virtually every platform anyone had ever heard of. Its playback performance across platforms is quite consistent, although slower devices may have difficulty playing back higher resolution and data rate streams. Flash's Spark codec offers better compression efficiency than MPEG-1, but is behind other proprietary codecs.

*AVI*   AVI is a commonly used format with a variety of players. AVI is a very simple (to a fault, many argue) file format, and so support has been easy to implement. The biggest problem with that support is that there is no common minimum set of audio and video codecs—Cinepak and IMA are the lowest common denominator. DivX AVI files are starting to use the extension .divx to define a new common set of features—MPEG-4 video with MP3 audio.

AVI files play in QuickTime, Windows Media/DirectShow, and RealOne, assuming the codecs are available on the client computer.

# Installed Base

Once a player is available for users to download and install, what you really want to know is whether and how many have downloaded and installed it!

Focusing on general install base numbers is rarely useful. All the metrics I know of are produced with flawed methodology (e.g., only counting video played in players, not embedded in a browser, and so massively understating QuickTime). Also, different demographics will have very different platform usages—corporate MIS directors will have a lot of Windows Media Players installed, but video editors overwhelmingly prefer QuickTime.

The best thing to do is to put a script in your Web site to detect which plug-ins your users actually have installed, and proceed accordingly.

Also, bear in mind that for compelling content, people are generally willing to install or update a media player. People prone to playing back media files likely have all of the Big Three installed on their machines. Also, both the OBJECT and EMBED HTML tags can automatically guide the user to the download of a player, so the old problem of users having to hunt for a compatible player has largely gone away, although they still have to choose to install it.

Note managed corporate desktops often don't allow users to install software, so it pays to be much more conservative when targeting corporate users than with consumer sites. If you're targeting a particular company, find out what media players they have preinstalled, and if they're willing to upgrade.

*QuickTime*   QuickTime is preinstalled on all Macs, and it is very commonly already installed on Windows machines. Its total penetration on Windows may have dropped some from the 85 percent or so it enjoyed back in the CD-ROM era when almost any CD you'd put in your machine would install or update QuickTime. CD-ROM discs still commonly install QuickTime, but there are fewer CD-ROMs being released these days.

QuickTime is still being widely installed. Apple claims about 100 million unique downloads a year of the Web installer, and many versions are also installed from CD-ROMs. A slight barrier to installation with QuickTime is its occasional requests to upgrade to the Pro version for $29.95. Apple has substantially improved this, with much less frequent nag messages, and a clear description on the download page for how and why to get the Basic player.

*Windows Media*  Windows Media's greatest strength is that it is preinstalled in all Microsoft OSs (Windows and PocketPC), and Windows Update includes Windows Media Player, so users often have the latest version available. For the corporate environment with managed desktops, Windows Media is the obvious choice.

Windows Media Player is much less likely to be installed on Mac OS, even though there is a good player available for both Classic and OS X.

*RealVideo*  RealVideo historically had the dominant installed base, but its lead has been slipping with the ascent of Windows Media. The launch of RealOne was also bungled, with many users being left with the impression that there was no free player, just the player with a $10/month subscription fee.

The difficulty in finding the free player has been a problem for RealNetworks through many versions. They're in need of revenue from as many paid users as possible, but don't want to reduce their installed base. They've recently responded to this criticism and made it much easier to find the free player.

RealPlayer does a good job of auto-updating components and versions. Most people with it installed should at least have RealPlayer v8, which includes the latest video and audio codecs.

RealNetworks has just announced their Helix open source effort, which includes a source code release of some components of RealPlayer by the end of 2002. This may encourage other vendors to start adding RealMedia support more broadly.

*MPEG-1*  Virtually everyone has MPEG-1 playback preinstalled on their computers.

*MPEG-4*  QuickTime 6 is the first architecture to ship with cross-platform MPEG-4 playback. RealOne for Windows is shipping with MPEG-4 playback support via plug-ins from Envivio with a version for Windows Media Player expected soon. However, different players may support different Profile@Level combinations.

*Flash MX*  Flash MX should quickly become part of the base install of virtually any Internet-connected media device. Macromedia has created an effective auto-update mechanism, so existing users should also upgrade in short order (although embedded platforms may not be able to upgrade).

*AVI*  While everyone has some kind of out-of-the-box AVI player, which codecs are available vary widely between platforms. The lowest common denominator of Cinepak and IMA offer much worse compression efficiency than even MPEG-1.

## Codec Options

*QuickTime*   QuickTime has the widest set of codec options, including delivery and authoring, free and proprietary, good and bad. From QT5 on, it has included an auto-update mechanism, where new codecs can be downloaded from Apple on the fly when content that requires it is encountered.

The best codec in QuickTime today, Sorenson Video 3.1 Pro, lags behind Windows Media Video v8, and offers lower compression efficiency but better performance than RealVideo v9. A number of promising third-party video codecs are also being developed for QuickTime, including ZyGoVideo and Streambox's ACT-L2. Historically, QuickTime has been worst in the audio codec department, with MP3 as the overall best audio codec, and the aging QDesign Music 2 Pro being the only general purpose audio codec for RTSP. However, QuickTime 6 includes the very competitive AAC as a standard QuickTime audio codec.

*Windows Media*   Windows Media has only a few default codecs, but they're the best in the industry today. Windows Media Video v8 couples industry leading compression efficiency with excellent performance, and it is available for almost all versions of the player. However, PocketPC 2000 is limited to WMV v7 and Solaris to MS MPEG-4 v3.

Windows Media Audio v8 is a great general purpose audio codec, although it lags somewhat in speech quality at 32Kbps and below. ACELP.net is a great speech codec, but it isn't available on PocketPC and MacOS X. For broad compatibility, you'll need to stick with WMA.

Not many details are yet available about the new "Corona" Windows Media 9 codecs. The only one discussed at length so far is Windows Media Audio 9 Professional—which will be the first audio codec to support six discrete channels of audio.

*RealVideo*   RealVideo, like Windows Media, has a few good codecs. RealVideo 9 is excellent, generally providing better quality than WMV v8 at the same data rate. The RealAudio 8 music codecs are quite capable and RealAudio Speech codecs provide a much broader selection of data rates than speech codecs on other platforms. Real just introduced new audio codecs that explicitly support preserving Dolby Pro Logic 2-channel matrixed surround sound information, which no other format currently supports.

*MPEG-1*   MPEG-1 has only one video codec (MPEG-1), and three theoretical audio codecs (Layers 1, 2, and 3). Most players don't include Layer 3 playback, so almost all MPEG-1 files user Layer 2.

The MPEG codecs aren't competitive in encoding efficiency, but can provide excellent quality at higher data rates (at least 1000Kbps for 320×240). The maximum resolution that is required to work is 352×288, but most players, including DirectShow and QuickTime, can handle higher resolutions, up to at least 640×480.

*MPEG-4* As a standard, MPEG-4 includes many, many combinations of codecs in its different profiles and levels. So far, the most important video codecs are Simple Visual (in ISMA Profile 0) and Advanced Simple Visual (in ISMA Profile 1). I'd expect Advanced Simple support to become the baseline for everything but the lowest powered mobile devices. On audio, the AAC and CELP codecs are becoming standard.

Future MPEG-4 profiles may include other features. However, the Profile@Level structure clearly indicates which features must be supported in a given player.

*Flash MX* Flash MX contains only the Spark video codec. MP3 or the Nellymoser speech codec are used for audio. Because Flash puts a premium on easy portability and light decoder requirements, its compression efficiency isn't as good as other modern codecs.

*AVI* There are innumerable codecs available for AVI, more than QuickTime even. These include VP3 and DivX on the high end, legacy options like Cinepak and Indeo on the low, and many proprietary authoring codecs. However, unless you can install codecs, you can't rely on much beyond Cinepak and IMA being installed.

To get around this problem, DivX has introduced .divx as a new extension for AVI files created with their codec and MP3, so those files can automatically be played in their player (named, sadly, "The Playa").

## Disc-based Playback

Disc-based playback includes anything played off a hard drive, CD-ROM, or DVD-ROM. I'm also including DVD-Video in this category.

*QuickTime* QuickTime has long been the leading technology for disc-based media. It has a good selection of codecs well suited to the task, and provides an excellent infrastructure for it. The unparalleled support for QuickTime inside of Director is also a big boon, as are the excellent interactive CD-ROM creation features of LiveStage Pro and iShell.

Sorenson Video v3 is a rather slow decoder for high-resolution playback, but On2's VP3 does quite well. There is an excellent, scriptable version of the QuickTime installer available from Apple for a free license.

*Windows Media* Although Windows Media was originally designed for Web-based playback, it works fine for CD-ROM as well, and can encode extremely efficient files with its 2-pass

VBR mode. Any DirectShow video player can embed Windows Media files. Alas, Director still uses VfW, which can't play back Windows Media. Xtras are available to get around this problem, but they only work under Win32.

*RealVideo*   RealVideo is not designed for local playback. It works, but Real doesn't have much in the way of advantages here—it lacks an unbuffered 2-pass mode, and has performance issues at the higher data rates CD-ROM can offer. RealVideo files, however, can be embedded in Director without any third party plug-ins.

*MPEG-1*   MPEG-1 works well for CD-ROM. Its relative inefficiency in compression is less of an issue on CD-ROM unless you need more than an hour or so of content, and its ubiquitous and undemanding playback are welcome. DirectShow and QuickTime can both play back MPEG-1 well. Playback in Director requires using QuickTime or a third-party Xtra for decode. It's best to encode in square pixel (e.g. 320×240 instead of 352×240) resolutions for CD-ROM playback, as square pixel offers better efficiency for both compression and decode, as well as more consistent playback.

*VideoCD*   VideoCD is an MPEG-1 on CD media format. Its biggest advantage is wide playability—pretty much all computers out of the box, DVD Players, and dedicated VideoCD players can play VideoCD discs. The standalone players are either NTSC or PAL, however, and can't play discs of the other format. Note some players may not be able to handle CD-R media, although all can play mass manufactured discs.

The drawbacks of VideoCD are a very inflexible specification (CIF resolution, fixed data rate), and very limited interactivity.

*MPEG-2*   MPEG-2 has long been considered a promising disc-based playback format. Like MPEG-1, MPEG-2 originally required hardware cards, but modern processors are fast enough to decode in real time (especially with modern video cards that can do some of the decoding work themselves).

Alas, licensing terms have kept software MPEG-2 decode from becoming a standard feature—typically, it requires a $20+ add-on. Better video playback codecs like DivX, VP3, and Rad Game Tools Smacker have made MPEG-2 less interesting for many applications.

*DVD-Video*   DVD-Video is also used for many disc-based applications, typically the DVD-ROM portion of a DVD disc. Since any machine that can play a DVD-Video disc has an MPEG-2 decoder, the question of decode goes away. Right now, good control of DVD-Video playback from an application requires DirectShow, and hence Windows, although kludgy solutions are possible by using AppleScript to control the DVD Player on Mac. Straight playback of a DVD-Video disc works on most platforms now, and the DVD-Video disc itself can contain a surprising amount of interactivity.

*MPEG-4*    It remains to be seen how well suited MPEG-4 is for disc-based playback. DivX v5, which is an implementation of MPEG-4 for AVI files, shows that MPEG-4 can offer good quality and decode performance. And QuickTime 6 can play back MPEG-4 with good performance inside Director and other QuickTime based apps. Eventually, MPEG-4 playback should be as ubiquitous as MPEG-1.

*Flash MX*    Flash MX is mainly designed for Web use, but should work fine for CD-ROM as well. The codecs in Flash MX are light to decode, so playback on the more powerful devices typical of disc-based players should work fine at moderate resolutions.

*AVI*    AVI is a commonly used file format for disc-based use. If circumstances allow for a codec install, there are many good options, such as DivX and VP3, for using AVI. If a codec install isn't an option, it's probably best to stick to another format.

*Bink*    Bink from Rad Game Tools is a very popular format for use in computer games. Bink offers okay compression efficiency and very fast decode, but its main strength is a very powerful API for using Bink inside applications. The API requires the C programming language, so Director presentations are out. Bink is also expensive; the licensing fee is $4,000 per title. That said, those integrating video into computer games say nothing comes close to the power, flexibility and tech support of Bink.

Bink replaced Smacker, an older 8-bit video technology with similar advantages. Smacker is still available, although very few new games make use of it.

## Progressive Download

Progressive download is a great way to deliver short, high-quality content over the web, especially to modem users. It is very different from real-time streaming, so formats that may be great for one might not be for another.

*QuickTime*    QuickTime was the format that pioneered progressive download, and it remains much stronger with PD than with real-time streaming. The default player and plug-in UI work well for progressive download, and both buffer intelligently. QuickTime has a good video codec (Sorenson Video v3.1 Pro) and a great audio codec (AAC) for doing progressive download. The 2-pass VBR mode used in Sorenson Video Pro with Cleaner and Squeeze is also optimal for progressive download use.

Older embedded QuickTime needs to be updated to support the hybrid OBJECT/EMBED tag required by recent versions of Internet Explorer. However, the hybrid tag eliminates the "format hijacking" problem, where RealPlayer or Windows Media Player would be assigned the .mov file type, even though they couldn't play modern QuickTime files.

QuickTime's Movie Alternates are an easy way to provide multiple versions of the same file for download—the standard MBR features in WM and RV don't work for progressive download, although there are other ways to implement similar solutions.

One caveat with QuickTime is that files must be Fast-Start for progressive download. Files are always encoded as Fast-Start, but any tweaking of the file after compression, even just changing metadata, requires the file be resaved.

*Windows Media*  Originally, Windows Media wasn't good at progressive download, but it has been a very viable solution since WM v7. WMA v8 is a great codec, as is WMV v8, especially with its 2-pass VBR mode (although it is only available in some tools).

The Windows Media progressive download experience closely follows that of QuickTime, down to the way the controller displays how much of the file has been downloaded. Windows Media's DRM is fully integrated for progressive download. Intelligent Streaming doesn't work for progressive download, as it would require some type of server-side scripting.

*RealVideo*  RealVideo is still focused mainly on real-time streaming, with less emphasis on progressive download. Doing progressive download with RealPlayer works fine; however, the architecture isn't tuned for it. For example, RealVideo has a maximum buffer size of 25 seconds, which is much less efficient for progressive download than an unlimited buffer.

SureStream doesn't work for progressive download. However, a SMIL document can do Quick-Time Alternative Movie-style selection among different streams at different connection speeds.

*MPEG-1*  MPEG-1 works fine for progressive download, although it will play in different players on different platforms. Its real drawback for progressive download is its relative lack of compression efficiency (although a good 2-pass VBR encoder can help quite a bit). While offering an MPEG-1 backup file for UNIX users may be a nice idea, it should rarely be the only format provided.

*MPEG-2*  MPEG-2 files are way, way too big for any real-world progressive download use.

*MPEG-4*  MPEG-4 is a fine progressive download format now that it is supported in QuickTime and RealOne, with other players forthcoming. How MPEG-4 will behave will depend on whatever player is assigned the .mp4 file type or is specified in an OBJECT tag. Excellent 2-pass VBR encoders are in development from companies like Envivio.

*Flash MX*  Flash MX only does progressive download, and it does it well. The basic encoding tool doesn't have a lot of depth, but Spark Pro inside Sorenson Squeeze supports an unbuffered 2-pass VBR and other useful features.

*AVI*   AVI is not a good choice for progressive download, since it's hard to predict which player and which codecs will be available to play it. The .divx variant doesn't have that problem, but isn't widely installed yet.

# Real-time Streaming

Real-time streaming is the flip side of progressive download Web video. It's more difficult to do right, and has significant variations between platforms.

*QuickTime*   QuickTime is the weakest real-time streaming platform, although it's a lot stronger than it used to be. Its biggest problem is its lack of any equivalent to SureStream or Intelligent Streaming that can dynamically switch among versions of the stream in real time. Recent versions of the QuickTime server are able to dynamically drop B-frames from Sorenson Video v3 Pro, but this only gives another 25 percent flexibility, and requires the files be encoded with B-frames in the first place.

There are also codec issues with RTSP QuickTime. Sorenson Video v3.1 works fine, but other video codecs like VP3 don't have native packetizers, and so won't stream reliably. While older versions of QuickTime lacked good RTSP codecs, the AAC included with QuickTime 6 is excellent. The low bitrate speech CELP codec is promised in a future version, but until then, PureVoice remains a workable alternative.

One big advantage of QuickTime is that the server itself is free, open source, and multi-platform. This means lots of folks just getting into streaming go with QuickTime.

*Windows Media*   Windows Media is very strong for real-time streaming. Its 2-pass CBR mode is optimal for streaming (although it isn't directly supported by mainstream encoding tools yet), and all its codecs work fine streamed. Intelligent Streaming can automatically switch between video tracks as bandwidth changes, but the file itself can only have one audio track.

Unlike all the other systems, which use the RTSP protocol, Windows Media uses MMS. This means that it may have more trouble getting through firewalls (although in that case it will fall back to the less efficient TCP/IP and then HTTP protocols).

Windows Media Server is included with Windows Server products, but isn't available for other platforms. However, RealNetworks' new Helix server supports serving Windows Media files via the full range of Real-supported OSs.

*RealVideo*   RealVideo was the original RTSP Web video technology, and it still has some significant advantages. Its SureStream technology can do both video and audio, making it the best available right now. Its codecs, while hard to decode at high data rates, are well suited to RTSP speeds.

RealServer is available on a variety of UNIX platforms and Windows. Its biggest drawback is that it's rather expensive, especially for large numbers of simultaneous users.

*MPEG-1*   Although it isn't normally thought of as a real-time streaming solution, MPEG-1 actually streams very nicely. QuickTime/Darwin Streaming Server can do this natively, as can some products from Cisco. Of course, much higher data rates are required than with competing formats.

*MPEG-4*   MPEG-4 is poised to become the fourth major real-time streaming format. Its streaming mode is based on QuickTime, with hint tracks, and is well supported in QTSS. The initial ISMA-based MPEG-4 solutions don't have any MBR options. Later MPEG-4 products based around the Scalable and Fine-Grained Scalability video profiles should offer excellent scalability, eventually.

*Flash MX*   Macromedia's new Flash Communication Center MX supports unicasting, multicasting, and videoconferencing—with the Flash 6 Player as a client. The product has only just been released, but it is receiving a lot of interest.

*AVI*   AVI doesn't support real-time streaming.

## Rich Media Support

Rich media support may be completely irrelevant to some projects, and absolutely critical to others. The Big Three formats support HREF tags (which can launch or modify browser windows). Beyond that, rich media support varies enormously. Make sure you understand your rich media needs before basing your platform decision on them.

*QuickTime*   QuickTime is the unquestioned leader in rich media support within a video format. Anything you can imagine making in Director or Flash, you can probably do in QuickTime, delivered as a QuickTime movie. QuickTime supports HREF triggers, movie-to-movie communication, user text input, and enormous depth. Interactive QuickTime languished without good authoring support for a while, but there are many excellent products now available, such as iShell and LiveStage Pro on the high end, and Totally Hip's LiveSlideShow, Interactive Solutions' MovieWorks, and even QuickTime Player Pro on the low end.

QuickTime also supports SMIL v1 for interactivity, which is less powerful, but offers automated authoring. All but the simplest rich media projects should strongly consider QuickTime.

*Windows Media*   Windows Media Player/plug-in itself doesn't have much rich media support, offering basic subtitles and HREF. Most rich media projects with Windows Media are targeted to Internet Explorer, with Windows Media as one of the component types. This works fine for projects that target compatible browsers, but doesn't offer the seamless all-in-the-player experience and cross-platform consistency of QuickTime and RealVideo.

*RealVideo*    RealVideo was the first on the market with a complete rich media solution, based around SMIL v1. RealOne now supports SMIL v2, which features very deep rich media support.

*MPEG-1*    MPEG-1 has no rich media support.

*MPEG-4*    MPEG-4 uses BIFS as its rich media format. Thus far, however, this is only available in the Envivio plug-in to RealOne and QuickTime. The QuickTime v6 MPEG-4 implementation only supports the ISMA profiles, which don't include any rich media. MPEG-4 could become a very powerful rich media delivery platform if BIFS-enabled players become standard. There is already a strong authoring tool from Envivio and iVast and Tribeworks are collaborating on one.

*Flash MX*    Flash MX, because it lives inside Flash, is the only true rival to QuickTime for rich media. Of course, QuickTime v6 can host Flash v5 files, so those wanting to integrate video and rich media can either put video in Flash or Flash in QuickTime.

*AVI*    AVI doesn't support rich media.

# Chapter 14

# QuickTime

## Introduction

Apple Computer's QuickTime is the granddaddy of all the media architectures. It started shipping over a decade ago, when a top-of-the-line computer ran at 25MHz. QuickTime's impact and legacy are hard to overstate. QuickTime laid the foundation for the entire desktop video authoring industry, as well as the practical delivery of video on desktop computers. And it's done this with a remarkable degree of backward compatibility. All the files on the QuickTime v0.9 developer's disk still play back in QuickTime v5.02.

QuickTime is far, far more than a simple system to play a rectangle of video and two channels of audio. It is a complete system for the creation and deployment of time-based media. Its name is an echo of Apple's pioneering QuickDraw API for 2D computer graphics, a core building block for the development of the modern GUI. QuickTime was intended to bring the same power to time-based media.

QuickTime is by far the deepest and most complex media architecture (even though it has some startling omissions), which is why this chapter is one of the longest in this book.

## QuickTime Format

The QuickTime file format has often been described as a "media container" and can hold almost any kind of media data. The basic unit of QuickTime from a user's perspective is the movie. A movie is built out of tracks. In a typical QuickTime file for distribution, all the tracks are inside the file. But they don't have to be. QuickTime supports a very useful technology called "Reference Movies" where different parts of the movie can be in different files. This way you can have multiple versions of the same content, all using the same underlying data, without having to duplicate the content. For example, you can trim the head and tail off a file, and instead

### History of QuickTime

|  |  |
|---|---|
| 1991 | QuickTime v1 announced, then shipped for Mac OS. |
| 1992 | QuickTime v1.5 included with System v7.1, QuickTime for Windows announced. |
| 1993 | QuickTime v1.6 released, includes Apple Compact Video (later renamed Cinepak). |
| 1994 | QuickTime v2 for Mac released, then (finally) v2 for Windows. |
| 1995 | QuickTime v2.1 ships for Mac and Windows. First with IMA. |
| 1996 | QuickTime v2.5 ships for Mac only. First implementation of interactive rich media with "QuickTime Media Layer," MPEG-1 playback. |
| 1997 | QuickTime v3 announced, wins best at show at NAB. |
| 1998 | QuickTime v3 finally ships for Mac and Windows almost a year after announcement. All new API for Windows (old applications require v2.1.5). First version with Sorenson Video and QDesign Music. |
| 1999 | *Star Wars* trailer gets 6.4 million downloads (with 2.2 million QuickTime downloads). QuickTime v4 released; first version with RTSP streaming, MP3 playback. QuickTime v4.1 released. |
| 2000 | QuickTime Streaming Server's source code released. |
| 2001 | QuickTime v5 ships, includes component download, MPEG-1 playback on Windows. QuickTime v5.02 ships, first version with Sorenson v3. QuickTime v5.03 released with new ActiveX component. |
| 2002 | QuickTime v6 ships, includes deep integration with MPEG-4. |

of having to export a whole new file, you can save a 4K reference movie that simply points to the original, unmodified file.

QuickTime movies also have a wide variety of metadata that can be attached to them, like name of the file, artists, and so on. QuickTime movies can also be programmed to do cool things, like automatically trigger the Present Movie full-screen mode when opened.

## QuickTime Tracks

The basis of many of QuickTime's unique features is the track. A track is a piece of media with a start time and end time. If it's a visual track, it appears on the screen. Those visual tracks can be smaller than the movie, and can overlap each other—including with transparency. Most track attributes can be changed by tween (for "in-between") tracks, which interpolate values like size and shape over time. And the user can interact with a file through Wired Sprites and Flash.

*Video*   The video track is the most basic type of track. This is your standard rectangular series of video frames. However, unlike some other formats, a given movie can have multiple tracks, all of which can have different frame rates, data rates, and even codecs. Each video track, however, has a single native resolution and codec, although the frame rate can vary throughout.

*Audio*   A QuickTime audio track is either mono or stereo, with no support for multichannel sound. Each audio track uses a single sample rate and codec. If there are multiple audio tracks playing at the same time, they will be mixed together on output. The audio, panning, EQ, and other properties of an audio track can be changed in the file or dynamically.

*Hint*   Content that is going to be served via RTSP needs a hint track, which is essentially an index to the media in the file that tells the server how to stream a file. The hint track can be pretty big, so you definitely don't want to have one in a file that isn't going to be RTSP'd, as it balloons file size.

**14.1**   The myriad of audio controls in QuickTime Player Pro.

You can hint any track type. However, optimal results require a native packetizer for the particular codec, which allows the codec to guess at missing content. Hinting with media types where the content must arrive generally requires sending multiple copies of the content, hoping at least one will arrive. This is a fine strategy for content that has very low data rates, hence a lot of material in a single packet, like music or text. However, for video and audio that doesn't support a native packetizer, you just need to cross your fingers and hope that the server can retransmit any dropped packets in time for playback. Otherwise, you get a white frame of video or audio drop-outs.

*MPEG-1*   When QuickTime imports simple file formats, like AVI, the content shows up as standard QuickTime audio and video tracks. MPEG-1 is a different case, as it has a very different structure. Among other things, QuickTime doesn't natively support MPEG-1 B-frames. So, instead of teaching QuickTime how to think like MPEG-1, MPEG-1 support was implemented as an import component.

This yields several limits to MPEG-1 in QuickTime, the biggest of which being you can't export the audio from an MPEG-1 movie with QuickTime. If you do an Export as QuickTime movie, audio won't even show up as an option. Thus QuickTime applications need to implement their own MPEG-1 reader if they want to extract the audio. Editing MPEG-1 in QuickTime Player Pro also doesn't work.

QuickTime includes a native packetizer for RTSP streaming of MPEG-1. While MPEG-1 doesn't offer competitive quality compared to modern codecs, other non-QuickTime tools know how to play back MPEG-1 through RTSP, so in cases where compatibility is an issue, you may want to consider using streaming MPEG-1.

QuickTime v6 implemented MPEG-4 as the standard QuickTime movie, not a component. You can purchase a quite capable MPEG-2 playback component from Apple for about $20.

*Text* A text track is simply a series of lines of text with a font, color, and location. Each line of text has a start and a stop time. Because many codecs don't compress text legibly, text tracks provide a very high quality, very low bandwidth way to provide perfect text on the screen. Providing subtitles as text instead of including them as part of the video can provide a much better experience. And, of course, you can have different text tracks in different languages, and not have to re-encode the video for each audience.

To author a text track, you can use a tool that supports making them, or write text files in a special format that QuickTime knows how to import.

*Streaming* A streaming track lets you embed a live stream into a QuickTime movie. You get a movie with one of these in them whenever you open any of the live streams from the QuickTime Player's Favorites, or open an rtsp:// address in QuickTime Player.

Going into the movie properties and picking Streaming Track>Bit Rate gives you a useful real-time status indicator of the stream (see Figure 14.2).

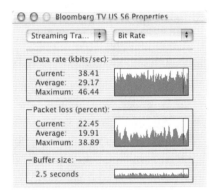

**14.2** QuickTime Player Pro's Streaming Track gives useful real-time information about the RTSP stream.

*Music* Music tracks use QuickTime as a synthesizer, telling it when and how to play a series of notes. While it sounds synthetic, the synthesizer (which uses the Roland General MIDI sound set) in QuickTime is pretty darn good, and typically requires less than 1kbp for a full score of music. You can import existing MIDI files into QuickTime to make a music track. QuickTime has a pretty full support for modern MIDI, including being able to import sound fonts. However, there's no way to convert normal audio files to MIDI—folks are working on the technology, but we've still got a long way to go.

By default, QuickTime music tracks use the built-in software synthesizer, but if you have a hardware MIDI synth, you can specify that should be used instead in the QuickTime control panel. Beware of doing that, however. Different synths have quite different sounds for different instruments, so you may hear something quite different than other users.

*QuickTime VR*   QuickTime took part in the virtual reality craze of the early- to mid-90s. QuickTime VR (QTVR) debuted in QuickTime v2.5 as a component of the QuickTime Media Layer (along with the less fortunate QuickDraw 3D). A QuickTime VR track enabled the viewer to navigate within and between panoramas. Unlike traditional QuickTime movies, which are temporal (time-based) but not interactive, early QuickTime VR panoramas had interactive navigation, but they didn't play from a start point to an end point the way noninteractive video would. In fact, user interaction was required to pan or zoom in/out of a QTVR panorama.

Since its initial release, QuickTime VR has become increasingly better integrated with the rest of QuickTime. For example, tween tracks can be used to do canned animation of a QuickTime VR panorama.

QuickTime v5 introduced the long-awaited cubic panorama. Where the original cylindrical panoramas couldn't allow the user to look too far up or down (no top to the cylinder), cubic panoramas model the scene by mapping the stitched images onto the six sides of a cube. This cubic approach allows you to create panoramas in which the viewer can actually look straight up and straight down.

There are a variety of tools from different vendors that allow you to "stitch" together a series of photographs into a panorama—a topic beyond the scope of this book. However, all those panoramas wind up being compressed as QuickTime pictures, using a QuickTime codec. Originally, Cinepak was the default, but most modern panoramas are done with JPEG. The Sorenson Video v3 codec has shown some promise for getting decent quality with very small files for QTVR and other still images, although it makes navigation somewhat slower. JPEG2000, incorporated in QuickTime v6, will probably become the codec of choice.

While QuickTime VR tracks originally didn't support any animation, the use of tween tracks makes animated VR now possible, although they are difficult to author.

## QuickTime VR Object movies

The QuickTime VR Object format is quite different from panoramic VR. With a QTVR panorama, the viewer is in the center of a scene that revolves around them. With an object movie, the viewer is presented with an object that they can examine by spinning it around. Internally, an object movie is simply a sequence of frames, compressed using one of same codecs you'd use for a panoramic VR movie.

## 3DMF

One of the great tragedies of QuickTime was the failure of QuickDraw 3D. In the same way QuickTime was a time-based expansion of QuickDraw, QuickDraw 3D was an enhancement of QuickDraw that included support for a good 3D model format (3DMF) and hardware acceleration. This acceleration was through the RAVE (Rendering Acceleration Virtual Engine) API.

However, Apple was in no position to compete against DirectX and OpenGL technologies, and virtually no PC software was written around RAVE.

QuickDraw 3D offered some very promising functionality within QuickTime, including support for embedding 3DMF objects inside of a QuickTime movie, and using tween tracks to modify their parameters over time. This presented some potentially amazing opportunities; for example, a 3D moving logo could be embedded into a movie as a moving 3D object, not as a series of bitmaps. This could have provided enormously higher quality at tiny data rates. Alas, no commercial products ever supported authoring full QuickDraw 3D movies in QuickTime. Apple has long since adopted the industry standard OpenGL as their 3D library of choice, which has certainly been a huge boost to the Mac 3D industry. Still, many of us old QuickTime hands long for what might have been.

Some of the QuickDraw 3D legacy lives on in the Quesa project (http://www.quesa.org), which is a reversed-engineered version of QuickDraw 3D's libraries that runs on top of OpenGL or Direct3D.

## Sprites

Sprite tracks had an inauspicious beginning in QuickTime v2.5. They were simply a means to make icon graphics move around the screen. Later, the ability to make wired sprites was added. A *Wired Sprite* is a sprite that knows how to do something. They form the basis for interactivity in QuickTime. If there is a button you can click in a QuickTime movie, it's either a wired sprite or Flash.

## Flash

QuickTime v4 added support for Flash v3 tracks, and QuickTime v5 upgraded to support Flash v4. Flash v5 was added in QuickTime v6. QuickTime supports most, but not all, of the Flash v5 specifications. QuickTime's support of Flash v4 allows it to do animations and Flash-based navigation of QuickTime movies. The launch of Flash MX has led to debates over whether it is best to put video inside a Flash file, or to put a Flash file inside a QuickTime movie.

## Skins

A Skin track replaces the standard UI of QuickTime Player with a new, custom one. This is useful for creating branded experiences. All the major architectures include skin features, but QuickTime's includes unique features like alpha-channeled edges for transparency effects. The leading application for Skin creation is Feelorium's Tattoo.

# Delivering Files in QuickTime

QuickTime not only predates the Web, it actually predates the CD-ROM as a standard feature in personal computers! However, the excellent work done on QuickTime's original design have enabled it to move into the world of the Web with relative ease.

## QuickTime for CD-ROM

QuickTime was the dominant file format of the golden age of the CD-ROM (with AVI a close second). Macs dominated multimedia production back then, and Macromedia Director, the leading authoring tool, has always had better native support for QuickTime than any other format.

QuickTime remains an excellent format for this kind of work. While Cinepak has been the historical format of choice because it decodes so quickly, SV3 and especially VP3 can offer much better quality at lower data rates and with acceptable performance on modern machines.

## QuickTime for progressive download

QuickTime pioneered the progressive download model for Web video, and it was used widely for Web video in progressive well before QuickTime v4 finally introduced RTSP streaming in QuickTime. QuickTime was pretty lousy at RTSP until Sorenson Video v3 and QuickTime Streaming Server v3 came out, so most QuickTime on the Web has been progressive.

Every QuickTime movie has a movie header, which is essentially an index to the file, allowing things like being able to drag the movie controller to a point on the timeline, and knowing where the data for that time is in the file. Because it wasn't known where all the information was until the file was completely encoded, the movie header used to be appended to the end of the file. As CD-ROMs can do random access, this structure wasn't a problem until the Web came along. However, when doing progressive download via FTP or HTTP, the file is always read front to back. Because none of the file can be shown until the movie header is available, this meant that progressive download flat out wouldn't work. So, in QuickTime v2.1, Apple introduced the Fast Start movie, which is a fancy name for a movie with the header at the start of the file. Originally, a special tool was needed to process the movie for fast start. However, making fast start movies is now the default mode of QuickTime.

However, if you do any editing in QuickTime Player, even simple things like changing annotations, the header gets moved to the back of the file again, and fast start no longer works. When this happens, you need to re-flatten the movie, either by doing a Save As of the file from QuickTime Player Pro, or by doing a batch flatten with Cleaner or a free, Apple-supplied AppleScript.

QuickTime v3 also introduced the compressed movie header. Because there was a lot of redundancy in the movie header, Apple found that applying a traditional file-based lossless compression to it made it quite a bit smaller. While this didn't reduce the size of the total file more than a

## MPEG-4 in QuickTime

QuickTime v6 has deep integration of MPEG-4 within QuickTime. The MPEG-4 file format is based on QuickTime, which enabled Apple to do more than make MPEG-4 a component within QuickTime. Instead, QuickTime treats an MPEG-4 file as a QuickTime movie. Thus, you can open an MPEG-4 file in QuickTime, and it will show up as a movie with an MPEG-4 video track and an AAC or CELP audio track. MPEG-4 codecs also show up as options inside the standard QuickTime export dialogs. This means a QuickTime movie can be made using those other codecs—especially welcome in audio because AAC is much better than both MP3 and Q-Design Music v2, and has a native packetizer for real-time streaming.

However, there are some differences between the QuickTime and MPEG-4 file formats, so a QuickTime file with the right codecs still isn't a legal MPEG-4 file. To make a straight-up MPEG-4 file, you'll need to use the Export dialog.

QuickTime doesn't natively support B-frames, which are part of Advanced Simple and ISMA Profile 1. Also, the Wired Sprite interactive movie structure of QuickTime is radically different than the BIFS system used in MPEG-4's Main and Core profiles. QuickTime's MPEG-4 support is ISMA-compliant and the ISMA profiles don't include any interactivity. However, the Envivio plug-in for QuickTime does support BIFS, although it is not yet available for Mac.

QuickTime Streaming Server from v4 on includes native MPEG-4 streaming as well. The open source Darwin Streaming Server version is the basis of MPEG-4 servers from other vendors like Envivio.

Lastly, Apple's QuickTime Broadcaster is a live broadcasting application that supports MPEG-4 as well as QuickTime streams.

few percent, because the header is before the video, this significantly shortened the minimum time before the first frame of the video could be seen. Some non-Apple QuickTime players and servers aren't compatible with compressed movie headers. But as long as the file is going to be played with QuickTime or streamed through the QuickTime/Darwin Streaming Server, they're fine.

## QuickTime for RTSP

QuickTime was doing video on the Web for years before QuickTime v4 finally introduced RTSP streaming. The RTSP implementation in QuickTime was rather elegant. Instead of introducing a wholly new file format and breaking compatibility with older file formats, QuickTime v4 introduced the hint track, which gives the server the information it needs to stream the file. QuickTime can hint anything that lives in a movie. However, codecs designed to be used in RTSP have native packetizers, which make it possible for the codec to guess what data is missing if packets are dropped (and some packets are always dropped).

For RTSP, you definitely don't want to use any audio or video codec that doesn't have a native packetizer. So if you just have to use VP3 or MP3, make it a progressive download project.

| Video | Sorenson Video v2, Sorenson Video v3, H.261, H.263, MPEG-4 |
|-------|-------------------------------------------------------------|
| Audio | QDesign Music v2, Qualcomm PureVoice, AAC |
| Other | MPEG-1, MPEG-4 |

Another important difference, at least so far, is that 2-pass encoding isn't available for QuickTime RTSP. Squeeze and Cleaner, the two programs that can do VBR encoding for QuickTime, both do an unbuffered VBR. This means you can have substantial data rate peaks, which can hurt streaming enormously. Using the Sorenson Video v3.1 1-pass VBR mode with streaming is fine, however, as the codec automatically tightens up the data rate as much as required if the streaming option is checked.

QuickTime's RTSP streaming support used to be infamously bad, and now it is merely mediocre. Older versions didn't support any retransmission of dropped packets, making pretty much any file watched over the public Internet prone to video and audio glitches. It also lacked any ability to respond to drops in the end-user data rate except dropping all delta frames and just playing keyframes.

This all got better with QuickTime Streaming Server v3, especially in conjunction with Sorenson v3 Pro. The critical new features of QTSS3 were described with the marketing phrase of "skip protection." Technically, they came through retransmission of dropped packets, allowing very large client-side buffering, and adding the ability to drop B-frames as an intermediate step between playing all frames and just keyframes. With the improved buffering, the server will keep spitting out packets to the client as fast as it can take them. Up to several minutes can be received in a short time, so you can actually unplug the network cable from the computer and the movie will keep playing for a while. Skip protection helps average out data rate fluctuations quite a bit. It also gives a lot more time for a dropped packet to be detected, requested, retransmitted, and to arrive before it is scheduled to play back. This functionality was improved still more in QTSS4.

QTSS4 is still behind Windows Media Server and RealServer, in that it has no way to actually switch between multiple versions of the encoded bandwidth to respond to the users' real world bandwidth.

However, QTSS has one big advantage—it's free and open source. Apple provides the branded QuickTime Streaming Server for Mac OS X Server only. However, you can also download binary versions of it under the Darwin Streaming Server name for FreeBSD, Red Hat Linux, Solaris, and Windows NT/2000/XP. And better yet, you can actually download the source code to compile, modify, or enhance it for other uses. While this obviously requires some technical savvy, a number of vendors are already building products around enhanced versions of DSS and add-in plug-ins for it. All versions come with good server Web administration.

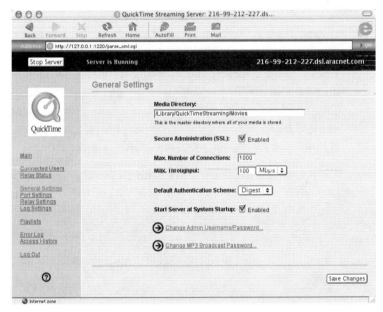

**14.3** QuickTime/Darwin Streaming Server has nice, simple web administration.

Overall, QTSS is a good, basic streaming server. It supports standard, useful features like play lists (making a virtual live broadcast out of series of prerecorded files), user authentication, fallback to HTTP to get through firewalls, using relays to local servers, and multicasting. Beyond RTSP streaming, it can also do Shoutcast streaming of MP3. RealNetworks Helix server also supports streaming QuickTime movies.

## QuickTime for Live Broadcasting

QuickTime has facilities for live broadcasting similar to those of other formats. Historically, broadcast was only possible on the Mac, and the Mac suffered from a lack of good capture cards with hardware preprocessing (ironically, the Mac capture cards were all too high-end to have videoconferencing-style features like hardware capture). However, some good USB video capture units came out, and eventually the combination of Moore's law and heroic optimization by Apple engineers made decoding the DV codec fast enough to leave lots of CPU power for compression.

The historic leading tool for QuickTime Live broadcasting was Sorenson Broadcaster. However, with QuickTime v5, Apple didn't publish the documentation needed to upgrade Broadcaster. Sorenson did a Windows version, but Broadcaster wouldn't work on modern Macs. This was a shame, because Broadcaster was a nicely designed, simple live encoder.

Live Channel Pro from Channel Storm filled in the gap. Live Channel is a unique, very cool broadcasting app with integrated switching and DVE, all running on a G4 without any additional hardware—another case of heroic optimization. This is overkill for many simple projects, so they also sell the much cheaper Live Channel Lite, which doesn't include most of the advanced features. Both versions are Mac OS v9.x only, but Channel Storm is working on porting them to OS X.

Apple provides QuickTime Broadcaster for free—a live compression tool for Mac OS X. It can broadcast via both MPEG-4 and QuickTime. It is simple and functional, although the initial 1.0 version doesn't have the depth of Sorenson Broadcaster or Live Channel Pro.

## Embedding QuickTime (and the Embed/Object tag fiasco)

The QuickTime plug-in for embedding QuickTime in Web pages was originally designed when Netscape was the dominant browser. Thus, it used the Javascript model for embedded objects standard on Netscape. Everything worked fine for many years. However, Internet Explorer eventually became the dominant browser. In the summer of 2001, Microsoft dropped all support for the old Netscape style of browser scripting with IE v5.5 SP2 and v6. This meant that all existing embed tags for QuickTime wouldn't work in current Microsoft browsers.

Apple released QuickTime v5.03 a few weeks later, which included an ActiveX control for scripting QuickTime, and produced new tools that generated an incredibly turgid nested Embed/Object tag, where every parameter had to be defined twice, once for each scripting model. However, there is a lot of legacy QuickTime code and tools out there which haven't been updated.

A typical tag now looks like this:

```
<OBJECT CLASSID="clsid:02BF25D5-8C17-4B23-BC80-D3488ABDDC6B" WIDTH="160"
HEIGHT="144" CODEBASE="http://www.apple.com/qtactivex/qtplugin.cab">
<PARAM name="SRC" VALUE="sample.mov">
<PARAM name="AUTOPLAY" VALUE="true">
<PARAM name="CONTROLLER" VALUE="false">
<EMBED SRC="sample.mov" WIDTH="160" HEIGHT="144" AUTOPLAY="true"
CONTROLLER="false"
PLUGINSPAGE="http://www.apple.com/quicktime/download/">
</EMBED>
</OBJECT>
```

Note how all the parameters like "Autoplay" have to be assigned their value twice. If they're given different values in both places, you'll get different behavior depending on whether or not the local system is using the Embed or Object format. A lot of cool parameters can be controlled here. The full list is available on www.apple.com. There are a few in particular I want to point out:

***AUTOPLAY*** If on, the movie will start playing when the amount of time remaining in the download is less than the total duration of the file (the standard progressive download buffering formula). If off, the movie will just sit there downloading until the user hits the play button.

***HEIGHT*** The Height tag controls how big a window to make for the file. If you're including the movie controller, you'll need to add 16 to this number to account for its height, or the top eight pixels of the movie and the bottom half of the controller are cut off.

***HIDDEN*** The Hidden tag makes a movie you can't see, and hence can't control. There was once an epidemic of Web pages that used hidden movies with music tracks to play background music for Web sites, offering no way to turn the music off. This was very annoying. If you absolutely must have music on your Web page (and you don't), make sure to offer some kind of control to turn it off.

***KIOSKMODE*** With Kioskmode="true", the pop-up menu on the extreme right is hidden. Thus, QuickTime Player Pro users lose the ability to save the movie from there. Also, the movie can't be saved through drag and drop. This is a simple, basic form of copy protection. However, a clever user can generally find the URL of the movie, and download it directly if you don't take other protection measures.

***Loop*** If Loop="true" the file will loop over and over. If you're making some kind of animated logo or such, this can be useful.

# Delivering Files in QuickTime   201

**14.4** The movie controller with (right) and without (left) kioskmode=true. You don't get the pop-up menu with kioskmode.

***QTNEXT***   QTNEXT specifies another movie to play after the current movie ends. Up to 256 next movies can be specified. Each additional movie adds its number to the end of the file, like this:

```
<EMBED SRC="nameof.mov" WIDTH="200" HEIGHT="240"
QTNEXT1="<http://www.apple.com/quicktime/movies/sample.mov> T<myself>"
QTNEXT2="<rtsp://www.apple.com/quicktime/movies/sample.mov> T<myself>"
QTNEXT3="<URL> T<myself>
QTNEXT4="GOTO0"
```

***QTSRC***   QTSRC is a clever hack that lets you make sure QuickTime will always play the file in question, even if another plug-in is assigned that media type. This helps keep the file from being hijacked by other players like Windows Media Player or RealPlayer; versions of both are prone to attempt to take over the .mov file type from QuickTime. QTSRC also allows other file types, like .mp3, to be specified to play in QuickTime. QTSRC works by first defining a dummy file that always plays in QuickTime, and then specifying the actual movie in the QTSRC. Note you'll need to have the dummy file actually exist at the specified URL, even though it won't be downloaded. A good choice is a QuickTime Image (.qti) file, which no other plug-in tries to use. You can make it a graphic that reads, "You need QuickTime v4" because earlier versions of QuickTime don't have QTSRC and it will default to showing the parameter defined by SRC. Note that this isn't necessary in an Object tag because it already defines what plug-in to use.

```
<EMBED src="dummy.qti" qtsrc="rtsp://www.apple.com/quicktime/sample2.mov">
```

*QTSRCCHOKE*   This turgidly named tag lets you specify a maximum data rate (in bits per second) at which the QuickTime plug-in will download the file. If a data rate isn't set, a single user on a very fast connection could potentially suck up 5000Kbps on a 20Kbps file. Typically, you want a limit around the peak data rate of the file; more than that doesn't do anything for the user, and less than that can cause more buffering than needed.

*QTSRCDONTUSEBROWSER*   QTSRCDONTUSEBROWSER="TRUE" tells the plug-in to use its own caching instead of the browser's. This generally offers better performance, and prevents the file from being stored in the browser's local cache, aiding copy protection. If the user has QuickTime Player Pro, they can still save the movie unless KISOKMODE="TRUE".

*PLAYEVERYFRAME*   PLAYEVERYFRAME="TRUE" simply does what it says. It'll play every frame of the movie in consecutive order. If the CPU isn't fast enough to play them all in real time, it simply drops the frame rate of any frame as much as needed. This means the frame rate can vary widely on different machines, and within a single file on a given machine. No audio will play because the sync would inevitably be very off. This can be a good option for audio-less content at high quality where spatial quality is more important than temporal accuracy.

For example, my wife and I announced the gender of our first child by posting an ultrasound video on the Web. Seeing the pertinent details required a lot of resolution, but the computing power owned by our members of our family varied enormously. Using PLAYEVERYFRAME meant that everyone saw a good version of our video, although the movie took a full minute to play on my 400MHz G3 and six minutes on my sister's 25MHz 68040.

*SRC*   SRC is the classic way to define what file should be played. Generally, the specific file should be assigned to a dummy file, and the real file should be referenced via QTSRC.

*TARGET*   The Target tag specifies where the movie should be played. If QUICKTIME-PLAYER is specified, the movie will open in the QuickTime Player app instead of playing within the Web page. This can be useful if you want it to play full screen via Present Movie.

If the Target tag is MYSELF, then the movie specified in SRC will be replaced with the movie specified in HREF when the movie is clicked. This is the "poster movie" functionality, where you have a clickable graphic that turns into the actual movie.

*TARGETCACHE*   TARGETCACHE="TRUE" specifies that the poster movie specified with TARGET doesn't get cached, enhancing copy protection.

# The Standard QuickTime Compression Dialog

The most basic way to encode video to QuickTime is through a QuickTime Export dialog, which is supported in QuickTime Player Pro and many other applications. While this isn't enough for a professional encoder (it especially lacks preprocessing), there is a lot of stuff it can be used for, including exporting a file for later processing in another application.

QuickTime has an extensive list of supported formats. Generally, you'll use QuickTime, MPEG-4, AVI, or MPEG-2 if you have DVD Studio Pro installed.

The QuickTime dialog lets you specify your audio and video settings, size, and Internet Streaming options. You'll want to pick Fast Start for files that are going to be used with applications like PowerPoint for Windows that don't use QuickTime v3 or better to play back files. For progressive and other non-streaming video, you'll want Fast Start + compressed movie header. This makes the file slightly

**14.5** The QuickTime Export components, as seen in QuickTime Player Pro. Most of these are standard, except for the MNG component I have installed.

smaller, but requires QT3 or better for playback. And for RTSP streaming, you'll choose the eponymous option. This automatically generates the hint tracks along with the audio and video files.

In the video option, you pick your codec and other settings. The Quality slider doesn't do anything in most codecs if you have specified a data rate—generally, you should leave the slider at its default. If you don't specify a data rate (or are using a codec like the JPEGs or Animation that don't have a data rate control) the quality slider controls the quality of the video, and hence the file size, yielding a 1-pass quality limited encode. Some codecs have two quality modes—spatial and temporal. The slider by default shows the Spatial mode. Hold down the Option (Mac) or Alt (Windows) key and drag the thumb around; if the codec has a temporal mode, the control will switch to that. The spatial slider typically controls the quality of keyframes and the temporal slider controls the quality of delta frames, although exactly how these are implemented is up to the codec.

If you leave Frames Per Second blank, the frame rate of your source is used. If the source's frame rate varies, the output will maintain that variation.

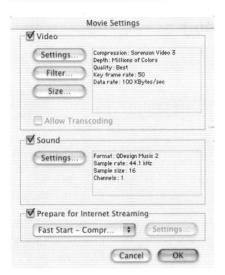

**14.6** The standard QuickTime Movie dialog, with some pretty typically progressive download settings.

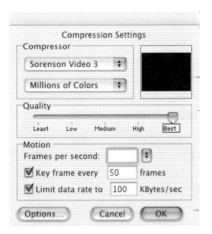

**14.7** The QuickTime video codec dialog. Note you'll have to click the Options button if you want to set codec-specific parameters.

Keyframe Every X refers to the minimum number of keyframes. A natural keyframe, which is normally inserted whenever there is a video cut, resets this counter, so the Keyframe Every value might not wind up meaning anything if your video contains a lot of rapid cuts. A good rule of thumb is to set a keyframe at least once every 10 seconds. If you expect the user to be doing a lot of random access in the file, you want keyframes to be more frequent. However, the more keyframes, the lower your average quality gets at the target data rate.

Limit Data Rate To specifies your target data rate. Note this is KBps (KiloBytes per second), although most other tools specify in Kbps (Kilobits per second). Also, this is the total data rate for the file. QuickTime automatically subtracts the audio data rate from this value to determine the video data rate. So if you type in 100, the total data rate will be 800Kbps. If you have 96Kbps of audio, QuickTime will target a video data rate of 704Kbps.

The options button opens up the special features of the codec, if any. The specific options are discussed with their codecs in the following sections.

The Sound settings are a lot simpler. They let you choose your codec, sample rate, and mono versus stereo. You'll never use 8-bit audio—the only codec that requires you to use 8-bit is MACE, which you'll never want to use. QuickTime has a decent audio rate resampler (much better than older versions), but it's still not the best. If you're making an intermediate file, pick the source's sample rate and resample with your final compression tool. There is no direct control of data rate—if data rate control is available for the audio codec, it'll be in its Options dialog.

The Export Size Setting lets you specify the output resolution. It doesn't allow any cropping or other features, so it is of limited utility for making final files.

## QuickTime Authoring Tools

QuickTime has had a powerful, open API for a decade now, and unsurprisingly a very healthy third-party industry has built up around using QuickTime as both an authoring and delivery technology.

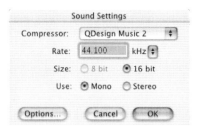

**14.8** The Audio codec dialog. If the codec supports audio data rates, you get to that with the Options button.

### QuickTime Player Pro

QuickTime Player Pro is the essential Swiss Army knife of QuickTime authoring. It's cheap at $29.99 (and as a bonus, you won't get that "Upgrade to QuickTime Pro" message again until the next major version of QuickTime), and the money helps finance QuickTime's development. If you use QuickTime, the time you save with Pro will more than cover its cost.

While you wouldn't use it much for final, professional compression, Player Pro is an amazing time saver. I won't go into its full depth here (check out *QuickTime Visual Quickstart Guide* by Judith Stern and Robert Lettieri). There are many tasks I use Pro for daily. It offers a simple way to change the annotations in a file. You can easily replace the audio or video track with a new one. You can trim the head or tail of the video, or paste in new head or tail video. You can hint a movie for streaming, or remove a hint track from a file to better optimize it for progressive download. You can fix the sync problems with Sorenson Video v3.1 B-frames (a full example of that process is given in Chapter 24, "Workflow Optimization"). And almost all of this can happen without recompressing the file—you're merely manipulating the already-existing data.

Whatever limitations QuickTime has as a delivery format, QuickTime Player Pro helps make it a vastly superior authoring format over all competitors.

### Discreet Cleaner 5

Historically, Cleaner and its predecessors have been the professional tools of choice for QuickTime compression since Movie Cleaner Pro v1.1. And while development of the product has lagged in recent years, Cleaner v5 still has a number of unique features for QuickTime.

First off, it was the pioneering tool that supported 2-pass VBR and authoring MP3 as a QuickTime track type. It also gives you the ability to mix and match which tracks are processed, copied without modification, or converted to audio and video (e.g. from text or music). Its EventStream technology also provides a simple way to do some basic rich media stuff with

QuickTime—such as adding hotspots for navigation and subtitles, although it's no substitute for a full interactive media authoring app.

One neat trick not often used is the "Flatten Only" option. This will flatten in-place all files in the current batch, making sure they're self-contained and fast start. With this feature, you don't have to export to a new file. Apple provides an AppleScript that can do this on Mac, but Cleaner is the easiest to use tool for doing this on Windows.

But the most invaluable feature of Cleaner v5 is its ability to make QuickTime alternates while it is compressing movies. This eliminates a labor-intensive step. Cleaner can even embed RTSP links automatically. And it supports important alternate features, like Control by Platform, not included in Apple's MakeRefMovie.

## Sorenson Squeeze

Sorenson's Squeeze was created during the half year between when Sorenson Video v3 was released and when Cleaner was finally updated to support it. The initial v1 release was obviously rushed, but the current v2 has matured nicely.

Squeeze for QuickTime is very focused, only supporting the Sorenson v3 video codec, and MP3, IMA, QDesign Music 2, and PureVoice for audio. Squeeze has a more flexible MP3 encode than Cleaner. It also has a better 2-pass VBR algorithm for Sorenson Video, giving it a slight quality edge. However, Squeeze's preprocessing isn't generally as good (although it has a better inverse telecine), and it obviously can't address as large a range of files as Cleaner. Still, it's quite a bit cheaper, and is easier to learn.

One unique feature of Squeeze is that it ties into Sorenson's Vcast video hosting network, so you can automatically publish your file as it renders.

## HipFlics

HipFlics is from Totally Hip software, makers of the essential Live Stage Pro authoring tool for interactive QuickTime movies. HipFlics is extremely innovative, pioneering a host of new features for compression tools. These features include an integrated timeline into which filters can be dragged to change the processing of different sections of the video. Different codecs can also be applied to different sections, so still images could use PNG or JPEG while video gets MPEG-4. All this is very slick and cool to demo. Alas, the current v1.1 version lacks some very basic preprocessing filters. It can't even deinterlace your video!

HipFlics is very frustrating, because Totally Hip got so many difficult things right, but missed some simple yet core features. I hope a future version will fill in the gaps and live up to its potential.

## Premiere

Before Movie Cleaner Pro, Adobe Premiere was the main tool used for QuickTime compression. Alas, Premiere's QuickTime authoring tools haven't been significantly updated since that version.

Like QuickTime Player Pro, Premiere also takes the data rate in kilobytes, not kilobits, and subtracts the audio rate from the total rate to determine video data rate.

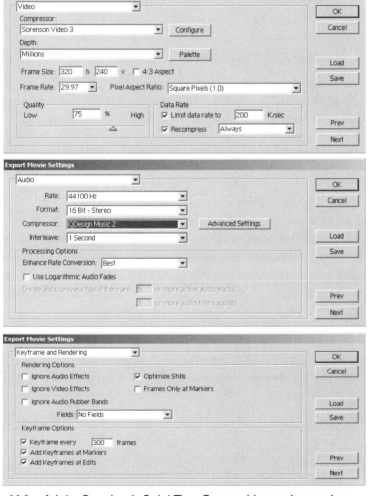

One great use for Premiere with QuickTime is to do assemble edits of existing content. If your input matches your output in Resolution and codec, you can set Recompress to "Maintain Data Rate". This only recompresses sections of the video where there has been rendering, or if the peak data rates are too high.

The Optimize Stills feature is also useful. It tells Premiere to let long frames (like a title card that appear on-screen for several seconds) to be counted as one long frame, not a series of identical frames. This helps keep quality up.

Frames Only At Markers makes a slide show where the only frames are where the markers are. You can then pretend it's 1982, and you're watching a filmstrip, but without those annoying beeps (anyone born since 1975 can ignore the preceding example).

Lastly, the Add Keyframes at Markers and Add Keyframes at Edits let you do just that. If you're

**14.9** Adobe Premiere's QuickTime Export video codec settings.

having trouble getting natural keyframes during transitions, at Edits will automatically drop one in at the cut. At Markers lets you specify exactly where you want keyframes to go.

## After Effects

Adobe After Effects was also frequently used for video compression in ancient times. Alas, its QuickTime support hasn't been updated for a long time either. The Output module in After Effects is really intended for interoperation with other video apps, not to make final deliverable files.

All After Effects does is provide the standard QuickTime effects dialogs from is own output window. Note you need to set the frame rate in the Render Settings Dialog, not with the Compression Settings dialog.

## MovieShop

Back in the paleolithic days of compression, Apple MovieShop was the tweaker's tool of choice. It was unsupported, buggy, and did incredible things with QuickTime codecs that have never been equaled. MovieShop was able to do manipulation of codecs to get consistent bitrates that nothing else could do. However, it hasn't been updated in ages, and its cool features don't apply to modern codecs. I doubt it'll even run on modern Macs. I only mention MovieShop to warn you away from it.

## Flash

Macromedia Flash MX has a couple of ways it can make a QuickTime movie. First, it can do a straight export via a custom QuickTime Export dialog. This turns the Flash movie into a standard QuickTime movie, although it lacks standard features like being able to pick an audio codec. Your best bet is to encode an intermediate file with PNG 24-bit and lossless audio, and then encode a final file from there. More interesting is Flash's ability to export a QuickTime movie with the Flash content turned into a Flash track. This preserves the vector graphics, object orientation, and small size of the Flash movie.

There are, however, caveats. Not all the features of Flash convert correctly. For example, the Load Movie action won't work within QuickTime. It's important to check Macromedia's current documentation on this, and thoroughly test playback. With these caveats, however, QuickTime is a powerful way to use Flash, and Flash is a great way to provide rich motion graphics in tiny files.

# QuickTime Integration Tips

## Macromedia Director

Macromedia's Director has been the CD-ROM authoring tool of choice since before CD-ROMs were standard equipment. Director has had good support for incorporating QuickTime on both Mac and Windows since its inception.

*Tip*

Director likes the resolution of all QuickTime files to be divisible by eight for both height and width.

## QuickTime Alternate Movies

QuickTime has a very unique Multiple Bit Rate system, called Alternates. Unlike MBR in Windows Media, RealMedia, or MPEG-4 scalable profiles, QuickTime doesn't bundle multiple versions of the data in a single file. Instead, authors create a number of different files, only one of which is played for a given user.

Alternates are supported in Apple QuickTime from v3 on.

### Master Movie

The key to alternates is the Master Movie. A master movie stores a list of the available files, the relative paths to them, and properties on which to base the decision of which to use.

### Alternates parameters

Each movie linked to the reference movie has a number of properties used to determine which file is played. Cleaner and MakeRefMovie support different subsets of the full range of features. Of the files that match these parameters, an Alternate will be played if

- the clip is at the target data rate,
- none are at the correct bandwidth, at the next lowest,
- none are lower, those at the lowest available bandwidth, or
- play the file with the highest quality setting matching the specified parameters.

*Connection* Connection refers to the user-specified connection speed. There were only a few bands available in QuickTime v4, so users who have upgraded to version v5 might still only have T1 or ISDN (Dual) selected. QuickTime v5 added a number of DSL and cable modem bands. There is no mechanism beyond user self-report to determine connection speed—a 56K modem connecting at below 28K still shows up as being at 56K. And a laptop configured to run on a corporate LAN would need to be manually changed to Modem for a modem connection to work.

Having the connection value be based on actual connection speed would be a huge improvement for QuickTime.

*Language*   This allows you to specify the language of the file. It should be left blank unless you're doing multilingual video.

*Platform*   The options are Mac OS only or any platform. This feature allows you to make alternate files to compensate for differences between Mac and PC gamma. Make the Mac gamma clip Mac OS Only with a higher quality value, and make the PC gamma file Any Platform with a lower quality value. This option is in Cleaner, but not MakeRefMovie.

*Quality*   Used to rank the priority of movie playback. If multiple files meet the requirements for playback, the highest quality clip will be played back. Cleaner gives a scale of 1 to 10; MakeRefMovie offers three levels.

*Computer power*   This is somewhat poorly defined. The values seem to roughly correlate to the MHz clock speed of 604e processors. Cleaner ranks them at 100-500, and MakeRefMovie 1-5 (just multiply by 100 to get Cleaner numbers).

*QuickTime version*   The version setting defines which version of QuickTime is required to view the clip. Note Cleaner only lets you select among major versions (v3, v4, or v5) of QuickTime, so you can't specify v4.1 for VBR MP3 support, or v5.0.2 to ensure Sorenson Video v3 support. MakeRefMovie can specify minor versions.

## Fallback movie

A QuickTime playback app will automatically skip past data it doesn't recognize. Because ancient versions of QuickTime and non-Apple .mov players don't know how to read the reference movie table, they'll automatically skip past it, and look for some kind of audio and video it can actually play. QuickTime v3 or later won't even get to that data because it will automatically skip to a reference movie as soon as the table is downloaded.

This allows you to make a fallback—additional video and/or audio in the stream that is only seen in a player that doesn't support a reference movie. This can be a complete version of the video using older, compatible codecs like Cinepak and IMA. Or it can be a simple still graphic telling the user they need to update to a current version of QuickTime, and where to download it (Cleaner automatically makes the latter for you if you don't specify a fallback movie).

## Authoring alternates

Alternates are a pain to create compared to the other formats. Each alternate requires its own unique encode, so targeting four bitrates means four different files need to be compressed. That process can take quite a bit of time. The files themselves can be encoded with any tool.

Making the master movie can be handled automatically by Cleaner, or manually with free tools like Apple's MakeRefMovie and Peter Hoddie's XMLtoRefMovie. MakeRefMovie provides an easy-to-use GUI, while XMLtoRefMovie uses XML files, and hence is easily scriptable.

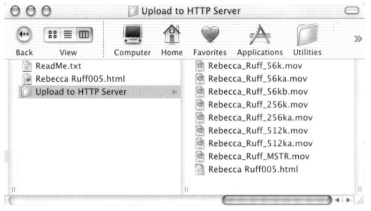

**14.10** The file structure of a Cleaner progressive download alternate movie output.

Cleaner is a dream to work with for reference movies, and is what most folks use. The reference movie generation is well integrated into the main UI. Cleaner can also automatically make a fallback file, and will even generate a fallback image for you with its own text or graphics if you don't actually encode a source file for fallback. Cleaner will also set references correctly for both progressive download and streaming files. It outputs a reference movie project into a complex directory structure, including both a ReadMe file explaining the project, and folders to copy up to each server type.

## Cool QuickTime Tricks

### Full Screen

The main playback-oriented feature of QuickTime Player Pro missing from the basic version is Full Screen. On Mac OS, this feature automatically lowers the resolution of the monitor to the lowest resolution equal to or greater than the resolu-

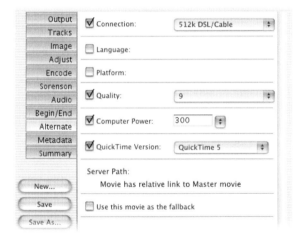

**14.11** Cleaner's Alternates tab.

tion of the file, turns off as many background processes as possible, and plays the file. The different modes—Normal, Double, Full screen, and Current—all affect what the resolution of the playback is before the resolution switches. For example, if a 320×240 file is presented in Normal, the resolution will typically switch to 640×480, and the file will be a quarter of the screen area. With Double or Full Screen, it'll switch to 640×480, and play back the file at 640×480. And with Current, it plays it at whatever the movie is currently set to. This enables movies with weird aspect ratios to play correctly.

Alas, on Windows the resolution switching aspect of Present Movie doesn't work. So, if a user has a 1600×1200 monitor, that 320×240 movie is going to be blown all the way up in Full Screen mode, causing seriously choppy playback on QuickTime versions before 6. The other options work just fine on Windows. I typically use Double for 320×240 and lower source to get close to the full screen experience while preserving decent playback unless the project is QuickTime 6 only.

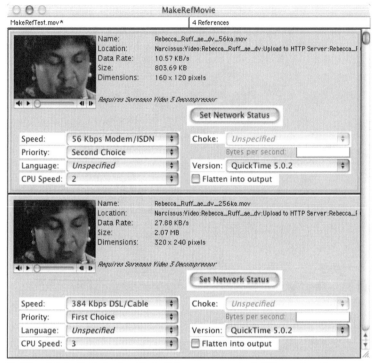

**14.12** MakeRefMovie. Note it has somewhat different options than Cleaner.

When saving a movie it can be set to automatically switch to Present Movie mode. The easiest way is via Cleaner v5's "Autoplay when opened in QuickTime Player" function, which supports the Full Screen and Normal modes. For Mac users, Apple includes an AppleScript to set this feature in its collection of free downloadable AppleScripts for QuickTime (Mac only, of course). The script supports Double as well as Full Screen and Normal.

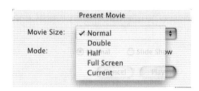

**14.13** The Present movie options available in QuickTime Player Pro.

## Anamorphic playback

Two common irritations of delivering at 640×480 from interlaced source are that deinterlacing leaves artifacts, and that playback requirements are high. *C'est la vie*, you may think. However, both irritations can be solved in a single, clever way. When we deinterlace and scale back up to 640×480, we're actually synthetically generating half of the lines of resolution. Instead of doing this during compression, you could just encode the video at 640×240, and let the video card do the scaling on playback. This would drop processor requirements in half, and should look just as good on a modern video card as does hardware scaling of the Y'CbCr overlay.

Because QuickTime files let the video tracks and movie be of different arbitrary sizes, this is actually quite simple to pull off. If you're in Cleaner, you can just type a Display size of 640×240, and you're done.

You can also pull this off in normal QuickTime as well. Make a file as you would normally, such as this anamorphic 640×240.

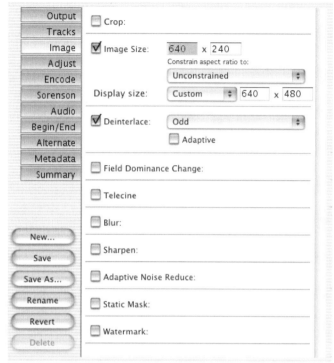

**14.14** Setting up Cleaner to do a half-height 640x240 encode with 640x480 playback

**14.15** The half-height video at its square-pixel resolution, and at full height.

Then pop it into QuickTime Player Pro and load up the Movie Properties Dialog. Go to the Movie>Size option. That will show you the current resolution of the file. The Normal value tells you the native size of the file. If the file has only a single video track, this movie size is the native resolution of that video track. Dragging the lower right-hand corner of the QuickTime movie resizes it. By default, QT keeps the aspect ratio. But if you hold down the Shift key, it'll let you drag to any resolution. Drag it so the Current value is 640×480. You'll need to be very careful—even being off by one pixel can hurt playback performance. Finally, Save As>Self-Contained (even adjusting just this one parameter will move the movie header to the end of the file, so it won't progressively download until you reflatten it).

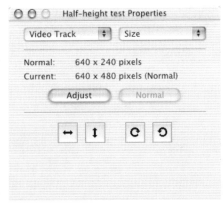

**14.16** How the half-height movie appears in the QuickTime Player Pro Size pane.

The file should work just like a normal QuickTime movie, and users shouldn't notice any difference. The one caveat is Present Movie will base the output aspect ratio on the Normal value unless you pick Current. Present Movie>Current preserves the modified aspect ratio.

## Media Keys

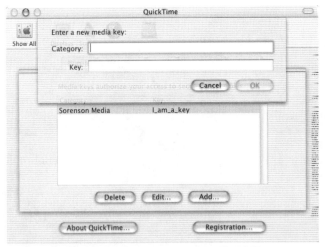

**14.17** The Media Keys control pane in QuickTime Preferences.

A Media Key is essentially a password for media. When you encode a file with a codec that supports Media Keys, like SV3 Pro or Indeo v5, you can specify the key along with the codec. To play the file back, the machine playing the media back needs to have the key. You can add the key manually through the Media Key tab in the QuickTime control panel.

However, the most common use of Media Keys is in locking the file so only a specific application can play it. iShell and LiveStagePro are the most often used apps, both of which can store internally an encrypted Media Key that is applied at runtime. This means the authorized player can unlock the media, but no other applications can.

## Fixed Audio Sync

When you encode with SV3 in bidirectional mode, the video is delayed by two frames, and hence doesn't match the audio. This is fortunately easy to fix in QuickTime Player Pro, although hopefully a later version of QuickTime or other authoring tools will support this automatically.

Load the movie into QuickTime Player Pro. Pick Edit>Extract Tracks. Extract the audio track, which becomes a new file. Do a Select All, and copy. Go back to the first movie, and Edit>Delete Tracks. Delete the audio track. Make sure you're at the beginning of the movie, and hit the right arrow key twice. This takes you forward two frames. Now, choose Edit>Add. This adds the audio in the clipboard to the movie at the selected point, which is a frame later than where it was before. You may need to delete a bogus last frame of the video. Go to the last frame of the video, and hit the right arrow key while holding down the shift key. This selects the last frame. Then press Delete. If there is still another blank frame, repeat. And that's it. Do a Save As and you've got perfect audio sync with SV3.

## Saving a Streaming reference movie

If you want to email someone a QuickTime movie, but can't actually send a huge attachment, you can send them a reference movie that links to the streaming file. This is dead simple to accomplish. With the movie to be streamed open, do a Save As. Note the file is only going to be 4K or so.

This technique will allow you to embed a QuickTime reference movie via HTML into a small email. However, unless someone is specifically expecting such a thing, it's rather rude to send it.

This is also how you put a streaming link into MakeRefMovie. You make a file with a streaming track, and include that as one of the references.

# QuickTime Delivery Codecs

QuickTime has always had the broadest selection of codecs for all kinds of tasks. I won't try to give a complete list of the codecs currently available and will focus on the ones you're most likely to use or encounter. QuickTime makes it very easy for third parties to create new codecs that automatically work in existing applications, which has led to a healthy QuickTime codec marketplace. Apple has been helping with this effort by opening its codec auto-update functionality to some third-party codecs (they need to pass Apple QA first). When a file is played that includes a codec not installed on the machine, QuickTime queries a server at Apple to determine if a downloadable codec is available. If there is, the user is asked for permission to download. If they say yes, the codec is installed, and the file plays without having to reboot the computer or even restart QuickTime Player. It's slick. The first two auto-update-able codecs were Sorenson Video v3 and VP3.

## Sorenson Video v3.1

Long-hyped and long-delayed, but now standard, Sorenson Video v3 is the default QuickTime codec. A completely new technology from the previous versions of Sorenson Video, SV3 still uses many of the same options and features as its predecessors. The Pro version is the most full-featured codec available today for any architecture.

SV3 is built into QuickTime from v5.02 on, and will auto-update with any version of v5.0. It isn't available for any earlier version of QuickTime.

QuickTime includes a Basic version of the SV3 encoder, and a complete decoder. The basic encoder has a good suite of default options, and is much better for CD-ROM and progressive content than the old Basic version of SV2. However, anyone doing compression professionally should have the Pro version, especially if they're doing RTSP streaming.

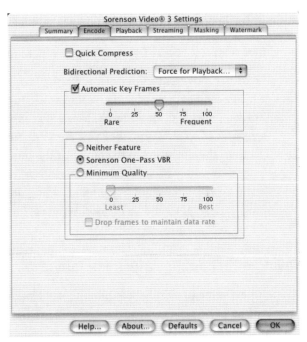

**14.18** Encode tab of SV3.1 Pro codec options, with typical RTSP settings.

The Basic version offers the standard QuickTime options of data rate, frame rate, and keyframe rate. Everything available via the Options dialog in Pro is assigned a preset in Basic. One difference from the old version of Basic is that this version isn't dog-slow, and it defaults to Image Smoothing on. This makes it a *much* more viable option for consumers to use for authoring CD-ROMs and progressive Web content.

Pro is currently at v3.1, which is backward compatible to the stock SV3 decoder. The Pro options are legion. I'll just give a quick summary of them here:

### Quick Compress

Quick Compress triggers a faster, slightly lower quality encode. It's mainly for live broadcasting, not video on demand. Quick compress is fine for making placeholder movies, but should never be used for final quality work.

## Bidirectional prediction

Probably the most significant single new feature in QuickTime, bidirectional prediction, enables B-frames. These allow about a 25 percent reduction (it varies according to content) in data rate at a given quality. Also, the B-frames can be dynamically dropped by QTSS during streaming if bandwidth isn't sufficient. Dropping B-frames dynamically causes the frame rate to drop by half. One drawback to bidirectional prediction is that it delays video playback by two frames. This can cause audio sync to be off—the lower the frame rate, the more off the sync. Bidirectional prediction also duplicates the first frame of the video, and cuts off the last frame. It's pretty easy to move the audio over to match the shift in video with QuickTime Player Pro, as described earlier.

If the frame rate is below 10, the bidirectional prediction option is automatically deactivated in "Allow" mode, but not in "Force for Playback Scalability."

## Automatic Key Frames

This option specifies how different a frame needs to be from the previous frame to trigger a keyframe. Automatic Key Frames should nearly always be on. The default 50 works well for most content, otherwise set this value in the range of 35–65. Greater sensitivity can help the video look better after cuts, and improve random access, but can also waste some data rate, reducing quality.

## 1-Pass VBR

Sorenson and Terran had a long productive partnership based around Cleaner's support for unique features in Sorenson Video, especially 2-pass VBR. However, after Cleaner's sale to Media 100 and Discreet, the Cleaner team was at least six months late releasing a version of Cleaner 5 that supported SV3, and so Sorenson took a series of moves to reduce their reliance on Cleaner. One-Pass VBR mode was part of this, so users of any QuickTime rendering application can get some of the benefits of VBR. However, it still isn't as good as 2-pass VBR for progressive download.

If you're not running Cleaner or Squeeze, both of which support 2-pass VBR, you'll want to use 1-pass VBR for progressive and CD-ROM content. Essentially, it loosens the data rate tracking enormously, letting the codec vary more over time.

You'll also want to use 1-pass VBR for RTSP streaming; if Streaming is checked in its tab, 1-pass VBR's tracking is tightened so that it's just tight enough for optimal RTSP.

## Minimum Quality

This specifies how poor the image quality for a given frame can be. If Drop Frames is selected, frames will be dropped as needed to maintain both the data rate and the minimum quality. If

Drop Frames is off, the data rate will just get higher. I almost never use this feature because I normally use 1-pass and 2-pass VBR and Minimum Quality is incompatible with them.

**14.19** SV3.1 Pro Playback tab.

### Image Smoothing

Image smoothing turns on a deblocking filter, which improves quality by making the video soft instead of blocky—the blockier the video, the larger the effect. Image smoothing is deactivated on slower processors above certain resolutions, depending on the processor:

| | |
|---|---|
| Pre G3/MMX: | Never on |
| G3/MMX/P II: | Only on below 320×240 |
| G4: | Only on below 500×375 |
| Pentium III: | Only on below 640×480 |

### Media Key

Media Key sets the track's media key, as described previously.

### Streaming

This option enables proper hinting for RTSP streaming. It should be on for doing RTSP streaming, and off for everything else. For the "Slice picture into packets of" option, don't change the default value for number of bytes unless you're doing something very unusual to account for an unusual network. If you are tweaking packet sizes, the value for number of bytes should be exactly 20 higher than the value specified in the QuickTime dialog to accommodate extra information added by QuickTime.

Force Block Refresh makes sure that every 16×16 block of video is refreshed at least every X seconds. This enables the frame of video to regenerate within the specified number of seconds if a packet was dropped or a user tunes into a stream in its middle. Force Block Refresh wastes some data rate (and hence lowers quality slightly) for low-motion content, but has no effect on high-motion content (which is constantly refreshing itself naturally). The default of five seconds is fine for most projects.

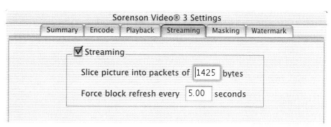

**14.20** The Streaming tab. Streaming should always be on for RTSP projects, and always off otherwise.

## Masking

SV3 leverages QuickTime's robust layering technology to allow real-time, alpha channel-based compositing of the frame. This is very powerful for times when you can segment your content into multiple layers, such as with a chroma key production over a static background. This is great stuff for those who need it, but it requires content produced for keying.

An example real-world use for masking is to mix very high quality background with lower quality foreground. For example, you could shoot a teacher in front of a blue-screen, and compress that video down to a 256×256 masked video. Then, you could make the background of the video a series of beautiful 640×480 PNG or JPEG images. Because the images are displayed for a long time, and the video is small and simple to compress, when combined they take only a fraction of the bandwidth that would be required to do the whole movie at 640×480 in SV3.

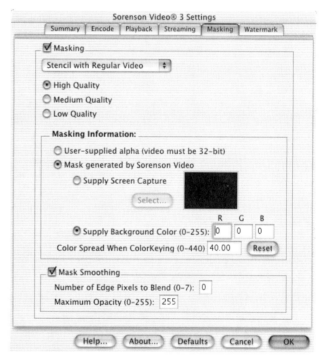

**14.21** And the Masking Tab. You can safely ignore this unless you are doing a chromakey project

## Stenciling

Stenciling is how the foreground and background objects are separated from one another. With Stencil Only, the foreground object's shape is extracted, but the content inside the stencil itself isn't saved. Typically, you'll use the Stencil with Regular Video option, where the foreground object will be encoded like anything else with Sorenson Video 3. The Stencil quality defines how much the stencil is going to be compressed. The High setting is lossless.

## Masking Information

Masking Information is where you define how the stencil is generated. Ideally, you'll have an alpha supplied, exported from a tool like After Effects into the Animation codec in Millions+ mode. Even if your original source doesn't have an alpha, you can use an AE plug-in like Ultimatte to generate a good key—professional tools do a better job than the built-in support of the codec. An 8-bit alpha is converted to the internal 1-bit alpha; all gradations of transparency are lost.

If you must use the codec, you can define a stencil based either on a screen capture or a color. The screen capture needs to be an image of a clean plate, that is, the shot is of the set with no objects in the foreground. This requires the video be shot with a locked-down camera on a tripod, of course. The more flexible option is to just pick the key's color. Note, however, this requires the video to be shot with very good, even lighting, and that foreground objects do not contain the key color. (Remember, if you choose blue as the color to key, blue eyed actors could prove problematic—their eyes may be keyed out.) Color spread sets how different a color needs to be from the specified key color to be considered part of the key. The better your lighting, the smaller this value needs to be, and the closer the color of foreground objects can be to that of the background.

## Mask Smoothing

Number of Edge Pixels to Blend specifies with width of the stencil. The bigger the number, the broader the area between the foreground and background elements. This can hide problems resulting from a bad key. Again, the better the lighting, the smaller this value needs to be. The Maximum Opacity value defines the opacity of the foreground. A value of 255 means that none of the background can be seen through the foreground; lower values make the foreground increasingly transparent.

## Watermark

SV3 allows a static user-created watermark to be composited in real time over the video. Because the complex graphics of logos are often severely compromised by compression, this approach can offer much higher quality than just compositing the watermark into the video stream before compression. Larger watermarks can entail a significant performance hit. Watermarks can now be in color, unlike watermarks created with an earlier version of Sorenson.

The file can be any still image format QuickTime can natively read, including PICT, TIFF, or PSD. Position defines the corner of the screen the watermark is in. Pixel Offset defines how many pixels there are from that corner of the video to the watermark. And Opacity defines how transparent the watermark is.

Note the resolution of the watermark is always the resolution of the source file—if you want to use your favorite 1024×1024 TIFF logo, you'll need to scale it down in advance. Also, the larger the file, the larger the CPU hit for playing the file. Sticking to 64×64 or less is a good rule of thumb.

## Sorenson Video v2

The original Sorenson codec, which uses the same backward compatible format in both SV1 and SV2, was the dominant QuickTime codec for three years. Of revolutionary quality when it was first launched, it had aged badly well before SV3 came out. Beyond its quality at a given data rate simply not being competitive any more, SV2 used the YUV-9 color space, where there is only one color sample per each 4×4 pixel block. This meant highly saturated content with sharp edges looked extremely blocky. SV2 didn't support B-frames, making it less suited for RTSP use. Finally, the Basic version of SV2 was extremely slow and limited, leaving QuickTime without a good option for a free consumer-grade codec.

Hopefully, folks won't need to continue to use SV2 for much longer, as SV3 offers such an improved experience. However, QuickTime's Movie Alternates feature makes it easy to create both SV2 and SV3 versions of the same clip. The user will then see the version appropriate to their version of QuickTime.

The SV2 decoder is somewhat faster than SV3, so folks doing CD-ROM content where high resolution is more important than quality or data rate might still use it. However, VP3 might be a better choice in those cases.

The only significant quality differences between SV1 and SV2 were the addition of Image Smoothing, and the repair of the Temporal Scalability option. Files from either version play in any version of the decoder, although the original decoder won't show image smoothing. There was also a ton of performance optimization for MMX, G4, and multiple processors. Alas, there is a serious bug that renders multiprocessor support on G4s unusable, although multiprocessors and SV2 on Windows get along fine.

The general structure of options in SV2 is quite similar to SV3. SV2 lacks B-frames, color watermarking, and masking. SV2 could set the keyframe size to a percentage of the size the codec would use by default. Because the codec almost always knew the right value, this option was mainly a way for novice users to get themselves in trouble, so I was pleased it was dropped from

SV3. Lastly, SV2 could set the data rate tracking sensitivity, which roughly controlled the buffer size. However, it wasn't documented or defined very well, so virtually everyone left it at the default of 50. This is replaced by the 1-pass VBR option in SV3.1.

Sorenson Video v2 Basic's encoder is included in QuickTime for all platforms. Sorenson Video v2 Developer Edition is no longer available, and wasn't ported to Mac OS X.

## MPEG-4

QuickTime can export MPEG-4 files or use MPEG-4 as a video codec inside a normal QuickTime file. Compared to SV 3.1 Pro, MPEG-4 offers lower compression efficiency, but faster encoding and decoding. And, of course, it's free. The only real option offered for MPEG-4 video is a choice between "Fast Compression" and "More Accurate Compression." The latter is still quite fast, and provides significantly better quality.

**Note**

You can use other MPEG-4 encoders to author a QuickTime movie, gaining access to features like 2-pass encoding. Just open the .MP4 file in QuickTime Player, extract the video track, and use as normal.

MPEG-4 is the first delivery codec to support automatic gamma correction in QuickTime. All MPEG-4 content is encoded internally with 2.2 gamma and is corrected to 1.8 when displayed on a Mac and 2.5 when displayed under Windows. Note other vendors of MPEG-4 decoders may not support this feature.

## ACT-L2

From Streambox, ACT-L2 is a very promising, though difficult to pronounce, new QuickTime codec. The encoder costs $499, and the decoder is free, and is available via Component Download.

Beyond QuickTime, ACT-L2 also targets set-top boxes, and has libraries available for playback on the TI C6000 and Equator MAP-CA processors as well as PowerPC and Intel. The Streambox Transport product claims to be able to transcode MPEG-2 to an ACT-L2 stream in real time, cutting data rate requirements in half without hurting quality significantly.

Streambox has several on/off options. Enable Drop Frames lets the codec drop frames to maintain data rate. Prefilter Image does some additional preprocessing (mainly smoothing sharp edges that would cause ringing on encode) to improve quality, but slows the encode down a little. Filter Residual also improves quality a little at the expense of speed. Auto Peak Bitrate Control tightens the data rate tracking. Fast Motion Search speeds up encoding quite a bit, with a potential loss in quality. I-frame and P-frame Minimum Quality set a quality threshold à la QuickTime's quality slider for keyframes and delta frames, respectively. These values are the opposite of the normal QuickTime convention, with 0 the highest value and

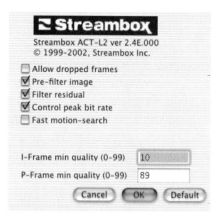

**14.22** The ACT-L2 dialog, tuned for maximum quality.

100 as lowest. Leaving them at their defaults is appropriate for most broadband Web video. Minimum Quality overrides the target data rate, but is overridden by Auto Peak Bitrate Control.

ACT-L2 can beat VP3 and SV3 for compression efficiency in some cases. At the time of this writing, playback and encoding are optimized for SSE, but not for AltiVec, meaning performance is quite a bit lower on Mac than Windows. AltiVec support is planned for a future version.

## H.263

H.263 is a videoconferencing standard, and also a commonly used codec (the MPEG-4 video codec is essentially an enhancement of H.263). Compared to SV2, H.263 dealt with fast motion much better, and had a much faster encoder, making it the default codec for broadband live streaming. However, SV3 encodes quite a bit faster and is higher quality. Because H.263 is a standard, some other non-QuickTime players can view via the proper media players streams produced by the H.263 codec. All MPEG-4 players also include baseline H.263 decode support.

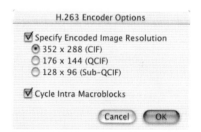

**14.23** H.263 options dialog.

If using H.263 for RTSP, turn the Cycle Intra Macroblocks option on. This achieves the same effect as Force Block Refresh. H.263 is limited in its internal resolution to either 352×288, 176×144, or 128×96. Encoding at other resolutions causes the codec to pick the nearest resolution and scale up on playback. It scales well, which doesn't normally cause quality problems.

## VP3

VP3, from On2 (formerly the Duck Corporation), is a close second to SV3 in the buzz arena. While not part of the default QuickTime install, VP3 was the second codec available through the auto-update feature. You can also manually trigger the auto-update via the QuickTime Updater application, which gets you the encoder as well as the decoder. Note that just having the decoder doesn't give you access to support—that requires the paid version, which is otherwise identical.

VP3 has also been released as open source, so the source code can be downloaded and tweaked. So far, we haven't seen much actual VP3-derived software, but this may change. Some improvements to the standard version of the encoder have already been made based on feedback from users working with the open source components. These haven't been rolled into the Component download version of the codec so far, so your best bet is to download the latest release version from www.vp3.com.

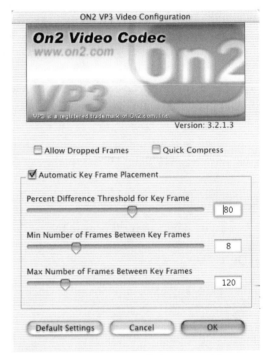

14.24  The VP3 options dialog, tuned for maximum quality.

Compared to SV3, VP3 offers better quality with some content, and a much faster decoder, enabling bigger movies on slower machines. However, it lacks 2-pass VBR and a native packetizer. VP3 is a good alternative to SV3 for high-resolution, progressive download, broadband content and CD-ROM. VP3 has a bad habit of substantially exceeding the target data rate if the data rate is below the optimum value for the codec, so be careful to check that the file size is what you expect. This has improved a lot in recent releases.

VP3 has a deblocking filter that improves quality when running on a processor with AltiVec (G4) or SSE (PIII, P4, AMD Athlon XP), but not on a G3, PII, or earlier.

VP3 is a fairly standard QuickTime codec with a few special options. It defaults to the Quick Compress mode, which entails less of a quality hit than most other Quick Compress modes, and speeds up encoding quite a bit. You can specify whether or not the codec drops frames to maintain image quality and data rate (turning this on can help hit data rate targets a lot more closely). And lastly, it has a pretty powerful Auto-keyframe mode. For most content, the default options are good, although I tend to change the Maximum Keyframe Distance to 10 times the frame rate.

If you use this, and I recommend you do, you should set the Keyframe Every value in the main QuickTime dialog to 9999 so it doesn't also try to insert keyframes.

On2 has announced VP4, aimed at the set-top box space, and VP5, a high-powered replacement for VP3. Neither have been announced for QuickTime.

## ZyGoVideo

The ZyGoVideo codec is a newcomer. It's based on wavelets, which are also the basis of Indeo 5 and JPEG2000. Wavelets offer a lot of promising features for Web video, like arbitrary scaling and dynamic thinning of quality bands to reduce bandwidth or CPU requirements. The current builds of ZyGoVideo don't support these features yet, although the company is looking into adding the functionality to a future build.

ZyGoVideo offers a free basic version and a $99 Pro version. The codec is also bundled with HipFlics. The Pro version adds a few new features, and a lot of behind-the-scenes quality improvements. The new features are configurable keyframe sensitivity, motion detection, and image smoothing. Increasing the motion detection value improves quality somewhat, dramatically increases encode time, and also increases decode requirements by roughly 10 percent.

One unique feature of ZyGoVideo is that a playback codec for Palm OS-based handhelds is being developed, so you'll eventually be able to use it to deploy content there. The handheld codec hasn't been released as of this writing, although developers were seeded with the components.

One significant issue with the codec is that it keeps its data rate very tightly controlled. Because of this constraint, keyframes are often starved of bandwidth, which causes the video to get blocky at keyframes, resuming quality over the next few delta frames. This is an odd design decision, as it doesn't have a native packetizer, and so isn't really suitable for tasks other than progressive download and local playback, where a loose data rate is fine. The quality slider in the main QuickTime dialog has an effect on keyframe size, so turning it up if you have this problem can improve quality substantially.

PDA playback aside, ZyGoVideo doesn't seem to be a particularly useful codec compared to SV3.

ZyGoVideo isn't installed with QuickTime, although it is available through Component Download.

## 3ivx

The 3ivx codec's name is a play on DivX. While it is based around QuickTime, the company is focused on ubiquitous playback of the QuickTime format on non-Apple players. They claim playback on Linux, BeOS, and Amiga. These files generally use MP3 for audio, because the standard QuickTime audio codecs are not available. And most of the groovy QuickTime features like sprites and interactivity won't work in any non-Apple tools.

There are three versions of 3ivx currently available. The Test version is free, and only supports adjustment of the Quality slider, without any data rate control at all. The $10 Personal edition adds deblocking and data rate control. Plus for $400 adds a deblocking filter, data rate tracking, and a high quality encode time that improves efficiency about 15 percent at the cost of doubling encode time. They've also announced, but have not yet shipped, a Pro version, which is intended to support 2-pass VBR and other features. Presumably this will be a standalone program.

In Personal and Plus, the Quality slider defines a minimum quality level. It will also raise the data rate above the target rate if necessary. This slider should be turned down to zero in most cases to guarantee the requested data rate. Artifact Reduction, as the deblocking filter is called, increases decode times roughly 200 percent, but with a significant boost in quality at lower data rates.

As of the v3.5 beta release, 3ivx is an extremely unappealing codec, with a slow encoder and decoder, and poorer quality than its competitors at all data rates. As it hasn't yet had a "final" release (although it is already being sold), perhaps it will improve dramatically by v1. Unless there are massive improvements, I expect it and other QuickTime-on-UNIX solutions will be eclipsed by straightforward MPEG-4 solutions.

## Legacy Delivery Codecs

### Indeo 5

The various Indeo codecs were originally produced by Intel Architecture Labs, and are now owned by Ligos. Intel's motive was to make a superior technology that would encourage users to buy new generations of Intel processors. Early versions were intended to be an enabling technology to grow the multimedia industry (and thus sell more processors). In an era where Macintosh computers dominated digital video production, it was also a way to encourage folks to use Intel's video capture and video playback cards in Wintel machines. In the early days of CD-ROM digital video, the Indeo v3.2 codec was second only to Cinepak in use, and had strong advantages for talking-head content. When Pentium machines came out, Intel wanted a new codec that would demonstrate the Pentium's power while still being usable on older machines. They started Indeo over from scratch with the renamed Indeo Video Interactive (IVI) v4, a scalable wavelet-based compression technology. The current version, IVI v5, is a further enhancement of that technology, originally optimized for the Pentium II processor. Indeo is available as a component download for QuickTime on Windows and classic Mac OS, but not for Mac OS X (although Indeo files play fine via the QuickTime Player in Classic).

Wavelet's promise is smaller files, better quality, less digital looking artifacts, and the ability to drop quality instead of frames on slower processors.

Like Sorenson Video 2, the Indeo codecs were hampered by their reliance on YUV-9 color sub-sampling. To help alleviate the YUV-9 problem, Indeo 5 running on MMX systems will interpolate between 4×4 color blocks, reducing (but not eliminating) the YUV-9 problem.

Indeo still has something to offer CD-ROM producers and game developers on the Windows platform. IVI supports a one-bit alpha channel, like SV3, and can do good real-time compositing with playback, either through an API or via a Director XTra. This is great for achieving a number of game effects, such as compositing an actor over a runtime–generated background in real time. The Indeo API can also do local decoding of a region of a movie. For instance, imagine a submarine game in which the user looks through a periscope. You could render a 1024×192 Indeo movie, representing a panoramic image around the sub. Programmatically, the Indeo codec could then be told to decode only a specific 192×192 region that the periscope is looking at, dramatically saving processing time.

One knock against Indeo back when it was being actively developed was its processor requirements for playback. But today, requiring a PII for 640×480 playback is nothing. However, playback requirements are much higher on Mac OS. Generally a hybrid project would be better off using a codec like VP3, which offers excellent quality, and has better decode performance on Mac.

## Scalability

Because Indeo is wavelet based, it's based around successive bands of increasing quality that are decoded in sequence, creating the final image (see Chapter 3). With scalability turned on, under high processor loads, the codec skips decoding the higher quality bands instead of dropping frames. This preserves motion at the expense of some image quality—generally a good trade to make. However, in the real world this is less useful than it sounds. The lowest quality decode is about 80 percent of the decode complexity as the high mode. Still, every little bit can help, and I recommend leaving this on in most cases. Note low end users are more likely to blame poor video quality on your encoding than on their slow processor.

## Transparency

Indeo was the first delivery codec with built-in support for transparency effects. The None mode has no transparency. First Frame uses a key color in the background of the first frame to define the alpha channel for the rest of the frames. Alpha channel attaches a 1-bit alpha channel to the entire movie. Note there is no edge feathering in a 1-bit alpha channel, so every pixel is either entirely in the image or entirely out of it, which can cause some ugly effects around the edge. If you know in advance what background is going to be attached to the image, you can premultiply the background into edge pixels.

### Indeo 5 Tips

Never use the Quick Compress mode for final rendering—the quality isn't as high. But it's great for rapid prototyping.

Indeo responds well to using the Quality slider instead of data rate controls. Setting a Q=85 instead of limiting data rate can often yield better results *and* a smaller total file size. Unfortunately, Indeo doesn't support explicit VBR, so the final data rate can be difficult to predict.

## Indeo v4.4

Indeo v4.4 was an earlier, lesser version of Indeo v5. Its features were very similar. The codecs are still available from Apple, mainly for playback of legacy content. If you need to make new files with Indeo, use v5.

## Cinepak

Cinepak was *the* CD-ROM video codec for many years. Originally created as Compact Video by SuperMac, brought to Radius when those companies merged, subjected to a last-minute trademark-infringement name change to Cinepak, licensed to everything from QuickTime and Video for Windows to the Atari Jaguar and the Java Media Layer, and split into an updated version by Compression Technologies, it is easily the most used, most ubiquitous codec ever. Free to end users, its combination of good overall performance and great price made it the dominant codec in CD-ROM applications until 1999. Virtually every personal computer shipped since 1994 has Cinepak preinstalled. Alas, progress has finally passed it by. There is little reason to use Cinepak anymore, unless you need to target sub-200MHz machines. MPEG-1 has similar ubiquity, and a great number of codecs provide better quality at lower data rates.

However, if you are targeting older machines, especially in the educational market, Cinepak can be a good choice. It is also useful for making QuickTime fallback movies for compatibility with older versions of QuickTime or UNIX QuickTime readers like Xanim.

### Cinepak tips

A favorite trick with Cinepak is to have very few keyframes. If a video requires no random access, and largely has a static background, Cinepak can have only a single keyframe, followed by literally hundreds of interframes. A keyframe will be automatically inserted if a frame is more than 85 percent different than the previous frame, so you'll generally get a keyframe with every camera cut.

For black and white content, the Cinepak 256 grays mode can provide beautiful results at reasonable data rates. Back in the VLB video card era, this mode could actually play back 640×480 15fps full-screen.

The 256-color Cinepak mode is rarely used, but can occasionally work well for animated content with too much motion for other codecs. However, you'll need to feed the codec 8-bit palletized source.

## Cinepak Pro

At Radius's booth at the National Association of Broadcasters 1995 convention, I saw an impressive demo of something called the Cinepak Toolkit, a compression program with a new version of the Cinepak encoder. I was a beta tester for it, and was thrilled to have a competitor to the venerable MovieShop. Alas, development stalled with Radius's financial implosion in 1996. But, finally, Radius's old Cinepak team worked out a deal, finished the product, and brought it to market. Cinepak Pro is available stand alone, or on the Mac OS with the Cinepak Toolkit, a simple, quirky, ancient compression tool notable mainly for a neat noise reduction filter. For those rare Cinepak jobs, I run Cinepak Pro through Cleaner.

The best thing about Cinepak Pro is that it makes standard Cinepak files—no user upgrade required. When you can't install software, Cinepak Pro provides an opportunity to still increase quality. Because it uses the same binary format as the old Cinepak, its improvements come from allocating that bandwidth more intelligently. First, it gives you *exactly* the data rate, frame-for-frame, you ask it for. This isn't a great way to conserve disk space, but does nicely eliminate data spikes. It also lets you select the keyframe/interframe size ratio and keyframe insertion sensitivity. And it is optimized to make things look generally better.

In general, Cinepak Pro does a superior job, especially at low data rates. One of the most obvious head-to-head comparisons is with text—at the same data rate, I've seen text look crystal clear under Cinepak Pro and completely illegible under the original Cinepak. However, there are clips where I think the original Cinepak looked better. For scenes in which the camera moves rapidly, the old Cinepak would smear, leaving streaks of color. Cinepak Pro doesn't smear. Instead, everything moving becomes very blocky. Much of the art of digital video is deciding which artifacts you like better, and in this case I like the old artifacts better.

But, in most cases, Cinepak Pro provides better quality than the old Cinepak. If you do use Cinepak professionally, you should have Cinepak Pro.

### Cinepak Pro tips

For footage with a static background, yank the keyframe-to-interframe ratio all the way to v5, reduce the keyframe rate, and you'll have better quality at low data rates.

When you have the Cinepak Pro codec loaded, some applications don't properly distinguish between Apple Cinepak and Cinepak Pro. They just list Cinepak twice, and selecting either yields Cinepak Pro. To use the old Cinepak with one of these programs, turn off Cinepak Pro in the Extensions Manager, then reboot.

## Indeo v3.2

Indeo v3.2 was Cinepak's biggest competitor in the mid-90s video codec industry. It was YUV-9, like all Intel codecs. Indeo v3.2 was freely available, but required an install for QuickTime on both Mac and Windows and in Video for Windows. Compared to Cinepak, Indeo v3.2 offered somewhat lower data rates and higher quality for low-motion, talking head-style content. It encoded somewhat faster than Cinepak, but had higher decode requirements.

Indeo v3.2 is still seen on some legacy CD-ROMs.

### Indeo v3.2 tips

The resolution of Indeo files absolutely must be divisible by four on both axes, or playback may be unstable on some systems.

The width of an Indeo file must be greater than its height, or playback will be unstable on some systems.

# QuickTime Authoring Codecs

Unlike the other major formats, QuickTime is also an excellent media authoring platform, and it includes a lot of codecs designed for capturing and editing content, not delivery. Thus QuickTime includes a wide variety of authoring codecs.

An authoring codec is a codec whose features make it useful for the acquisition, editing, and storage of full quality content. An authoring codec's output can be recompressed with minimal loss. Some codecs, like MPEG-2 and Apple Animation, can be used for both authoring and delivery when different settings are applied.

## DV

QuickTime supports DV as a native QuickTime media type, which means you can open a raw DV file in the QuickTime Player, and it'll behave just like a normal QuickTime movie.

In the past, many different vendors produced DV codecs for QuickTime. Originally this was because Apple didn't have one, and later because vendors felt that their codecs were faster and/or of higher quality than Apple's DV codec. However, the modern DV codec in QuickTime v5 is tremendously better than the older ones and wicked fast (it can decode at 300fps in memory on a 1GHz G4).

By default, DV is shown in QuickTime in a preview mode, which only displays a single field, although the data for full quality of the movie is still in the file. If you want to see the file as it actually is, you'll need to set the High Quality flag for the video track.

The biggest limitation of DV comes from its fixed 25Mbps data rate. While 25Mbps is generally adequate for content originally produced in DV, and hence cleanly captured and compressed in-camera, it often isn't enough bandwidth to provide adequate quality with noisy source, like a VHS dub. For this reason, I generally don't recommend transcoding existing analog source to DV before capture.

QuickTime v6 added a DVCPRO-PAL codec, so it can work with that normally incompatible 4:1:1 PAL format.

## Motion-JPEG

Motion-JPEG is the granddaddy authoring codec in QuickTime. Essentially, it takes JPEG, makes it 4:2:2 instead of 4:2:0, and allows storing each field as its own bitmap (although you can do progressive scan M-JPEG as well).

The M-JPEG A and B codecs were invented by Apple to provide a common file format for exchange of media between the many varieties of digital video capture hardware available in the mid 90s. While there were many different vendors, and many different proprietary codecs, only two main chipsets were used in all these systems, so it was theoretically possible to create a single file format that would allow lossless interchange among different systems. In a heroic effort, Apple was able to get the vendors to open their kimonos, and provide the information necessary to make universal file formats. Alas, most vendors still require the use of their own formats, although they provide the ability to losslessly transcode to one of the transition formats. The A and B flavors of M-JPEG are based around the chipsets used by different cards. One welcome exception is the Aurora cards, all of which use the standard Motion-JPEG A format.

One limitation of the basic codec inside QuickTime is that it only supports quality limiting, not data rate limiting. A value of 85 is good enough to be visually lossless for most uses. Ideally, you'll have the storage space to encode the files at 100 percent, which is mathematically lossless.

## Animation

Animation is the only QuickTime compressed codec that can contain an alpha channel, useful if you're taking advantage of QuickTime v3's transparency effects.

Using run length encoding (RLE), Animation's compression depends on long horizontal lines of identical pixels, and as many as possible identical lines between frames. Reducing the Quality slider for better compression largely works by flattening out these lines. Superior results can often be gained doing this before the file is compressed.

Note Quality settings aren't available in 256-color mode, only in Millions and Thousands. Typically, you'll want to use the Graphics codec for 8-bit graphics, which generally provides the same quality at half the data rate.

Animation is an ancient technology, and is very much behind the times. Modern codecs like Window Media Screen can provide equal quality with the same content at a data rate a fraction of what Animation requires. Because it is so inefficient, you can generally get an Animation file to about half the size with Zipit or Stuffit.

One promising alternate is MNG (Motion Network Graphics, pronounced "ming," and based on PNG), which is an open standard. Hopefully Apple will include a MNG codec for QuickTime soon, so we can finally stop using Animation as a delivery codec.

## Graphics

The Graphics codec is similar to Animation, in that it does an excellent job of compressing flat-color synthetic source. The big differences are that Graphics only does 8-bit graphics, and offers about twice the compression with the same content. Graphics doesn't have any data rate or quality controls, so it's lossless in 8-bit color, and its file size is very sensitive to the complexity of the content presented. Like Animation, you want to give it content with long lines of exactly the same color, and have most parts of the screen identical between frames.

For rendered graphics without many different colors, I've attained pleasing results by converting to 256 color. However just saving the movie as 256 color causes horrible dithering. Instead, make a superpalette of the source with Equilibrium DeBabelizer, then remap the video to the new palette in DeBabelizer using the Set Palette and Remap Pixels command, with dithering off. This yields a custom palette tuned towards the content. Leaving dithering off maximizes long lines of identical color. This technique is discussed in Chapter 22, "Miscellaneous Formats."

It's generally easier to create content in 8-bit in the first place than to try and take true color images and convert them to 8-bit. The operating system typically draws simplified elements when in 8-bit mode.

Again, Graphics is ancient technology, and ripe for replacement by a MNG codec in QuickTime.

## PNG

The PNG (pronounced "ping") codec is Apple's implementation of the Portable Network Graphics lossless RGB format as a QuickTime codec. PNG can do black and white, 4-color, 16-color, and 256-color indexed color, and 8-bit-per-channel RGB with and without an alpha channel. PNG is always lossless, and offers much better compression efficiency than Animation for stuff not amenable to RLE. However, PNG is keyframe-only, so Animation can offer better compression efficiency for content in which most of each successive frame is unchanging, as with screen animations.

PNG is ideal for intermediate files of RGB content like screen shots (almost all the graphics in this book were PNG files at one point in their life), and for interoperation with RGB-based applications such as Cleaner, Premiere, and After Effects. Its only real drawback is that it's a little slow for real-time capture or playback. PNG is best suited for storage.

In the Options dialog, you have different choices for compression efficiency versus speed. Although they're not so labeled, they're roughly in order of quality versus speed. I normally just park it on Best.

**14.25** The Microcosm codec, tuned for smallest file size.

### Microcosm

Microcosm, from Theory LLC, is another lossless, keyframe-only RGB codec. It offers file sizes and performance in the ballpark of PNG. However, it has an optional 16-bit-per-channel mode, the first codec to support that. This is very useful for storing intermediate generations of high-end authoring projects, especially when targeting 10-bit-per-channel video formats like D1. Of course, you'll need to use Microcosm with applications like After Effects v5.5 that support reading, processing, and writing 16-bit-per-channel.

Microcosm is currently a free public beta. It isn't part of Apple's Component Download program, at least not yet, so for the time being you'll need to make sure it is manually installed on all machines.

## Component

The Component codec is an uncompressed 4:2:2. It's a good alternate to Motion-JPEG when encode/decode performance is more important than conserving drive space. Because it's 4:2:2, it takes two-thirds the data rate of None, and should be used instead of None with Y'CbCr apps like Final Cut Pro and ProCoder.

## None

The None codec is an uncompressed RGB in the full variety of color spaces. I rarely use it except when I need to make a file that is interoperable with DirectShow apps—None is the only authoring codec standard for both QuickTime and DirectShow.

## Video

The Apple Video codec was in the original version of QuickTime. Its code name was Road Pizza because it left the video looking vaguely like a flat, crushed version of the original.

Video is a weird duck compared to modern codecs. It was quality-limited, not data rate-limited, and encoded in 5-bit-per-channel RGB. It had an extremely fast encoder and decoder because it was designed to run on 25MHz computers, so it was used for doing comps and that sort of thing. But even DV and JPEG are fast enough for real-time comps on a modern computer.

I mention Video because it shows up in the codec list as "Video," which causes some people to pick it assuming it'll be useful for compressing video. There is absolutely no reason why you'd want to encode to Video for delivery—it was largely replaced by Cinepak!

# QuickTime Audio Codecs

QuickTime, and the Mac OS before it, has had audio codecs since the late 1980s. None of those codecs has been removed since. A number of the older formats are listed in the next section: "Legacy Audio Codecs."

## MPEG-4 Audio

Like MPEG-4 video, QuickTime can use MPEG-4 audio as a standard audio codec. Advanced Audio Coding Low Complexity (AAC-LC) is the standard general purpose audio codec from MPEG-4 and is the only one available in the MPEG-4 audio mode (the CELP speech codec is listed, but grayed out and unavailable). AAC is easily the strongest audio codec in QuickTime—providing much better compression efficiency than MP3 and QDesign Music 2 and with a native packetizer for RTSP streaming.

Beyond the normal channels and sample rate control, the codec also lets you pick a data rate between 16 and 256 Kbps. 128 and beyond should be indistinguishable from the source and quality can be surprisingly good even at 16.

One complication is that Apple's AAC-LC implementation automatically reduces the sample rate of the audio depending on the combination of data rate, sample rate, and input number of channels. This isn't a limitation of AAC-LC itself and other implementations offer other combinations. AAC-LC is the first VBR audio encoder in QuickTime. It requires a change in how it is called by software. Apps may crash encoding to AAC if they are not updated to support it. Table 14.1 shows the maximum sample rate Apple's AAC codec will use with particular combinations of bitrate and sample rate.

## QDesign Music 2

QDesign Music 2 was an enhancement to the original QDesign Music codec that added RTSP support and improved quality. Originally introduced in 1999, it was the best general purpose codec of its era. It is still excellent for instrumental music, although it is quite prone to phasing anything with speech. AAC is a much better all-around codec, and should eclipse QDesign Music 2 quickly.

Table 14.1  Apple's MPEG-4 AAC-LC audio codec automatically converts the sample rate to a specific one based on source sample rate, data rate, and number of channels.

Input sample rates (KHz) to output sample rate (KHz)

| Data rate (Kbps) | 44.1 mono | 44.1 stereo | 48 mono | 48 stereo |
|---|---|---|---|---|
| 8 | 8 | na | 8 | na |
| 16 | 16 | 8 | 16 | 8 |
| 20 | 16 | 8 | 16 | 8 |
| 24 | 22.05 | 11.025 | 22.05 | 11.025 |
| 28 | 22.05 | 11.025 | 22.05 | 11.025 |
| 32 | 32 | 16 | 32 | 16 |
| 40 | 32 | 22.05 | 32 | 22.05 |
| 56 | 32 | 22.05 | 32 | 22.05 |
| 64 | 44.1 | 32 | 48 | 32 |
| 80 | 44.1 | 32 | 48 | 32 |
| 96 | 44.1 | 32 | 48 | 32 |
| 112 | 44.1 | 32 | 48 | 32 |
| 128+ | 44.1 | 44.1 | 48 | 48 |

QDesign has continued to enhance this technology, and is developing its new QDX platform for portable device music around it. However there has been no word about getting QDX into QuickTime.

There is a basic version of the QDesign encoder bundled with Quick-Time, but it only goes up to 48Kbps. The Pro version ($399) can go up to 128Kbps, and offers much faster encoding, slightly higher quality, and some controls to tweak quality slightly. As of this writing, the Pro version wasn't available for Mac OS X, although the Basic one was.

**14.26** The QDesign Music 2 Pro dialog.

## Qualcomm PureVoice

Qualcomm PureVoice is a QuickTime implementation of their PCS digital phone audio codec. As would be expected, it produces audio that sounds like a cell phone—speech is intelligible, anything else sounds lousy. Like other speech codecs, it fails to reproduce much non-speech content, so it can even be used in a pinch for extracting intelligible voice from noisy audio.

PureVoice has two data rate modes: Full Rate and Half Rate. At full rate 8KHz is 14.2Kbps, and at half rate it is 6.7Kbps. These data rates go up linearly with the sample rate, so 44.1kHz Full is 78.4Kbps and 44.1kHz Half is 37.1Kbps. PureVoice generally shouldn't be used at these high sample rates for this reason. A 48Kbps mono MP3 file will almost always sound much better than the 78.4Kbps Full, even with pure speech content.

PureVoice only works in mono.

PureVoice is natively VBR, in the sense that it can reduce (it never increases) its data rate during quiet or silent passages. In essence, easy to encode packets of video wind up with a bunch of zeros at the end. Alas, this feature isn't supported in the file itself, just via RTSP streaming. Turning on Use Streaming Format tells the server not to bother to transmit those zeros. Just leave it on—you can't predict when it'll help, but it never hurts.

Before QuickTime 5/QTSS v3, PureVoice needed to be encoded at 8kHz for the native packetizer to do loss recovery. Under QTSS3 or later, there is packetization for other sample rates, although it is still most effective at 8kHz. Of course, if the file isn't being sent via RTSP, any sample rate is fine.

## MP3

QuickTime supports MP3 as an audio codec for playback, although it doesn't include an MP3 compressor directly. MP3 files encoded in other tools can be added to a QuickTime movie using QuickTime Player. Cleaner v5 and Squeeze can directly use MP3 as a QuickTime audio codec. MP3 can provide better quality for mixed voice and speech than QDesign, and decodes much faster. However, it doesn't include a native packetizer for RTSP streaming, and so should only be used for progressive or local playback.

While the Fraunhoffer codec in Cleaner is a good one, its VBR mode makes it impossible to predict its final file size, and the CBR mode isn't as efficient as possible. Lastly, it is limited in the combinations of sample rate and channels. For example, 96Kbps is the minimum data rate that can do 44.1kHz stereo. Squeeze's Fraunhoffer codec has more flexible bitrate/sample rate combinations.

If absolute maximum quality VBR is required, I recommend using the LAME MP3 encoder, which lets you specify a target data rate for the VBR file. It also allows a broader range of sample and data rate combinations. LAME is available as a free command-line tool, or integrated into a number of standalone MP3 encoders.

QuickTime v4.1 and higher supports playback of MP3 files recorded in VBR mode. QuickTime v4 will crash when presented with VBR MP3, so make sure it sees never such a file by using the Movie Alternates feature.

## IMA

The IMA audio codec (created by the Interactive Multimedia Association) was the dominant CD-ROM codec of the Cinepak era. It offered a straight 4:1 compression over uncompressed, so where a 44.1kHz mono track would be 705.6Kbps, an IMA version of that would be 176.4Kbps. Stereo required twice the data rate, so full 44.1kHz stereo in IMA would take 352.8Kbps. Audio quality was quite good at 44.1kHz, but a loss of frequencies in the higher ranges of the sample rate quickly degraded quality at lower data rates.

Today, there are two big reasons to use IMA: performance and compatibility. IMA is dead-simple to decode, and so takes up a negligible amount of the available CPU time even on legacy machines. This makes it a good choice for local playback when low performance requirements are more important than compression efficiency.

IMA is generally playable by non-Apple .mov players, like XAnim and Windows Media Player. Its support is somewhat more common than MP3, so IMA is typically used as the audio track in QuickTime Movie Alternate fallback files.

## None

As with the other formats, QuickTime's None mode is uncompressed PCM. As such, it's mainly used for authoring. However, a lot of older QuickTime files that predate IMA (before QuickTime v2.1) will use the None codec, either in 16-bit or (heavens forbid) 8-bit.

## Legacy Audio Codecs

*QDesign Music*  QDesign Music (no version number) was the original implementation of QDesign. QDM2 isn't backward compatible and requires QuickTime v4, so QDM is still included for legacy playback, and for creating content that targets QuickTime v3. The original QDesign Music Pro options were much more complex than its QDM2 equivalent, which had *six* parameters to control. Needless to say, this made it very hard for anyone who wasn't an audio engineer to get optimal results out of it. Fortunately, quality improvements in the audio support of QuickTime v4 allowed the elimination of some of the controls, and several of the remaining ones were made automatic, allowing dynamic change over the file.

*MACE*  MACE (Macintosh Audio Compression and Expansion) 3:1 and 6:1 are 80s-era speech codecs from the Mac OS, that predate even QuickTime. They're natively 8-bit, and provide strikingly bad quality. Of course, they were optimized for real-time compression and playback on 8MHz computers. MACE is included for legacy playback in QuickTime for Windows from 3.0 on and in all versions of QuickTime for Mac. Never use either MACE codec for new content.

*μ-Law*  μ-Law, (that's the Greek letter, pronounced "mew" or "moo" depending on whom you ask) is an old-school telephony codec, offering 2:1 compression, and hence 16-bit quality over 8-bit connections. It sounds okay for speech, but this poor compression efficiency makes it an unusual choice for any modern applications. It is also called "U-law" (for those who can't easily type the 'μ' character) and G.711. It has the distinction of being the best compressed audio codec supported in older versions of Sun's Java Media Layer. The current versions support IMA.

*A-Law*  A-Law is quite similar to μ-Law, being a 2:1, 16-bit telephony codec. It is less commonly used, and offers no significant advantages.

## For Further Reading

*QuickTime for the Web: for Windows and Macintosh* by Steven W. Gulie

*QuickTime 5 for Macintosh and Windows* (the Visual Quickstart Guide) by Judith Stern and Robert A. Lettieri

# Chapter 15

# Windows Media

Windows Media is Microsoft's digital video technology. Like many a Microsoft technology, the first few versions were weak. But over time, it has matured into a true powerhouse. Windows Media does a great job of handling live broadcast and on-demand video, as progressive download and streaming files on both Windows and Macintosh computers as well as mobile devices. Windows Media was the first architecture with integrated digital rights management (DRM). And because Microsoft sees control of digital media infrastructure as a critical part of their mission, they've hired a large number of enormously talented, driven people to push the technology forward.

## The Advanced Streaming Format

The file format for Windows Media is called Advanced Streaming Format, or ASF. It's quite a bit more complex than AVI, although ASF has some surprising limitations. For example, while an ASF file can have multiple video tracks to switch among in MBR, current implementations must have one and only one audio track.

ASF is the weakest format for supporting rich media types. Microsoft's general thrust has been that animated text, graphics, and such should be handled by the browser through HTML+Time, not in the media player. ASF files can also include scripts, markers, and some other metadata functions, which are mainly used to trigger the HTML+time events.

An ASF file requires indexing to enable random access. In most cases, indices are dynamically generated by the Windows Media SDK, but if a file doesn't have an index (as would be the case with, say, a recording of a live broadcast in which an index wasn't generated after the broadcast), random access can only skip to markers.

The original file extension for ASF files was, unsurprisingly .asf. However, because Windows solely uses file extensions to discriminate among file types, there was no way to assign different icons or default actions to video files (which can't play on music devices) and audio files (which can). So, with the introduction of Windows Media v7, .wmv (video) and .wma (audio) extensions were created. This lets different file types use different tools (such as Windows Media Player for video and RealOne for audio). These don't represent any change to the underlying file type, just to the extension, and so you can generally mix and match extensions with no fundamental problem. For content using codecs other than WMA and WMV, the .asf/asx extensions can still be used.

Like RealVideo, Windows Media for true streaming uses metafiles that reside on Web servers. These metafiles link to the actual media. The original extensions for those metafiles was .asx, but currently the preferred options are .wax (audio) and .wvx (video).

Windows Media can also use customized player skins, which redefine the appearance of the player's UI. Content providers are thus able to specify particular skins to complement particular content to, for example, emphasize a client's brand. Uncompressed skins files use the .wms extension and compressed skins (preferred for delivery) use .wmz.

# Windows Media Player

A big issue with Windows Media Player (WMP) is that Microsoft is quite aggressive about dropping support for older platforms. This can make targeting features for a player quite complex. While dropping Windows 95 support in WMP v7.x wasn't a big problem, dropping it for NT was, as a large number of corporate users are still on NT. The WMP v6.4 version for NT can auto-download the WMV v7 and v9 codecs. However, the user needs to have administrative access for the auto-update to work, and corporate NT users often do not have such access privileges. Also, the scripting model for the ActiveX object that embeds video in Web pages changed substantially, meaning complex projects either have to implement v6.4 objects or create two different versions of the script. Lastly, proper playback of VBR-encoded WMV files requires at least version v7 of the player.

Microsoft now provides the WMV v7 and v8 codecs in a package that can go into an NT disk image, so administrators can provide the latest codecs by updating the corporate-wide disk image.

## Windows Media for CD-ROM

Windows Media isn't as widely used for CD-ROM as other formats. However, its modern codecs provide excellent quality and compression efficiency at typical CD-ROM resolutions like 640×480.

CD-ROM content that needs a fixed data rate should use 2-pass VBR. 1-pass VBR can be used to set a target quality instead of file size; however, file size won't be known in advance, so in general it isn't an appropriate option.

Table 15.1  Windows Media Player version and features.

| Operating system | Most recent player | DRM version | Latest video codec in default installation |
|---|---|---|---|
| Windows 95 and NT | 6.4 | v1 | MS MPEG-4 v3 |
| Mac OS 8.1-9.x | 7.1 | v1 | WMV v8 |
| Mac OS X | 7.1 | v1 | WMV v8 |
| Solaris | 6.3 | v1 | MS MPEG-4v3 |
| Windows 98/ME/2000 | 7.1 | v7 | WMV v8 |
| Windows XP | XP | v7.1 | WMV v8 |

## Windows Media for progressive download

Although Windows Media was originally targeted at streaming, it has become arguably the strongest technology for progressive content, beating out the long-time champ, QuickTime. This is due to WMV v8's excellent decode performance and quality, the 2-pass VBR mode, good cross-platform support, and the integrated DRM.

There aren't any particular tricks to encode for local playback of WM files, except to only use a single video track instead of the Intelligent Streaming multiple bitrate mode. For progressive download and other file-based playback, the ideal encoding mode is 2-pass VBR.

## Windows Media for on-demand streaming

Unlike all the other streaming formats, Windows Media still uses its own proprietary Microsoft Media Server (MMS) protocol instead of RTP/RTSP. Thus a direct URL to a streaming ASF file starts with mms://. In a practical sense, MMS isn't much different than RTSP—they're both based around UDP. Microsoft has indicated that they'll add RTSP support in a future version.

The Windows Media Server (WMS) comes with Windows NT/2000/XP Server, and isn't available otherwise. Some vendors may eventually provide MMS streaming from UNIX-based server operating systems. RealNetworks' new Helix Server includes Windows Media support on a variety of operating systems.

Windows Media Server is a full-featured streaming server, familiar to administrators of Microsoft's Internet Information Services (IIS). It offers all the appropriate features of a modern server architecture. By default, it will fall back to TCP delivery if MMS is blocked by a firewall, and it can be configured to fall back to HTTP. The Intelligent Streaming MBR mode will correctly switch data rates even when being used in this mode.

Although WMS isn't available for big-iron UNIX machines, Windows Media supports Microsoft's clustering model, where multiple servers can be brought together to operate as one large server, providing scalability and fault-tolerance.

The ideal encoding mode for on-demand streaming is 2-pass CBR (known in other formats as buffered VBR, or in RealVideo as VBR). You can specify a buffer size in this mode. The larger the buffer, the higher the average quality, but the greater the latency.

## Windows Media for live streaming

Windows Media provides a good, mature system for doing live streams. Most folks just use the default Windows Media Encoder for capture, but other tools like Discreet Cleaner Live can also do it.

WME can use any DirectShow-compatible capture device for input, including OHCI 1394. However, it's always best to use a hardware-based system that can do much of the preprocessing in hardware, leaving as much CPU power as possible available to the codec itself.

WMV can use up to four processors for video compression and two for audio compression (for stereo audio), so if you're doing live broadcasting, a hefty multiprocessor machine can substantially increase the maximum resolution, frame rate, and data rate you can target.

WME can also simultaneously capture and serve video from the same computer. However, unless you're going to have more than a handful of viewers, it's best to send a single copy of the stream to a dedicated server, so CPU cycles don't get gobbled up by serving instead of being used for encoding.

## Windows Media for portable devices

One of the best features of Windows Media is its support for a variety of platforms, including mobile devices—specifically those running PocketPC. You can download Windows Media Player v7.1 (which supports WM v8 codecs) for anything running at least Windows CE v3 and PocketPC 2002, which comes with an enhanced player installed. PocketPC 2000 is limited to WMP v7 and WMV7. These devices support playback from local storage, of course, and streaming if the device has a (typically wireless) Internet connection.

Of course, mobile devices are very limited in processing power, memory, storage, connection bandwidth, and other features compared to desktop computers. While pretty much any WMA file will play on a mobile device, you have to severely constrain your video resolution and frame rate for video to work on such devices. Also, the limitations of different devices vary quite a bit, both in performance and in maximum video quality. For best results, a machine running at least at 150MHz with a 16-bit display is recommended. Because mobile devices follow Moore's law, expect the playback performance of a PDA of 2004 to perform like a desktop computer of 1999.

Microsoft provides downloadable profiles for encoding to different PocketPC devices. However, these only target 28.8 playback. For downloadable files, 150Kbps total should work on the higher-end devices like iPaq. Older systems like the HP Jordana and Casio Cassiopeia should target more like 100Kbps. 320×240 15fps and lower are appropriate targets. Note that PocketPC devices use portrait orientation, so they are 240 pixels wide and 320 tall.

## Embedding Windows Media in a Web page

Like all Web video formats, Windows Media files can be embedded in a Web page. This works for all the Windows players. Earlier versions for Mac could not be embedded in a player. Instead, embedded content launched the Windows Media Player on Macs. This has been fixed in current versions.

Windows Media doesn't have the same depth of features as QuickTime does as a plug-in, but it does offer some control of things like playback resolution via scripting. This scripting is handled via WAX/WVX metafiles.

# Windows Media DRM

Digital Rights Management is essential to establishing some level of comfort among owners of valuable copyrighted content (such as recent movies and television shows) so they'll feel safe releasing their content in digital formats. Microsoft identified DRM as a feature that would give Windows Media an important competitive advantage.

Version 1 of the MS DRM was released in 1999, well before similar offerings from competitors. Unfortunately, v2 and 7 were cracked in the fall of 2001, with a readily available tool that anyone can use to convert any copy-protected WM file to an uncopy-protected file. Updated versions have since come out that provide the same features, but with a fix for the vulnerability.

DRM v1 is still secure, and it has been the most used DRM to-date, especially for downloadable music. Also, the Windows Media Players for Mac (v7.1) and NT (v6.4) only support v1. You can use both DRM models with a given file, so it'll play in any player, but you'll only have access the rules in a given DRM.

The basic model of WM DRM is similar to other DRM solutions, like those in RealVideo and (eventually) in MPEG-4. The content is encrypted so the file can't be played without a license. It doesn't matter if the file is on a computer or a mobile device, or being streamed, or on a CD-ROM. With v7, you have a lot of flexibility in those rules. For example, you can specify that a file can only be played for $x$-number of days.

You need to specifically license the DRM technology from Microsoft—it's not part of the encoder or the public SDK.

There are a variety of rules you can apply to the media, with a lot more in v7 than in v1.

## DRM rules

Like all DRM technologies, striking the proper balance between protection and flexibility for the consumer is difficult. Windows Media left such business decisions in the hands of the content provider, so you can be as open or as strict as you desire.

Table 15.2   DRM rules in Windows Media.

| Version | v1 | v7 | v7.1 |
|---|---|---|---|
| Available Rules | Expiration Date<br>Unlimited Play<br>Transfer to SDMI/non-SDMI<br>Burn to CD | v1, plus:<br>Start time<br>End time<br>Duration<br>Counted operations: plays, transfers | v7, plus:<br>Application Exclusion<br>DRM Exclusion<br>Expiration after First Use<br>Expiration on Store<br>Allow saving of protected streams |

Most of these rules are fairly straightforward:

***Expiration date***   The media won't play after this date.

***Unlimited play***   The file doesn't have any restrictions.

***Transfer to SDMI/non-SDMI***   Determines whether the file can be copied to SDMI and/or non-SDMI devices. The Secure Digital Music Initiative was an early attempt to build a secure digital music system. Its proposed encryption and watermarking schemes were cracked pretty quickly, so there isn't much in the way of SDMI-compatible devices out there.

***Burn to CD***   Determines whether the media can be burned to a standard audio CD. The files themselves could be burned to a CD-ROM, of course, but you'd still need the license to play them.

***Start time***   A particular date before which the content can't be played.

***End time***   A particular date after which the content can't be played.

***Duration***   Determines how long the file can be played after its initial use.

***Application exclusion***   Prevents the file from playing on specified players.

*DRM exclusion*  Excludes the use of particular DRM models. This can be used to easily modify the license to prevent playback on compromised players.

*Expiration after first use*  Allows the file to be available only for a defined amount of time after it is first played.

*Expiration on store*  Allows the file to be played only for a defined amount of time after it is stored on the user's computer.

*Allow saving of protected streams*  The use can save a protected stream into a local file. That file remains packaged, and still requires a license.

## The DRM process

*Packaging*  The Windows Media Rights Management application encrypts (packages) the digital media file with a key. The key itself is also encrypted, and stored in a separate license file. Information such as the URL of the license can also be added. This packaged media file is saved as a WMA or WMV.

*Distribution*  The packaged file can be distributed as a file for progressive download or local playback or via an MMS server for streaming. If it's done as a file, that file can be distributed to other users and can be played anywhere as long as the license can be verified.

*License server*  A license clearinghouse stores the license complete with its business rules, and implements the Rights Manager services. The clearinghouse brokers requests for the licenses and validates the user's right to access the content. Except for the largest media companies, the clearinghouse will be different than the content provider.

*Getting the license*  Once users acquire the media file, the player needs the key to decrypt the file. This process automatically begins whenever the user downloads the packaged media file, acquires a pre-delivered license, or plays the file for the first time. If the user already has access to the license, everything works automatically—they hit Play and the media plays. If they don't have or can't get a valid license, they're sent to a page at the license clearinghouse site and are instructed on how to buy the media or whatever. This is one reason why timed licenses are a good idea. The customer doesn't have to always have an Internet connection to use the content.

*Playback*  For all this to work, the user needs to have a media player that supports the DRM mode used by the file. Once both the license and media file are available, the user can treat the file like a standard file, as long as the license is valid. Depending on the license, users may be able to copy the files to a different computer or device (such as a portable music player). Or the file may be locked to a specific machine.

The audio part of playback is run through the Secure Audio Path. This encrypts the data on the way from the drive to the sound card, so it cannot be extracted in memory. This requires a Microsoft-certified sound driver, or the file won't play. Of course, the analog audio itself isn't protected.

# Windows Media Encoding Tools

## Windows Media Encoder

Microsoft gives away their Windows Media Encoder for free. It's a pretty simple GUI encoding tool that handles live broadcast, live capture, offline encoding, and recording screen activities.

WME is designed for novice users, and can be frustrating for more experienced users. For example, you can't simply define the target data rate you want to use. If a given data rate isn't a default, you have to specify, a process that is frustrating (more on that follows).

WME's preprocessing isn't deep, but it's useful. You get the usual cropping and scaling, as well as deinterlacing and a decent, but not fabulous, inverse telecine.

Your two initial options are to go through an encoding wizard, or jump into the tabbed session interface. It's probably best to work with the wizard until you get a handle on the feature set.

**15.1** The Windows Media Encoder New Session Wizard walks you through a series of questions to determine appropriate compression settings. Once created, you can edit them via Properties.

First, you define the basic mode: live broadcast, live capture from a capture device or from the screen to a file, or conversion from a file. One neat feature of the live capture to a file is that it'll actually buffer the capture if the CPU isn't able to encode in real time. This prevents dropped frames while providing most of the speed of a real-time capture. For example, the computer might capture an hour-long file, but still encode for 20 minutes after the capture is completed. A fair amount of drive space is required for the buffer.

If you're capturing from the screen, you can specify the whole screen or a region. You can also define a named window to capture. This is a great way to record a specific application without having to carefully align the capture window.

For typical file-to-file conversion, pick the Convert option. You're then asked to specify whether the file will be streamed from a WM server or will be for local/progressive playback. The only real difference between these is that you only get one video track with the local/progressive option. You then select or create a profile. This part frustrates me, as there is no way to just specify on the fly the settings you want. First you have to create or modify an existing profile. And then if it doesn't have the data rates you want, you'll have to create a new data rate option as well.

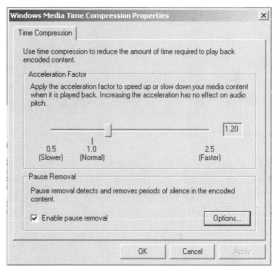

You'll also see options for a script data rate. This allows you to save a certain amount of the total bandwidth for any scripts going into the file. This and audio are subtracted from the total bandwidth to determine the output bandwidth. For a typical script, 2Kbps is a good start. You specify the buffer size for a streaming file with the Advanced button. As always, the bigger the buffer, the higher the average quality, but the worse the latency.

Frame rate in Windows Media can go up to well past 60fps. You can do 60-frame output with 60-field source, as the deinterlacing can extract two progressive frames from each interlaced frame's two fields. Lastly, you can specify keyframe interval in seconds, and image quality versus frame dropping with a 0 to 100 slider. Even with a value of 0, WMV 8 does an excellent job of maintaining the target data rate.

**15.2** The time compression dialog. (Those who have heard me speak in public have suggested that I implement this feature personally with a setting around 0.5.)

One odd feature of WME is time compression. This allows you to speed up or slow down the speed of the content. This is done with pitch correction, so you don't get the Mickey Mouse effect. You can also do Pause Removal, which will cut out any portions of the video where there isn't audio (as occurs when a presenter stops talking). These features work as advertised, but aren't of practical use for most content.

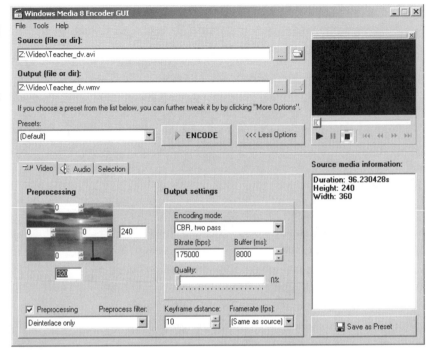

**15.3** The invaluable WME8EGUI front-end to the Windows Media Encoding Utility. Note: it thinks that DV is 360×240, so you'll manually have to enter the proper square pixel resolution.

## Windows Media Encoding Utility

The Windows Media Encoding Utility (WMEU) was originally the only way to encode to Windows Media in anything other than 1-pass CBR and is still the only free tool to do so. But this should change with Corona.

For those encoding Windows Media on the Mac, using the encoder through the command line on VirtualPC is also a surprisingly viable option. On a fast G4, the encoder is pretty quick even emulated—especially when working with preprocessed source.

First, get a GUI front-end for the WMEU. WM8EGUI.EXE is the best I've found so far. Also, know the only formats WMEU accepts are .avi and .wav (and there's no reason to use WMEU for .wav files). It won't take other DirectShow-compatible files like MPEG-1 or QuickTime files (even if they're DirectShow-compatible). Because of this, many users preprocess the video in another application and do the final encode in the WMEU. The Huffyuv codec is typically the best choice for this preprocessing.

## Windows Media Resource Kit

The Windows Media Resource Kit is a free collection of tools for working with Windows Media. These vary from obscure command-line only tools to a decent GUI batch encoder. There are several dozen programs in the set. A few of the most useful ones for compression are mentioned here.

*Advanced Script Editor*   This GUI tool can be used to set in and out points, properties, markers, and scripts for an ASF file. The GUI tool provides a simple timeline with which you can sync the events. In many ways, it feels like an early version of Microsoft Producer.

*Attribute Editor*   The Attribute Editor lets you edit metadata attributes relevant to WMA files, such as album titles.

*Batch Encoder*   The batch encoder is a very simple, but useful batch encoder for Windows Media. You can select a number of source files, and assign pre-existing profiles to each along with a few metadata items. Alas, you can't create a new batch file from within the batch encoder's interface.

*Stitcher*   Stitcher lets you combine several files into one new file, concatenated. Alas, stitcher doesn't allow mixing audio and video from different sources. It works on the WM formats as well as AVI, WAV, MPEG-1, and MP3.

*Windows Media Author*   Windows Media Author is also known as TAG, from Digital Renaissance (now ExtendMedia). It was a promising early attempt at the kind of graphical tool to integrate rich media with video, à la Microsoft Producer. It hasn't been updated since 1999, and ExtendMedia appears to be defunct.

*Windows Media On-Demand Producer*   From Sonic Foundry, this tool dates back to circa 1999. It served as the basis for Sonic Foundry's anemic StreamAnywhere product. The standard Windows Media Encoder is a much more capable tool.

## Discreet Cleaner 5

Cleaner v5.1.2 for Windows is, after a rocky start, a decent Windows Media encoder. Cleaner's strengths are its superior preprocessing and batch capabilities. Alas, it lacks a few core features for Windows Media encoding. First, you can't specify a different frame rate for each Intellistream output—they must all share the same frame rate. Second, it doesn't offer the full 0 to 100 range for the spatial quality versus dropped frames slider. It simply offers three discrete levels. With 5.1.2 on Windows, Cleaner now implements 1-pass and 2-pass VBR and CBR.

Cleaner v5.1.2 for Mac is a lot less capable than the Windows versions, due to limitations in Microsoft's Windows Media SDK. First, the Mac SDK only supports a single video track, not the Intelligent Streaming MBR mode. Second, it features only the WMV v7, MPEG-4 v3, and WMA v7 codecs. Thus it lacks the WM v8 codecs, and the ACELP low-bitrate speech codec. Due to these limitations, anyone trying to do professional-quality WM encoding needs to do that encoding on a Windows platform.

## Adobe Premiere

While Premiere 6.0 doesn't support WM encoding through its normal Make Movie dialog, it does have an Advanced Windows Media module, which can work either on the current clip or on the timeline. This module is a custom GUI on top of the standard WM export module. It has the same features and limitations, and uses the same profiles, as other WM applications. You can also create a new profile through the standard WM UI, and those profiles will be usable by other applications. The module doesn't have any preprocessing options—preprocessing should be done in Premiere first.

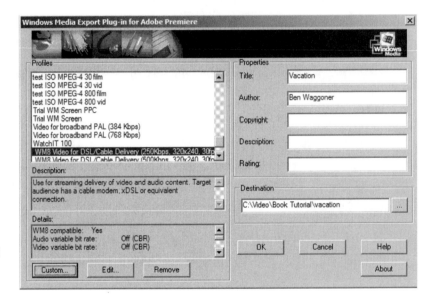

15.4  Adobe Premiere's Advanced Windows Media export command. It lets you render straight from the Timeline to Windows Media, with access to most of the settings.

Because the Premiere option contains all the features of the normal WM SDK, it's a fine idea to export straight from the Premiere timeline if you don't need any preprocessing beyond what Premiere can provide. This can save time and cut down on the storage required for intermediate files.

## Microsoft Producer

Producer is a rich media authoring tool for Windows Media, tightly integrated with Microsoft's PowerPoint. In many ways, it's the first step in building a dream tool that integrates video capture, editing, and compression with rich media markup and authoring. Producer isn't as powerful as some of the more focused tools, but the combination of functions in Producer is compelling. And its price (free for PowerPoint users) is hard to beat.

The current main use for Producer is for the integration of PowerPoint with Windows Media. Producer lets you easily import an existing PowerPoint presentation, and convert it to high-quality,

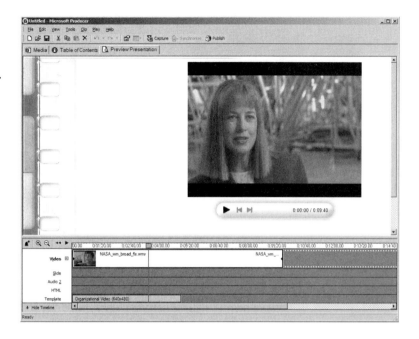

**15.5** The interface for Microsoft Producer, for integrating PowerPoint presentations with video.

bandwidth-light vector graphics and client-side effects. You can then record a live video or audio stream, or import an existing file, and synchronize the presentation with the media. Then you turn the whole shebang into a presentation, either for local or online playback. While doing all of this was previously possible, Producer does a decent job extremely quickly. Its biggest drawback is a lack of decent preprocessing. You can either create a preprocessed .avi for source or give Producer pre-encoded media to work with.

The biggest drawback of Producer is that the presentations only play back in Internet Explorer for Windows v5.5 or later, ruling out other platforms and browsers.

# Windows Media Video Codecs

Windows Media has a limited number of available codecs, none from third parties. The current MS SDK includes these video codecs: MS MPEG-4 v3, Windows Media Video v7, Windows Media Video v8, ISO MPEG-4, and Windows Media Screen.

## Windows Media Video v8

Windows Media Video v8 is the main Windows Media codec for video content. It is all-around excellent. Compared to the leading codecs in other formats, WMV v8 provides very competitive quality across the full range of data rates, from modem to above 1000Kbps. It also offers very CPU-efficient encode and decode (although its raw encode speed is quite a bit slower than WMV v7). WMV v8 also scales well to provide optimum playback across a wide range of CPUs and speeds, from 100MHz PocketPC devices, to iMacs, to the latest P4 screamers.

It will auto-update with WM players from v6.4 on, and is preinstalled from v7.1 (Windows) and v7.01 (Mac). An Enterprise Deployment Pack allows system administrators to install the latest codecs with older players.

WMV v8 supports four basic data rate modes. These have only been added to the SDK, but should be broadly available in products by the time this book comes out.

Table 15.3  The Windows Media Video v8 encoding modes.

| One-pass CBR | This is the traditional encode mode, and the only one available in the older v7.1 SDK. A buffer size can be specified. The larger the buffer, the higher the average quality, but the worse the latency. |
| --- | --- |
| Two-pass CBR | Also buffered like One-pass, but it first does an analysis pass. Offers higher average quality at the same bitrate. This is the optimal mode for real-time streaming. |
| One-pass VBR | This mode specifies image quality, not data rate. The data rate changes, sometimes radically, as needed to maintain the target image quality. This is only for files where relative quality is more important than a predictable file size. Requires WMP v7+ for reliable playback. |
| Two-pass VBR | This full-file VBR is great for progressive download and local file playback, but shouldn't be used for MMS. Two-pass VBR is effectively the same as doing a 2-pass CBR with a buffer size equal to the file size. It requires WMP v7+ for reliable playback. This is the optimal mode for progressive and local file playback. |

During playback, WMV v8 can apply four increasing levels of post-filtering, because greater surplus CPU power is available in the client machine after the base decode.

Table 15.4  Postprocessing levels with WMV v8.

| Level 0: | Nothing is on. |
| --- | --- |
| Level 1: | Light deblocking, deringing off. |
| Level 2: | Strong deblocking, deringing off. |
| Level 3: | Light deblocking, deringing on. |
| Level 4: | Strong deblocking, deringing on. |

## Windows Media Video v7

WMV v7 was pretty darn good. It isn't much different than v8, except it is less efficient in terms of quality at a given bitrate. WMV v7 was also available via auto-update for WMP 6.4 or greater, and was built into WMP v7.0. The current WM SDK v7.1 now includes the WM v8 codecs. However, only SDK v7 is available for Mac OS, so folks encoding with Cleaner on a Mac currently only have access to WMV v7. WMV v7 was quite a bit faster to encode than WMV v8.

## MS MPEG-4 v3

This was the last Microsoft codec to be labeled MPEG-4. Microsoft has since renamed their codecs Windows Media Video, skipping v4, v5, and v6 in the process. MS MPEG-4 v3 isn't compatible with real MPEG-4 in any way. However, it was the subject of the original DivX v3.x hack to enable its use in AVI files.

The Windows Media Video codecs offer substantially higher quality at a given bitrate than MS MPEG-4 v3. The main reason that codec is still used is because it is the last version pre-installed with Windows Media Player v6.4, the most recent available for Windows NT and 95. While WMP v6.4 can auto-update to the latest codec, this requires Administrator privileges, which many Windows NT users in corporate environments do not have. Thus, this codec is still used quite a bit for content distributed in managed corporate environments.

MS MPEG-4 v3 is also the last version available for the WMP v6.3 for Solaris.

## ISO MPEG-4

After catching a lot of flack over having a non-compatible MPEG-4 codec in previous versions of Windows Media, Microsoft introduced a real ISO-compatible codec in WM v7. However, this made little practical difference, as WM doesn't support the file format, stream, or audio codec from MPEG-4, so the codec can't be used to interoperate with other MPEG-4 solutions.

While many perceive Microsoft as anti-MPEG-4, Microsoft has played a leading role in developing the MPEG-4 video codec and has contributed technology for it.

Theirs is a basic implementation of Simple profile MPEG-4 Video, and isn't competitive against WMV v8 or even WMV v7. I hope future versions of Windows Media Player will include complete MPEG-4 playback.

## Windows Media Screen

The Windows Media Screen is an absolutely wonderful codec that hasn't received its due buzz. As its name suggests, Screen is designed to record screen activity. This is similar to what the Animation codec from QuickTime is often used for, but Screen offers enormously improved compression efficiency, delivering file sizes only a few percent of what Animation can do with the same source. Screen can also do real-time streaming.

Unlike the WMV codecs, which are natively Y'CbCr, Screen is natively RGB, and can function at 24-bit, 16-bit, and 8-bit.

The easiest way to record screen content is through Windows Media Encoder itself. Like all lossless codecs of this sort, the content on the screen can have an extreme effect on how efficient the compression is. Simply dragging a cursor around a static screen is very easy to encode. Recording a large window with video playing is extremely difficult, and requires a data rate many times that of the video playing in the window. In general, keeping all extraneous motion to a minimum, turning off any animation, and using a simple background will produce the best results. Also, 8-bit will typically compress quite a bit more easily, both because there are only a third as many bits per pixel, but also because a lot of difficult to compress gradients are automatically deactivated in 8-bit mode. You can also convert screen capture RGB sources in codecs like Camtasia.

The data rate of Screen is very dependent on the keyframe rate, because with typical screen motion, a keyframe can be hundreds of times larger than the average delta frame. Unless random access is needed, start with the canonical rate of one keyframe every 10 seconds, and consider increasing that if data rate/frame rate targets are proving hard to hit.

## MS MPEG-4 v1 and v2

These older codecs were the initial versions of Microsoft's MPEG-4–derived codecs, and predate the final MPEG-4 standard. They offered substantially lower quality than even MS MPEG-4 v3. Encoders for these haven't been in the standard WM SDK since 1999, so only very old tools are going to be able to encode to these formats. MS MPEG-4 v2 was the last version of the codec that also functioned as an AVI codec, although the DivX hack makes MS MPEG-4 v3 work in AVI files.

# Windows Media Audio Codecs

## Windows Media Audio v8

There have been several iterations of the Windows Media Audio encoder. However, Microsoft has been able to make the last few generations of the encoder backward-compatible with older decoders, all the way back to WMA v2. This was important, as it has made WMA a stable competitor to MP3 for mobile devices. But this doesn't mean quality has stalled—Microsoft has made continual enhancements that improve quality at a given bitrate and add new options.

Like RealAudio, WMA only supports certain combinations of data rate, sample rate, and content. The complete list follows. WMA is a stellar music codec, but even the latest versions can produce noticeable phasing with voice, even at 32Kbps. The voice is always intelligible, but this phasing can be distracting. 64Kbps and above generally sounds excellent, and can be difficult to distinguish from uncompressed audio in typical consumer listening environments. At 128Kbps and above, WMA is capable of maintaining the two-channel matrixed surround sound information of Dolby Pro-Logic, delivering a surround sound experience when played through a Pro-Logic system.

One odd feature of WMA is the 0Kbps codec. Because current implementations of the ASF file format always requires an audio track, adding this audio track without any audio was the only way to support video-only content without wasting bandwidth. This feature has been in and out of the SDK, so not many tools support it yet.

Table 15.5   Available Windows Media Audio v8 options by data rate and sample rate (m=mono and s=stereo).

| Kbps | 8,000 | 11,025 | 16,000 | 22,050 | 32,000 | 44,100 | 48,000 |
|---|---|---|---|---|---|---|---|
| 0 | | | | na | | | |
| 5 | m | | | | | | |
| 6 | m | | | | | | |
| 8 | m | m | | | | | |
| 10 | | m | m | | | | |
| 12 | s | | m | | | | |
| 16 | | | ms | m | | | |
| 20 | | | | ms | s | | |
| 22 | | | | s | | | |
| 32 | | | | s | ms | ms | |
| 40 | | | | | s | | |
| 48 | | | | | | ms | |
| 64 | | | | | | s | |
| 80 | | | | | | s | |
| 96 | | | | | | s | |
| 128 | | | | | | s | s |
| 160 | | | | | | s | s |
| 192 | | | | | | s | s |

# ACELP.net

ACELP.net (Algebraic Code-Excited Linear Prediction), like the other low-bitrate voice codecs, provides intelligible speech at low bitrates. Microsoft licenses it from Spiro Labs. The mathematical models behind the codec mimic the behavior of the human vocal apparatus, reproducing the sounds of the human voice very well, and other sounds barely at all. For speech-only content, ACELP at 16Kbps can sound better (due to less phasing) than WMA at 32Kbps. At 8kHz it can do 5, 6.5, and 8.5Kbps, and at 16kHz it only does 16Kbps. ACELP is mono only.

As of v7 of the Windows Media SDK, this codec was only available in the Windows version of Windows Media, not in Mac OS, so Mac tools like Cleaner v5 can't encode to it. This will hopefully be changing in future versions of the Mac SDK.

Due to licensing reasons, ACLEP.net is also not available for all platforms. PocketPC 2002 and MacOS X don't support it.

## VoxWare MetaSound and MetaVoice

These were the pre-WMA/ACELP audio codecs for Windows Media. As the name implies, they were general-purpose, low-bitrate speech codecs, respectively. While decent for their era, they were not competitive with WMA and ACELP in quality, and took substantially more CPU power to decode.

They aren't in the current Windows Media SDK, so only old versions of software (such as Terran Media Cleaner Pro v4.03) are able to encode to them. There is no longer any reason to do that, though decoders are included in all current versions of WMP.

# Chapter 16

# RealSystem

## Introduction

RealNetworks (then Progressive Networks) created the streaming industry. RealPlayer v1 was launched in 1995. It was the first real-time streaming application. The initial dominance of the company was in streaming audio, but they added video to RealPlayer v4 in 1997, making it the first real-time video streaming solution. While there were other formats for distributing audio and video on the Internet, RealNetworks was the innovator and the market share leader for real-time solutions throughout the history of the industry. Of course, other vendors (especially Microsoft) identified streaming media as a strategic area of interest worth competing over, and have been pushing hard into the market.

RealNetworks is also unique in that they're a standalone vendor, competing against products from companies like Apple and Microsoft that can afford to subsidize their video distribution technologies with revenue from other sources. This means RealNetworks needs to see more direct revenue from their products and services than their competitors. Historically, the main sources of Real's revenue were purchases of server software and the Plus version of the player software. However, pressure from other vendors has forced Real to put more functionality into the free version of the player, reducing the number of truly useful features in the upsell version. Prices on the RealSystem Server have remained high, but it isn't clear how long RealNetworks will be able to maintain those prices as their once-dominant market share declines.

As a corporation, RealNetworks has been de-emphasizing the revenue they get from tools, players, and servers, and is increasingly focusing on revenue from their content distribution network, Real Broadcast Network, and content syndication through the $9.95/month RealOne SuperPass.

In the summer of 2002, RealNetworks announced thier "Helix" line of products, for which details were just becoming available as this book went to print. The Helix Server is capable of hosting MPEG-4 and Windows Media files as well as the RealMedia and QuickTime content it handled before.

## RealMedia Format

The RealMedia format, often called a RealVideo file even if it only contains audio, has the extension .rm. RealAudio, or .ra, is an older version of the format, and is no longer in use. Real's metafile format is .ram. As with other metafile systems, you normally put the .ram file on an HTTP server, and it contains the actual link to the .rm media on a RealServer.

The RealMedia format includes the best multiple bitrate implementation, SureStream. With SureStream, up to eight baseline copies of the media can be made, all varying in data and frame rate. RealProducer and other tools also often add additional streams of lower data rates than the requested streams to provide additional fallbacks.

## RealOne Player

The current version of the RealVideo Player app is RealOne released for Windows and in beta for MacOS X. It replaced RealPlayer v8 (which is still the most recent GM version supported for non-Windows platforms). Media players and their revisions are typically contentious matters, and the Real players have certainly had more than their share. Historically, the two biggest objections have been to the hard upsell for users to purchase the paid version of the player (which used to be $29.95, and is now $10/month), and the cluttered interface. The good news is that RealNetworks has done a much better job of showing where the free player can be downloaded (a common user complaint was that people couldn't find the free player's link on Real's site). The RealOne interface is somewhat less cluttered than the default mode of RealPlayer v8 (although the Compact mode of RealPlayer v8 was quite nice). RealOne introduced a new three-pane style of working with video and linked HTML navigation.

Formerly, the big upsell for the non-free player was access to special features. However, most of these features (like video level controls) were actually more likely to get users in trouble. The one important RealPlayer Plus feature was support for PerfectPlay, which enabled a large buffer for playing files encoded at a higher data rate than the connection speed. This only worked with files encoded with PerfectPlay enabled (which I recommend you do by default).

RealOne SuperPass's main feature is access to subscription content, like radio broadcasts of major league baseball games and NASCAR races. It also includes the relatively minor video and audio EQ as well as other tweaks from RealPlayer Plus. But its big new feature is called TurboPlay. It speeds up buffering on fast connections, by buffering the initial video to the player as fast as the connection can handle. If the connection speed is a lot faster than the clip's data rate, latency can be improved substantially. The current version of QuickTime Streaming Server and the forthcoming .NET version of the Windows Media Server include a similar feature.

All the RealPlayers support a variety of other formats besides RealVideo, including MPEG-1, AVI, WMA, Liquid Audio, DirectShow-compatible QuickTime, and so on. Some are built-in,

**16.1** The GUIs for RealPlayer v8 and RealOne Player are quite different, but both focus on integrating rich media with video.

and many others are available through a good component download mechanism. Currently MPEG-4 support is handled through the licensed Envivio plug-in, which so far is only available under Windows.

## RealVideo for CD-ROM

RealVideo is not designed for CD-ROM use. Certainly, .rm files can play from a CD-ROM. Director includes native support for Real playback (both for CD-ROM and Shockwave movies), so integration is fairly easy. However RealVideo doesn't have much to offer for CD-ROM use. Its codecs require faster machines for playback and the advantages of SureStream don't apply.

## RealVideo for progressive download

Earlier versions of RealVideo weren't well suited to progressive download. However, current versions do a fine job with it. RealVideo files encoded for progressive download shouldn't use Sure-Stream, of course. A file played locally will only play the highest data rate audio and video stream, so including more than one just makes the file larger without adding much. One exception to this is if the file was encoded for RealPlayer v5 compatibility. If so, the compatible portion is at the front of the file, so you'd see that portion of the file instead of the stream with the highest data rate (which is not good).

The new Helix Producer supports unbuffered encoding, although other applications that use the RealMedia SDK don't yet. This new mode only works in RealPlayer v8 and above, requires a component update, and uses a new file extension, .rmvb. Initial tests look promising.

## RealVideo for real-time streaming

RealVideo and RealAudio before it were designed for real-time streaming, and this is still the area at which it excels. RealVideo has always been the best solution for video being delivered to modem users. Modem data rates vary enormously, so a good MBR is required. QuickTime doesn't have any explicit MBR, other than B-frame dropping, and Windows Media can reduce video data rate but not audio. RealVideo's SureStream allows multiple data rates for both audio and video, up to eight each. The video bands can vary by data rate and frame rate, but not resolution. The audio bands can vary completely. Each band is given a target connection type, and when the RealPlayer connects, it starts with the stream that matches the specified connection type in the user's player. Once the stream starts, the server will dynamically adjust it up or down to the largest streams that fit within the target data rate. Although there need to be equal numbers of audio and video bands, SureStream may end up playing the audio track from a higher quality band than the video, if it can fit within the target bandwidth.

SureStream can also include a single fallback stream for RealPlayer v5 compatibility. At this point the installed base is overwhelmingly G2 or better, so this feature is much less important than it used to be. The fallback uses the same target data rate, frame rate, and so on as the lowest SureStream data rate.

The biggest drawback to the Helix Universal Server is its price. There is a free version (Helix Universal Server Basic), but it's limited to 10 simultaneous users, noncommercial use, plus it times out after 12 months. It's mainly useful for evaluating the technology. Pricing and versions for Helix Universal Server haven't been announced yet. The three versions of the Helix Universal Server announced are: Enterprise, Intranet, and Mobile. Enterprise targets those putting video on the public internet; Intranet stays within an organization. And Mobile supports the panoply of mobile device streaming formats.

One of the biggest advantages for many shops is that Helix Server runs on most major server platforms, including Windows NT/2000/XP, Linux, Solaris, HP/UX, AIX, and Tru64.

## RealVideo for live broadcast

RealVideo is a fine system for live broadcasting. There are a number of tools for it, including Helix Producer and third-party live broadcast software packages like Cleaner Live. Helix Producer supports a variety of video capture cards, with Viewcast's Osprey line being the "official" partner.

## RealVideo for rich media

RealPlayer itself has good rich media support via SMIL. Rich media in RealPlayer includes the usual skinning features, synchronized HTML pages, clickable hot spots, and so on. However, these features are all handled through SMIL instead of inside the stream itself. The authoring process is to create the compressed .rm file and the rich media interactivity as separate tasks in the authoring process.

There are a number of good SMIL authoring tools. The one used most with RealOne is the GRiNS editor for RealOne (Mac Classic and Windows, licensed by RealNetworks from Oratrix). As you might gather from the name, its focus is on making rich presentations for the RealOne player, especially content in its three-pane interface.

For simpler presentations, and those targeting the older versions of RealPlayer, there are RealPresenter and RealSlideShow. RealPresenter (in a free Basic version and the $199.95 Plus version) converts PowerPoint presentations into rich media that can be streamed from a RealServer. RealPresenter Plus even supports doing live broadcasting, so an existing presentation can feature live narration, video, and/or audio.

RealSlideShow (also in a free Basic and a $99.95 Plus edition) takes existing still image and audio assets and coverts them into a slide show presentation.

## Embedding RealVideo

Installing RealPlayer also installs Netscape-style plug-ins or an ActiveX component, as appropriate to the browser.

# RealVideo Encoding Tools

One of the best things that RealNetworks has done is to fully open up all the features of their encoder in an SDK for third party development. Unlike QuickTime and Windows Media, this includes full support for 2-pass encoding, which has been included in third-party RealMedia encoding products for some time. The SDK also integrates scaling, deinterlacing, and even inverse telecine—features that may not otherwise be available to other formats in some tools. RealNetworks is also the only company to offer cross-platform live broadcast API, supporting it on all four targeted platforms (Windows, Mac classic, Linux, Solaris).

The nice thing about this is that the baseline of RealVideo encoding support is very high and consistent. The choice of video compression tool can be made based on features other than format support.

## Helix Producer Basic

Helix Producer Basic (formerly RealProducer) is the free version of the RealVideo encoder, available for classic Windows and Linux. It's the only mainstream commercial video compression app that runs on Linux. Like Windows Media Encoder, Helix integrates real-time compression—including live broadcast and file-based encoding in the same software.

My biggest frustration with Helix Producer is that it has sporadic support for reading from many file formas I care about. For example, I use a lot of QuickTime movies and Huffyuv-encoded .AVI files as source. The initial version of Helix Producer has trouble with some but not all of these files. Workarounds are documented in the ReadMe, but hopefullly won't be needed in a future version.

Helix Producer Basic is easy for novices to use, but it doesn't support fine tweaking of encoding settings. For professional results, you need to get Helix Producer Plus.

Helix Producer hasn't been announced for either Mac OS or Solaris. For the time being, users of those platforms will need to use the older, kludgy RealProducer 8.5 and forego support for the RV9 codec.

## Helix Producer Plus

Helix Producer Plus is the paid version of RealProducer. Unlike previous versions of RealProducer Plus, Helix Producer Plus is *a lot* more capable than the free version.

First, it can create more than two SureStream data rates per file, which is critically important for files that are going to be played on the public Internet. Second, it adds a very useful bandwidth simulator feature to RealPlayer. This allows you to simulate what the stream would look like at different actual data and packet loss rates. The bandwidth simulator is the only easy way to check the quality of the lower speed bands in a SureStream file. Note the bandwidth simulator file only works with files streamed off a RealServer.

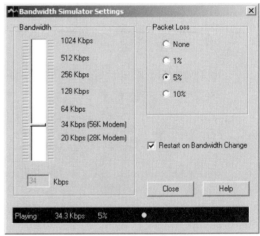

**16.2** The bandwidth simulator, included with Helix Producer Plus, lets you specify both a connection speed and packet loss to test a stream.

Helix Producer Plus is also much easier for professionals to use than RealSystem Producer Plus. It allows you to make and save new audience settings on the fly and vary preprocessing settings on a per job basis. This may not sound like much, but is certainly a big deal for anyone who ever had to use the older version!

## RealMedia Editor

The RealMedia editor is a simple tool for trimming and tweaking already compressed RealVideo files, distributed with RealProducer. It allows you to cut and paste files, trim ends, and change metadata settings. It's used for many of the simple tasks QuickTime Player Pro is great for,

RealVideo Encoding Tools **263**

although it doesn't have the same depth. Because all processing is done without recompressing the files, changes are almost immediate. RealMedia 9 wasn't available as we went to print, but it expected shortly.

## Cleaner

Cleaner, like other tools that use the RealVideo API, does a good job of supporting the SDK. The speed of the RealVideo codec itself is so high that it accentuates Cleaner's lack of speed compared to other tools—RealProducer will generally run the same task at least three times as fast as Cleaner on a single-processor machine, and faster yet on a multiprocessor machine.

Cleaner's EventStream offers some support for RealVideo, through automatically generating SMIL files and adding interactivity to the .rm file.

The RealVideo 9 codec isn't in the current Mac OS SDK from RealNetworks, nor is it available on Mac OS X. This means Mac users for the time being must encode in Classic and with older codecs.

## Premiere

Adobe Premiere, like other systems based on the RealVideo SDK, does a fine job of compressing RealVideo files. Premiere includes a plug-in that can export to RealVideo from the timeline, supporting all the SDK features. This includes an inverse telecine, which Premiere doesn't otherwise support.

**16.3** The Advanced RealVideo export dialog from Premiere. Note it gives you inverse telecine as an option, which isn't available for other formats from Premiere.

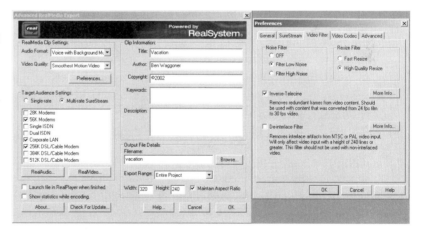

# RealVideo Codecs

## RealVideo v9

In April 2002, RealNetworks announced the RealVideo v9 codec and a command-line encoding utility to make files with it. It is now supported in the current versions of Helix Producer and the RealMedia SDK for Windows and Linux. A number of vendors including Anystream and Discreet also have support for the codec.

RealVideo v9 is available as an automatic download for RealPlayer v8 and later, for both Mac and Windows. RealVideo v9 has significantly improved compression efficiency over RV8, providing the same quality at perhaps a 30 percent lower data rate on average. At a given data rate, processor requirements are perhaps 20 percent higher. This means, however, that the quality you can deliver to a given CPU is slightly higher than that of RV8.

Overall, RV9 appears to be a good, incremental improvement on RV8. Functionally, it doesn't have any new features, just much improved compression efficiency. All the details described in the following (under RealVideo v8) are still valid.

## RealVideo v8

As this book is being written, RealVideo v8 is the mainstream codec, although the industry should transition rapidly to RealVideo v9. While the older RealVideo G2 and G2+SVT are still available for encoding, there isn't really any decent reason to do so. RealOne and RealPlayer before it come with RV8 preinstalled. Their robust auto-update feature means that the vast majority of RealVideo customers are using the upgraded version.

RV8 is a solid, modern codec. It offers a number of configurable parameters, like emphasis on image quality versus frame rate. It also supports a 2-pass VBR mode with a buffer size up to 25 seconds. RealVideo v8 isn't at all aimed at the CD-ROM market, and so doesn't have a progressive/local mode.

One unique RV8 feature is its decoder. It has a widely scalable post-processor that offers increasingly better quality on faster machines. For slow machines, audio and the basic frames are shown. For machines that exceed the minimum system requirements of a given clip, RV8 can actually play the clip back at a *higher* frame rate than that at which it was encoded. RV8 does this by interpolating the motion vectors in the codec to generate a new frame between existing ones. This generally works a lot better than you might suspect. And when there's a lot of CPU power available, a deblocking filter is applied. To see if it's working, check the status of the "Post Filter" in the Statistics window in Streams->Video. On the flip side, the codec is able to ratchet down frame rate somewhat on lower-end machines. The number and pattern of dropped frames isn't predictable, as it gets much worse with data rate spikes. There's no equivalent to Sorenson

Video v3's B-frame–based temporal scalability, where the lower-end experience preserves smooth motion. This frame dropping is particularly common under Mac OS v9.x.

All this makes it difficult to predict the user experience, as different machines can display the same content in radically different ways. RealVideo v8's biggest drawback is its relatively high CPU requirements, especially on Mac OS systems. Even the fastest Macs struggle with a 250Kbps broadband stream, especially when running the Classic-only RealPlayer under Mac OS X.

RealVideo offers a number of useful features. First, it has a 2-pass buffered VBR mode, accessible both within RealProducer and third-party tools like Cleaner.

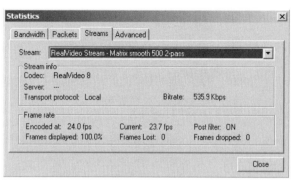

**16.4** This info dialog inside RealPlayer tells you whether or not the PostFilter is being used, and what the playback frame rate is compared to the file's frame rate.

The buffer has a maximum value of 25 seconds, and can provide quite improved quality. It also offers four levels of emphasis on image quality versus frame rate. In order of frame rate, these are: Smoothest Motion, Normal Quality, Image Quality, and Slide Show. For most content, I find Smoothest Motion offers the best quality. Note the frame rate synthesis feature means that, on fast machines, the actual playback frame rate may be much higher than the encoded rate.

## RealVideo G2+SVT

RealVideo G2+SVT was the Scalable Video Technology version of RealVideo G2. Using technology licensed from Intel (originally developed for their Indeo and ProShare projects), SVT was designed to address the performance issues of G2 to allow adoption of RealVideo in broadband environments. This introduced the technologies used in Real v8 to dynamically optimize for quality on higher speed CPUs, specifically the deblocking filter and frame rate interpolation. G2+SVT also included substantial general optimization from Intel for playback on MMX machines.

## RealVideo G2

RealVideo G2 was the first "modern" version of Real's technology, and offered a codec much improved from the RealVideo v5 and RealVideo (Fractal) codecs it replaced. G2 was introduced with the current RealMedia file format, which also introduced SureStream.

G2's biggest problem was its heavy CPU requirements. Doing even 300Kbps was risky on the fastest machines of the day. G2 had the deblocking post-filter of SVT, but not the frame rate upsampling.

## RealVideo

This was the original version of RealVideo, introduced with RealPlayer v4—the first version of Real with video support. It was based on H.263, and did a good job of providing then-good quality at modem data rates. RealVideo wasn't updated until the release of G2 several years later. RealVideo is still included in the SDK, and it is possible to include a RealVideo v5 fallback file inside a SureStream file.

## RealVideo (Fractal)

The RealVideo (Fractal) codec was a licensed implementation of Iterated's ClearVideo codec. It could, in some cases, provide better quality than the original RealVideo, but at the expense of incredibly long encode times. While a decoder is available to RealPlayer through component download, the encoder hasn't been included in the Real SDK for years, and no modern tools can author to it.

# RealAudio Codecs

RealPlayer was an audio-only solution until v4, and so it has always had a strong focus on audio. Instead of offering a few codecs with different settings, Real has always presented audio codecs as a list of content type, data rate, and channel pairs.

Some are marked "High Frequency." These options support higher frequency response, but may increase artifacts when handling complex audio content. You may need to experiment to determine which version is better suited for your particular content.

One nice feature of RealAudio is that SureStream allows a single file to provide multiple audio tracks, so a RealServer can automatically provide the optimum data rate version to a given user.

There are four basic audio codec families still in use.

## Voice

This is a speech codec. It's a fine example of the genre, and is unusual in that data rates go up to 64Kbps. That data rate is overkill for most applications, but the 32Kbps is an excellent choice for audio books and the like.

Table 16.1   Voice codec options.

| | |
|---|---|
| 5Kbps | Voice |
| 6.5Kbps | Voice |
| 8.5Kbps | Voice |
| 16Kbps | Voice |
| 32Kbps | Voice |
| 64Kbps | Voice |

## Stereo Music—RealAudio v8

The Stereo Music—RA8 codecs were introduced in RealPlayer v8.5. They were a substantial enhancement in providing cutting-edge compression at low data rates. There are actually two different sets of algorithms under the RA8 name: A version of Sony's ATRAC-3 codec from minidisc players (enhanced for RTSP streaming) is used for 96Kbps and higher data rates. This provides quality from excellent at 96Kbps to archival at 384Kbps. Below 96Kbps, a system developed in-house at Real is used that is very competitive with other general purpose codecs across its entire data rate range. In most apps that use the RealSystem SDK, the in-house codecs are referred to as "RA8" and the ATRAC-3–derived codecs are designated "RA 8" (with a space between the 'A' and the '8').

Although I usually recommend using mono over stereo for lower data rates, the superior quality of the RA8 Stereo codecs over the mono Music codecs is enough that RA8 should be used instead.

Table 16.2   Stereo Music RA8 options.

| | |
|---|---|
| 12Kbps | Stereo Music – RA8 |
| 16Kbps | Stereo Music – RA8 |
| 20Kbps | Stereo Music – RA8 |
| 20Kbps | Stereo Music High Response – RA8 |
| 32Kbps | Stereo Music – RA8 |
| 32Kbps | Stereo Music High Response – RA8 |
| 44Kbps | Stereo Music – RA8 |
| 44Kbps | Stereo Music High Response – RA8 |
| 64Kbps | Stereo Music – RA8 |
| 96Kbps | Stereo Music – RA8 |

Table 16.3  Stereo Music RA8 (ATRAC-3) options.

| | |
|---|---|
| 66Kbps | Stereo Music – RA8 |
| 94Kbps | Stereo Music – RA8 |
| 105Kbps | Stereo Music – RA8 |
| 132Kbps | Stereo Music – RA8 |
| 146Kbps | Stereo Music – RA8 |
| 176Kbps | Stereo Music – RA8 |
| 264Kbps | Stereo Music – RA8 |
| 352Kbps | Stereo Music – RA8 |

## RealAudio Surround

The Surround codec was announced in April 2002, along with RealVideo v9. As of this writing, the only tool that supports RA Surround is Helix Producer, but tools from other vendors should be available soon.

RealAudio Surround is a tweaked version of the normal RealAudio v8 codecs that maintains the high frequencies and phase relationships required to maintain Dolby ProLogic matrixed surround sound. Note this isn't a true 5.1 solution. It just preserves existing surround information encoded in two channels. You'll need to use an authoring tool like Digidesign ProTools to add the surround information.

Because this codec preserves otherwise inaudible phase relationships and other high-frequency information the non-surround versions wouldn't use, files encoded with this codec require higher data rates and have lower compression efficiency than the normal RA8 codecs. You should only use RealAudio Surround when you have explicitly surround sound content you want to preserve.

RA Surround uses the in-house RealAudio codec at 96Kbps and below, and ATRAC-3 at 132Kbps and above.

Table 16.4  RealAudio Surround.

| | |
|---|---|
| 44Kbps | Surround Audio - Preview |
| 64Kbps | Surround Audio - Preview |
| 96Kbps | Surround Audio - Preview |
| 132Kbps | Surround Audio - Preview |
| 146Kbps | Surround Audio - Preview |
| 176Kbps | Surround Audio - Preview |
| 264Kbps | Surround Audio - Preview |
| 352Kbps | Surround Audio - Preview |

## Music

Music is Real's mono audio codec. At data rates above 8Kbps you'll usually use the RA8 codec, but for 6- and 8Kbps general-purpose audio, Music is still the best option, because RA8 doesn't go down to data rates that low. Of course, music reproduction at those data rates is extremely difficult. When at low data rates the speech component is more important, consider using the Speech codec instead. While 8Kbps Music may enable a user to recognize the song, they aren't going to dance to it.

Table 16.5    Real Audio Music options.

| | |
|---|---|
| 6Kbps | Music |
| 8Kbps | Music |
| 11Kbps | Music |
| 16Kbps | Music |
| 20Kbps | Music |
| 20Kbps | Music High Response |
| 32Kbps | Music |
| 32Kbps | Music High Response |
| 44Kbps | Music |
| 64Kbps | Music |

## Stereo Music

The pre-RA8 Stereo Music codec is no longer considered adequate. RA8 offers superior quality at similar data rates, and at a much broader range of data rates.

Table 16.6    RealAudio Stereo Music options.

| | |
|---|---|
| 20Kbps | Stereo Music |
| 32Kbps | Stereo Music |
| 44Kbps | Stereo Music |
| 64Kbps | Stereo Music |
| 96Kbps | Stereo Music |

# Chapter 17

# MPEG-1

## Introduction

MPEG-1, released in October 1992, was the Motion Pictures Experts Group first standard. It offered revolutionary quality for its time, although it has since been eclipsed by later technologies, many of which built upon ideas first introduced in MPEG-1.

The initial hype for MPEG-1 was its application as a compression format for VideoCD and interactive movies on desktop computers. The interactive movie angle never really caught on—it required users to buy a $200 card to get full-motion video and there were never enough compelling games available that utilized MPEG-1 video. In fact, it quickly became clear that users were much more interested in fully interactive real-time 3D games than in interactive cinema. Most CD-ROM projects made use of QuickTime or AVI, both of which offered greater flexibility and lower decode requirements. By the time computers sported CPUs fast enough for MPEG-1 playback to be ubiquitous, other formats like Indeo v5 were available that offered superior compression efficiency. However, MPEG-1 eventually gained an important role as the format of choice for universal playback.

Initially, VideoCD itself looked to be an abysmal failure. The quality it produced was barely that of VHS, let alone the laserdiscs that were then the darling of the videophile community in the U.S. and Europe. However, once hardware costs dropped enough, VideoCD became enormously popular in Asia, due to the prevalence of very cheap pirated movies and ubiquitous, inexpensive players.

## MPEG-1 Video

The MPEG-1 video codec reused a lot of work already done in JPEG and H.261. Given the extreme limitations of the encoding and decoding power available back in 1992, MPEG-1 was designed as a straight-up implementation of DCT without a lot of the special features offered by modern codecs. MPEG-1 is a 4:2:0 color space codec, with 8×8 blocks and 16×16 macroblocks. It doesn't support partial macroblocks, so resolution must be divisible by 16 on both axes.

Compared to modern Web codecs, MPEG-1's largest deficiency is its lack of a deblocking filter. In general, MPEG-1 does well at the data rates and resolutions it was designed for: 1150Kbps at 352×240, 30fps (NTSC) or 352×288, 25fps (PAL). This data rate was chosen to match the transfer rate of an audio CD, so VideoCD players could be built around the physical mechanism of audio CD players.

Modern high-end MPEG-1 encoders with such enhancements as 2-pass encoding, better preprocessing, and more exhaustive motion search algorithms offer much higher quality than their predecessors. These modern encoders output better quality at lower data rates. However, the lack of a deblocking filter means quality drops off sharply once bandwidth falls below a certain point. Depending on how complex and thus how difficult the video is to encode, the video can become extremely blocky, with none of the post-processing modern codecs use to hide those artifacts. As we learned in Chapter 2 regarding vision, the human eye is much more attuned to sharp edges, so these blocky artifacts are quite distracting.

**17.1** MPEG-1 sample at 800Kbps *(left)* and at 500Kbps *(right)*.

## I, P, and B-Frames

MPEG-1 supports I-frames (keyframes), P-frames (predictive or delta frames), and B-frames (bidirectional frames). Strings of the letters I, P, and B are used to indicate how many B-frames are included in a given file. The most common configuration for high-quality compression is IBBP—you get two B-frames per P-frame. In a typical file, you might see a pattern like this:

IBBPBBPBBPBBIBBPBBPB. . . .

Bidirectional frames are essential for getting maximum quality out of MPEG-1. Some older, cheaper encoders didn't do B-frames because they take longer to encode. No B-frames meant less quality. Most modern encoders support B-frames, which offer better encoding efficiency and can be dynamically dropped at low data rates. The number of frames in the BPP sequence is called the reference interval. A value 3 is typical (and required for delivery on VideoCD).

MPEG-1 also tends to have many more I-frames than Web content: two per second. This is because random access was more strongly emphasized for MPEG-1 than in Web formats where latency is traded for compression efficiency. Additional I-frames can be injected into the MPEG-1 stream at scene cuts and the like.

## GOP

One critical bit of terminology in MPEG-1 is the Group of Pictures or GOP. A GOP is the minimum unit that can be decoded in its entirety. It always begins with an I-frame, and continues until it reaches the last frame before the next I-frame. Random access in MPEG-1 can typically only go to the beginning of a GOP. In other formats, a GOP is "a keyframe and all interframes based on it." GOPs are of fixed size in MPEG-1—typically 12 frames for PAL or Film, and 15 frames for NTSC. The number of frames in a GOP must be a multiple of the reference interval.

A subtle but important distinction in MPEG needs to be made between Open and Closed GOPs. In a Closed GOP, no frame can reference any frame in any other GOP. In an Open GOP, they can. For example, the last frames in a GOP are typically PBB. In a closed GOP, the last B-frames can only reference the P-frame before them. In an Open GOP, they can reference the first I-frame of the next GOP, improving their quality. This makes random access slower, but can significantly improve compression efficiency. Files for local playback should always be Open GOP.

MPEG-1 is progressive scan only. Adding support for interlaced encoding was one of the major features of MPEG-2.

## Video Buffer Verifier (VBV)

Because MPEG-1 was designed to play back on hardware devices with limited memory, MPEG-1 specifies a minimum amount of buffer RAM. This buffer size is measured in 16K blocks. The default for MPEG-1 playback is 20. Higher values allow the codec to spread the data rate out more, improving average quality, but reducing compatibility. Leaving it at the default of 20 is appropriate unless you know you're targeting a player (typically a software player) that can handle higher values.

## MPEG-1 Aspect Ratios

One continual source of frustration with MPEG-1 is that its default resolutions are non-square pixel 352×240 for NTSC and 352×288 for PAL. With 4:3 source, this means just displaying the pixels would result in NTSC being shown 10 percent too wide and PAL 9 percent too tall. A flag in the MPEG-1 file specifies its pixel shape, so this is supposed to be corrected on playback. All hardware devices do this correctly. Most software playback devices do it correctly, too. For example, QuickTime scales the video down to 320×240 by default (but still can be pixel doubled or whatever). However, Windows Media Player and other DirectShow-based players just display the actual pixels in question. Worse, you sometimes see MPEG-1 files in which the aspect ratio was pre-corrected for Windows Media Player distortion, so those files will play back incorrectly on legal MPEG-1 players.

The best solution is to encode video for computer playback at square pixel, and at a square pixel aspect ratio. Thus it will play correctly both on players that support the pixel shape flag and those that don't. It's also more efficient to do this—if the video is going to wind up 320×240, it's pointless to encode pixels that aren't going to be displayed. The typical choice is 320×240. However, if you have widescreen content, you can go up to the full 352 wide and still be MPEG-1-legal. For example, a 1.85:1 film could be encoded at 352×192 square pixel.

If you know you're targeting playback on a device that honors playback ratio, you can use some other aspect ratios, especially the widescreen ones. For example, doing 352×240 at the 0.8055 ratio is very close to the standard 1.85 film ratio.

Table 17.1 MPEG-1 aspect ratios.

| Aspect ratio | 352 wide plays as |
|---|---|
| 0.6735 | 524 |
| 0.7031 | 500 |
| 0.7615 | 462 |
| 0.8055 | 436 |
| 0.8437 | 416 |
| 0.8935 | 396 |
| 0.9157 | 384 |
| 0.9815 | 360 |
| 1.0000 | 352 |
| 1.0255 | 344 |
| 1.0695 | 328 |
| 1.0950 | 320 |
| 1.1575 | 304 |
| 1.2015 | 292 |

## MPEG-1 Audio

MPEG-1 has three audio codecs. All are based on similar technology with psychoacoustic modeling. The higher the level, the better the compression efficiency, but the higher the CPU requirements for encoding and decoding. While this isn't such a big deal these days, it was when the format was created. MPEG-1 audio can only be at 32k, 44.1k, or 48kHz.

*Layer 1* MPEG-1, Layer 1 audio sounds pretty good, but offers very low compression efficiency. Back in the day, it was used by radio stations as a transport format because it sounded

**17.2** MPEG in NTSC *(above left)*, PAL *(above right)*, and with correct *(lower left)* aspect ratios.

excellent at high data rates and could be encoded and decoded in real time with just software. There is no reason to use it today.

Layer 1 data rates run from 32 to 448Kbps.

*Layer 2*   MPEG-1, Layer 2 audio has eclipsed Layer 1 for use as a codec inside MPEG-1 files because it offers substantially better encoding efficiency and puts an insignificant hit on modern processors. Layer 2 audio is also supported in all MPEG-1 hardware playback devices, and is mandated as the audio codec for VideoCD. The general rule of thumb is that Layer 2 sounds like an MP3 file at twice the data rate, so 192Kbps sounds about as good as 96Kbps MP3. Thus, VideoCD typically has much better audio than video quality.

Layer 2 is the same technology as Sony's MUSICAM.

Layer 2 was also used in audio-only files before the MP3 explosion. MPEG-1, Layer 2 audio files use .mp2 extension. This was the high quality format of choice on the pioneering Addicted to Noise music Web site.

Layer 2 data rates run from 32 to 384Kbps, with 32 to 64 being mono only.

*Layer 3* MPEG-1, Layer 3 audio was never widely adopted in MPEG-1 authoring or playback tools due to higher licensing requirements. However, it became an enormously popular technology in its own right, albeit under its shortened name, MP3.

Unless you're specifically targeting an MPEG-1 player that supports MP3 (which is unlikely), you should use Layer 2 instead. For more details on MP3, see Chapter 22, "Miscellaneous Formats."

Layer 3 data rates range from 32 to 320Kbps.

## MPEG-1 for VideoCD

The VideoCD format was a precursor to DVD. VideoCD combined MPEG-1 and limited interactivity delivered on a standard audio CD. While VideoCD's quality limits kept it from being widely adopted as a format for paid content, it has seen renewed popularity as a format for video pirating, especially in Asia. There were already a huge number of pirate CD replicators in the region, so it was a relatively trivial process for them to start making pirate VideoCD discs. While decent quality is possible with VideoCD, most discs are pirated and suffer abysmal quality. It's common for such VideoCDs to be shot with a video camera in the back of a theater, so the heads of patrons going for popcorn can be seen à la *Mystery Science 3000*. Don't be fooled by these and believe VideoCD can't do a good job. With a good encoder and proper care, a VideoCD can be made without much in the way of visible artifacts.

VideoCD uses a very precise version of MPEG-1, with effectively no room for different parameters. This format is called "White Book." The data rate is exactly the same as that of audio on an audio CD, so for every minute of disc capacity, one minute of VideoCD can go on—80 minutes maximum as of late 2001. VideoCD requires precisely 1150Kbps for video, 44.1kHz stereo, and 224Kbps audio.

Table 17.2  VideoCD formats.

| Format | Resolution | FPS |
|---|---|---|
| NTSC | 352×240 | 29.97 |
| Film | 352×240 | 23.976 |
| PAL | 352×288 | 25 |

When authoring an MPEG-1 file to burn for VideoCD, you're almost always going to be making a White Book disc. The one exception is if you're using a version of Toast (the leading Mac OS CD mastering program) that's earlier than v5.0. Toast was made by Astarte, who also made MPEG-1 encoding software. Toast required a specific format of MPEG-1 that only their software could produce. Tools like Cleaner were made to jump through a lot of hoops to get their "Toast Compatible" formats to work, generally just in time for Adaptec Toast v5 to come out. Toast v5 supports White Book just fine. Normal White Book VideoCD files play back just fine in any normal MPEG-1 player software, but pre-v5.0 Toast Compatible files made won't play anywhere.

MPEG-1 for VideoCD **277**

**17.3** 720×480 *(top)* and 720×486 *(middle)* compared to 352×240 *(bottom)*.

There are three VideoCD formats: NTSC, Film, and PAL. Note the Film rate format uses the telecined NTSC frame rate, and plays in NTSC compatible VideoCD players. PAL transferred film uses the PAL format. Computers can play back either type, but VCD and DVD players normally handle only PAL or NTSC.

## Preprocessing for VideoCD

Like DVD, you typically don't do any image processing with VideoCD, as it's meant for playback on a video device in the end. Therefore, gamma correction, contrast, and luma range should be maintained or modified to fit video range if the source doesn't already. The video needs to be deinterlaced, of course, as MPEG-1 is progressive scan only. It's also appropriate to do noise reduction on troubled source material to make the file easier to compress.

The default pixel shape of MPEG-1 is the same as with 720 wide formats. The MPEG-1 resolutions are cut down by 50 percent from their ITU-R.BT601 equivalents. However, 352×2=702, not 720. This is because 720/2=360, which isn't divisible by 16. However, the active part of the image with 720 wide is only supposed to be 704, and thus the 720 source grabs a wider part of the raster than 352×2 does. Crop eight lines off the right and left edges of your source before scaling down to 352 to keep the aspect ratio accurate, and everything will be perfect. Don't crop from the top or bottom if it's 480- or 576-line source. If it's 486, you'll need to crop six lines out of the source to grab the 480 line raster that 240 is doubled to fill (refer to Figure 17.3).

## VideoCD authoring

Once you have your White Book MPEG-1 files, you'll need to use a VideoCD mastering program to actually make the disc image and burn it. This step allows you to set up the chapters and markers in the file, add some graphics for menus, and so on.

## VCD variants

*eXtended Video CD (XVCD)*   XVCD is a widely supported variant of VideoCD that supports double the video data rate, and hence higher quality, especially with lower-end and real-time encoders. This of course halves the playback time.

*Super VideoCD (SVCD)*   SVCD is a VideoCD variant that actually uses MPEG-2. It still uses CD-ROM media. Durations are very short, around 20 minutes, due to MPEG-2's greater data rates. SVCD is covered in detail in Chapter 18, "MPEG-2."

## Why VideoCD?

So, why make a VideoCD in the modern era? Because CDs are the most ubiquitous digital playback medium in the world and they are by far the cheapest to manufacture (they currently run about $0.05). A VideoCD will play in any VideoCD player, DVD Player, or personal computer made since 1995.

VideoCD also offers better interactivity than VHS tape, although it offers much less than a DVD can provide.

VideoCDs don't support any meaningful copy protection, so they should only be used for content you're not worried about being pirated.

## VideoCD Players

The canonical VideoCD Player is a standalone device a lot like a DVD Player. New ones also support the XVCD and SVCD formats. DVD players also play VideoCD discs. However, many older DVD players can't play VideoCDs or audio CDs on recorded on CD-R media; they can only play discs produced at commercial CD stamping facilities. Playing CD-R on DVD drives requires a second laser running at a different frequency, which all modern DVD players have.

There are also VideoCD software players available for all major operating systems. Windows Media Player can do it on Windows, as can QuickTime on both Mac and Windows. In both cases, you'll need to load the .dat file from the VCD into the player. For a more DVD-like experience, there are lots of commercial and shareware players as well. Note that most VCD software doesn't correct for video gamma, so VideoCD will typically look a little bright on Mac and a little dark on Windows.

## MPEG-1 Playback

Most MPEG-1 software playback is done either with the QuickTime or DirectShow API. Both of these are quite capable, and can easily handle resolutions up to 640×480 on any G3, PIII, or better CPU. MPEG-1 for CD-ROM use should be encoded as square pixel.

There are a number of MPEG-1 players for other operating systems, especially Linux and other flavors of UNIX. These vary radically in performance and stability, and many are developing rapidly. If you're targeting playback on a wide variety of systems, it's best to stick to the canonical MPEG-1 resolutions of 352×288 or less.

### MPEG-1 for CD-ROM

MPEG-1 was originally designed for CD-ROM playback, and remains an excellent choice in that role at the data rates and resolutions it was originally designed for. Basic MPEG-1 playback can be done in most CD-ROM authoring environments by treating it as either a QuickTime or DirectShow movie. Both can easily support MPEG-1 up to 640×480, so you don't have to be limited by VCD resolutions.

 **Note**

QuickTime for Windows prior to v5 didn't support MPEG-1 playback, although the Mac version of QuickTime has since v2.5.

### MPEG-1 for progressive download

MPEG-1 can be used for progressive download, and QuickTime, Windows Media, and RealPlayer all support MPEG-1 playback embedded in a Web page. MPEG-1's big advantage is that it can be played *anywhere*, including on Linux and UNIX systems that typically don't have access to the modern codecs found in other formats. If you need to provide a version of your content that everyone can see, MPEG-1 fits the bill. Of course, much higher data rates are required to get competitive quality. Also, there is no standard syntax for Embed/Object for MPEG-1 files. Typically, you define the player you want in HTML, and use that player's syntax, though this won't work on Linux/UNIX browsers.

### MPEG-1 for streaming

Odd as it may seem, there are some fairly robust implementations of MPEG-1 for RTSP streaming. MPEG-1 was the basis for Cisco's initial line of Web video tools, and MPEG-1 is supported in the Apple QuickTime/Darwin Streaming Server. While MPEG-1 doesn't support huge MBR flexibility, these servers are able to dynamically drop B-frames to thin bandwidth before dropping down to I-frame only transmission.

MPEG-1 streaming is typically used for unusual circumstances, such as in a heterogeneous computing environment with a fast network of UNIX workstations that don't support other formats. The compression efficiency of MPEG-1 just isn't competitive against other solutions; WMV 8@500Kbps can beat MPEG-1@1500Kbps in quality.

## MPEG-1 Authoring and Tools

Because MPEG-1 files can be processed by almost any computer, and can offer good quality at higher data rates, they can be useful as intermediate files. MPEG-1 is never truly lossless, but it can get close enough to be acceptable for many tasks.

When using MPEG-1 for authoring, you're balancing encode time versus file size. File size can be a significant issue, as many MPEG-1 encoders have a file size limit of 2GB, so high data rates can reduce file duration. Typically, you don't have to worry about resolution for authoring files—they generally don't need to be decoded in real time, and DirectShow and other architectures that most applications use to decode MPEG-1 can support higher resolutions.

When file size is more important than encode speed, you simply make a normal MPEG-1 file, but at higher data rates. Put the Layer 2 audio to its max of 384Kbps, and target a total data rate of at least 4000Kbps for 320×240. Raise data rate accordingly for higher resolutions—640×480 should use around 16,000Kbps. The higher quality modes with better motion search helps. Because MPEG-1 doesn't support interlaced video, it's only appropriate for storing progressive content.

When you're not so worried about file size, you can make an I-frame only MPEG-1. This is equivalent to using JPEG for an intermediate file. However, unlike JPEG, you have to give a data rate for MPEG-1, not just a quality setting. MPEG-1 encoders typically pad out the file to whatever data rate you request, so if you request a data rate higher than is needed to store the video uncompressed, the file will still encode at the absurdly high data rate. For 320×240, 20,000Kbps is as high as you should ever need to go.

QuickTime, alas, doesn't support MPEG-1 as an authoring format for audio, just video. This is due to limitations in QuickTime's MPEG-1 support. Hence Mac OS tools like Cleaner tend to have their own MPEG-1 decoders to get both audio and video out of a file.

Because MPEG-1 is an international standard, myriad groups and companies have tried to create better MPEG-1 encoders, each offering a different blend of speed, quality, and ease of use. MPEG-1 encoding is very well suited to SIMD architectures like AltiVec and SSE, so an encoder optimized for those architectures can be many times faster than non-SIMD optimized encoders.

Quality differences can be substantial between encoders. Quality can also vary within a single encoder based on the mode in which it is run.

There are tons of MPEG-1 encoders out there, with more coming all the time. Some typical examples:

## Heuris MPEG Power Professional

One of the longest-lived MPEG encoding packages is Heuris MPEG Power Professional (MPP). It comes in several versions, but all have complete MPEG-1 support. MPP earned its reputation as a professional tool with deep support for the myriad features of MPEG, lots of verification tweaks, and so on. However, with a starting price of $349 for the MPEG-1 only version, it's quite expensive for its market.

The Heuris interface is rather complex, as its controls reside in a series of tabs. With some tweaking, you can get enormous changes in encoding time and significant changes in video quality.

For most content, leaving the core options at "auto" will strike a good balance between quality and data rate. For maximum quality, set Mode to Two-pass, and have Auto Analyze on.

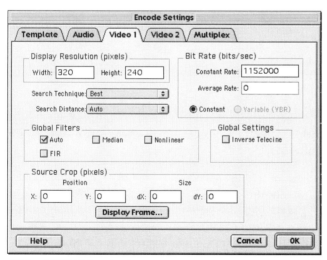

MPP has decent preprocessing with basic cropping, scaling, deinterlacing/inverse telecine, and an odd Global Filters option. Global Filters lets you specify video filters to make the video easier to compress. These are only on/off, and so lack fine control. If you don't already have preprocessed input, you're best off checking "Auto" where it will dynamically adjust the settings as appropriate while compressing.

**17.4** MPP Video 1 tab.

MPP offers an impressive degree of control if you really want to drill down. It uses Edit Control Lists (.ecl) to configure encoding settings on a frame-by-frame basis. These store useful information such as where I-frames should be injected, which filters to run, and the cadence for inverse telecine. At a minimum, you want the Auto Analyze mode turned on, which automatically generates an ECL based on an initial analysis of your video before the file is encoded (the analyze pass is usually a lot faster than the actual encode pass). You can also do an analysis-only file, and manually tweak the ECL. This is generally only necessary when it's critical to get maximum quality at a constrained data rate, but it's a great way to spend a weekend for the obsessive compressionist.

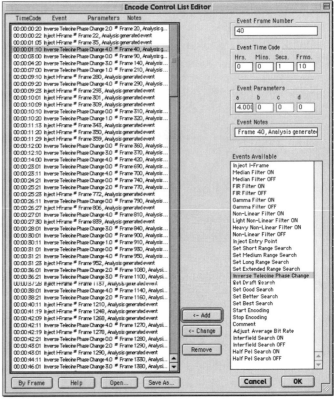

**17.5** ECL tweaking interface.

## Cleaner (and Charger)

Cleaner has a decent, simple MPEG-1 implementation, licensed from PixelTools. It's slow and somewhat inflexible, but can provide very high quality. It's integrated with Cleaner and so has a lot of preprocessing and workflow advantages.

However, there are annoying limitations in Cleaner's MPEG-1 support. For one, Cleaner doesn't allow encoding to VideoCD for NTSC film format, just NTSC video and PAL. Also, there's no way to control mono versus stereo audio modes.

Cleaner has an enhanced MPEG authoring option called Charger. It mainly provides new features for MPEG-2. It also adds 1-pass and 2-pass VBR encoding modes. $500 seems a lot to pay for added VBR modes.

The important control in Charger is Encode quality, which offers three quality/speed levels. At the Slower (Higher Quality) mode, Cleaner provides some of the best MPEG-1 encoding out

**17.6** Cleaner 5 MPEG-1 encode tab.

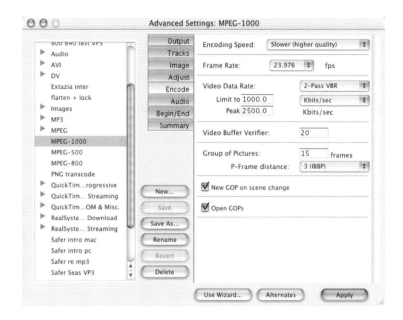

there, without having to wade through a lot of encoder dialogs. Charger isn't particularly swift, though. At its maximum it runs at one-quarter the speed of TMPGEnc, discussed next.

## TMPGEnc

The TMPGEnc encoder is theoretically commercial, but it's been in free public beta for ages. It offers an excellent suite of highly configurable MPEG-1 options including the usual GOP options, several speed versus quality modes, and so on. It also has enormously flexible options for encoding the video, including some preprocessing options that are very specific. While the defaults are typically fine for most content, tricky clips that need to fit into very narrow bandwidth are best compressed with TMPGEnc.

TMPGEnc is wicked fast. Even in its highest quality mode, it encodes at more than half real time on a dual 1GHz PIII with preprocessed source.

## Canopus ProCoder

Canopus's ProCoder is the new kid on the video compression block. Its initial 1.0 version is focused on the MPEG formats and it includes excellent MPEG-1 support. ProCoder has four different quality/speed bands and does CBR, 1-pass, and 2-pass VBR. The highest quality mode is called "Mastering Quality," and though the UI suggests against it with dire warnings of slow encoding, it is still much faster than Cleaner and offers superlative quality.

ProCoder's UI is much easier to use than TMPGEnc.

## Adaptec Toast Titanium

On the opposite extreme from TMPGEnc and MPP is Adaptec's Toast Titanium. This tool is mainly known as an excellent CD-ROM mastering app. However, it can also author VideoCDs and encode QuickTime files to VideoCD format.

Toast Titanium offers no options for the encoding process—it just does it. The quality is, as you'd expect, not particularly good. You can import other White Book MPEG-1 files into Toast for authoring.

Titanium also provides an iMovie plug-in so you can directly export to White Book from iMovie. It doesn't work with any other applications.

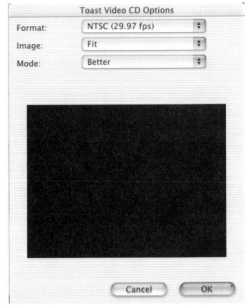

**17.7** Toast Titanium VCD authoring window.

# Chapter 18

# MPEG-2

## Introduction

By almost any standard, MPEG-2 is the most ubiquitous and important video standard in the world today. While the Web gets all the press, MPEG-2 gets the eyeballs, through DVDs and digital satellite and cable and soon HDTV. MPEG-2 is a fairly linear extrapolation from MPEG-1. The main new features are

- improved compression efficiency, and
- support for interlaced video.

## The MPEG-2 Format

There are two basic MPEG-2 formats: program streams and transport streams. Program streams are for standard file-based uses, such as on a CD-ROM or other media. Transport streams are designed for transmission through lossy environments, and so include a lot of error resiliency and recovery functions.

As a format, MPEG-2 is quite similar to MPEG-1. Structurally, MPEG-2 uses the same profiles and levels descriptors as MPEG-4, although it has many fewer versions, most of which are never seen in the wild. Almost all standard definition content uses the MainProfile@MainLevel option, which is 720×480, 29.97fps in NTSC or 720×576, 25fps in PAL, 15Mbps maximum.

## MPEG-2 Video

The MPEG-2 video encoder is an enhancement of MPEG-1, carrying on its features, often enhanced, and adding a host of new ones. The basic structure of macroblocks, GOPs, IBP frames, and so on, is the same.

For typical standard definition content, MPEG-2 with a good VBR encoder is well served by average data rates between 2.5Mbps and 6Mbps. Much below that, and quality degrades with most source, and it's hard to provide significantly higher quality that a high-end encoder can provide at 6Mbps (although lower-end encoders can attain better quality at higher data rates).

## Interlaced Video

The biggest addition to MPEG-2 is native support for interlaced video. Interlacing is handled in a more subtle way than just splitting the video into two half-height bitmaps. Instead, a video can be progressive or interlaced at three different levels: (1) A given stream can be progressive or interlaced. A progressive stream must only have progressive frames, but (2) an interlaced stream can include both interlaced and progressive footage. (3) Frames themselves can be interlaced or progressive. A progressive frame must have only progressive macroblocks, but an interlaced frame can have a mixture of progressive and interlaced macroblocks. This flexibility is extremely useful. For example, in a frame where part of the image is moving and other parts aren't, the static sections can be encoded as progressive (for better efficiency), and the moving sections can be interlaced (which is required as the fields contain different information).

Because an interlaced macro block actually covers 32 lines of source instead of 16, the zigzag motion pattern normally used in MPEG for progressive scan must be modified. The alternate scan pattern attempts to go up and down two pixels for every one it goes left and right. (This is shown in Figure 3.13 on page 44 of Chapter 3.)

## High Definition (and what happened to MPEG-3)

MPEG-2 was originally targeted at SD resolutions, although the spec itself allows for absurdly high resolutions well beyond the theoretical maximum 4096×4096 of MPEG-1. However, it wasn't initially obvious how well MPEG-2 could scale, so the plan was to produce an MPEG-3 that handled HD. Extending MPEG-2 to support HD turned out to be fairly easy, so MPEG-3 was dropped. And so HD became another profile of MPEG-2. You'll see this profile referred to with the initials "HL" for High Level.

## Scalability

Another important feature of MPEG-2 is scalability, wherein a decoder decodes only part of the bitstream and still presents an image albeit of lower quality scaled to the available bandwidth. Scalability is based on enhancement layers, where a reference image can be improved by adding additional data. This enables bandwidth scalability, as the enhancement layer can be dropped if there is insufficient bandwidth.

Scalability is interesting for end-to-end transmission systems, but isn't commonly seen in file-based encoders, and it isn't used in either DVD or in the USA's ATSC (Advanced Television Standard Committee) standard for digital and high definition televisions.

# MPEG-2 Audio

Oddly enough, most MPEG-2 applications don't use MPEG audio. Instead, they use another audio format that provides better quality or compression efficiency.

All DVD audio codecs sample at 48KHz, slightly better than the 44.1KHz of audio CD.

## PCM

Consumer-level DVD authoring systems typically just provide the audio as a pair of uncompressed tracks. This is due to the expense and complexity of the Dolby Digital licensing arrangements. PCM (Pulse Code Modulation) also allows 96KHz recording at 16-, 20-, and 24-bits, making it potentially much better than CD as a recording technique, although most or all of this detail is lost with typical listeners, systems, and environments. For example, broadcast SD television audio is 32KHz audio.

Because PCM has no compression efficiency, the only other MPEG-2 application it is used for is authoring.

## Dolby Digital (AC-3)

Most professional DVDs use the Dolby Digital audio compression standard, also known as AC-3. It is also mandated in the ATSC standard for digital broadcast, and is used by many digital cable and satellite systems. Dolby Digital provides very good quality at decent data rates, and does an excellent job providing multichannel support. It provides substantially better compression efficiency than multichannel MP2, which is why Dolby Digital is almost always used on DVDs. Also, most modern feature films have soundtracks created in 5.1 Dolby Digital surround, which can be included on a DVD without having to be converted to another audio format.

A 2-channel Dolby Pro Logic matrix encoded stream is almost always provided on the DVD, which is turned into multichannel audio by the receiver. For DVD, and increasingly with the other digital transmission systems, a full 5.1 mix may be provided, especially with recent films that were released in 5.1. The full 5.1 channels are Center, Front Left, Front Right, Surround Left, Surround Right, and a Low Frequency Effect (LFE, for the sub-woofer in the system). With a maximum data rate in the spec of 448Kbps, quality is very good. The old standard was 384Kbps, from AC-3 laserdiscs, but for DVDs, I recommend sticking to the full 448Kbps for the main audio track of multichannel sources. Stereo is typically encoded at 192Kbps.

Dolby Digital also supports "Dialog Normalization." This allows the dynamic range of the audio to be compressed when the range between, say, dialog and explosions, would be too wide to allow intelligible speech on one end and keeping explosions to a neighbor-approved level on the other.

# Chapter 18: MPEG-2

Table 18.1  AC-3 supported channels.

| Label | Channels | LFE? | Data rate range |
|---|---|---|---|
| 1/0 | Center | No LFE | 64-448Kbps |
| 2/0 | Left, Right | No LFE | 96-448Kbps |
| 3/0 | Left, Center, Right | LFE optional | 128-448Kbps |
| 2/1 | Left, Right, Surround | LFE optional | 128-448Kbps |
| 3/1 | Left, Center, Right, Surround | LFE optional | 192-448Kbps |
| 2/2 | Left, Right, Left Surround, Right Surround | LFE optional | 192-448Kbps |
| 3/2 | Left, Center, Right, Left Surround, Right Surround | LFE optional | 224-448Kbps |

The Dolby Digital encoding process can be quite complex, with a number of different preprocessing and compression modes. If 5.1 or other multichannel source is being used, I suggest consulting a sound recording engineer with experience in Dolby Digital to get the correct encoding settings. For typical compression projects, where the source is 2-channel audio, it's important to tell the encoder whether or not the source is matrix encoded for the older Dolby Surround format, or the wrong mode could be used during playback.

## DTS (Digital Theater Systems)

DTS, from Digital Theater Systems, is often positioned as the high-end DVD alternative to AC-3. It is also sometimes used for theater sound, but never for broadcast. DTS's big advantage and big limitation in the marketplace is its low compression. Originally designed as a theater standard with higher quality and ease of use than Dolby Digital, it used CD-ROM as its data standard. The base version of DTS only does 5.1, but there is a new 6.1-variant called DTS-ES that offers backwards compatibility to current players. In DVD, DTS can use a data rate of 1,536Kbps or 768Kbps. There was also a big movement to implement DTS as an audio format, but it didn't get much mass-market traction, and is likely to fall by the wayside with the competing DVD-based audio standards.

DTS's higher data rate requirements and lower market share often means that DTS discs may have fewer features and cost more than AC-3 versions. However, aficionados say the superior audio makes this a small price to pay.

While most professional DVD authoring tools include AC-3, many do not include DTS. DTS is usually only used in high-end Hollywood productions. For most applications and listeners, 448Kbps AC-3 is more than good enough.

## MPEG Layer 2

Ironically, the one audio codec you're unlikely to ever see alongside MPEG-2 video is MPEG audio. While the initial PAL DVD spec supports 7.1-channel MPEG Layer 2 (MP2) audio, it was dropped for the final version and was never included for NTSC at all.

There were a few additional MP2 sample and data rate combinations added with MPEG-2, which are often called Layer 2.5. These allow higher sample rates at lower bitrates.

## Advanced Audio Coding (AAC)

AAC, originally designed as a sort of Layer 4, offers what is probably the best compression efficiency of all the MPEG formats. However, while AAC was originally designed for MPEG-2 applications, it isn't widely used with MPEG-2. It lives on in MPEG-4. AAC was also the basis for audio formats such as Liquid Audio.

# MPEG-2 for DVD Video

The best known use of MPEG-2 is in DVD authoring. It's the application that is most likely to require a compressionist's touch. MPEG-2 on DVD uses a tight subset of MPEG-2, albeit one subset for NTSC and another for PAL. These are, however, very capable of producing exceptional quality. DVD can provide the highest quality of any standard definition consumer video format. DVD provides a maximum bitrate of 9.8Mbps for video and peak MPEG-2 data rates of 8Mbps are quite common. In practice, experts recommend using a peak of at most 9.6Mbps for manufactured discs and below 8 Mbps for replicated formats like DVD-R. DVD also supports VBR, and high-end discs almost always use 2-pass VBR to produce optimal quality within the disc space requirements.

Because DVD's high bandwidth is more than adequate to yield superlative quality with MPEG-2, there isn't any real pressure to upgrade to MPEG-4 or another, more efficient standard. DVD will remain essentially what it is today, until the release of some future HD system, which will presumably use MPEG-4, perhaps in conjunction with MPEG-21 for interactive and rich media.

MPEG standard resolution and frame rate for DVD is 720×480 29.97fps for NTSC and 720×576 25fps for PAL. Note that unlike VideoCD there is no NTSC film mode. Because MPEG-2 supports progressive blocks inside interlaced video, telecined film is encoded straight into an NTSC stream.

Quality expectations for DVD are quite high, especially as compared to other MPEG-2 applications. Typically, any visual artifacts are considered unacceptable. "Good enough" data rates with professional encoders are typically 6Mbps for CBR encodes and 4Mbps for 2-pass VBR (with much higher peaks, of course).

**Note**

The DVD spec supports using lower resolution MPEG-2 and MPEG-1 video. This can be useful if you need to put dozens of hours of content on a single disc. You can also mix and match, so the feature film can be MPEG-2, with supplementary information in MPEG-1.

## VOB files

DVD video doesn't actually use MPEG-2 files, but incorporates the MPEG-2 video and audio information in whatever format resides in Video Object (VOB) files. This means you can't just read a DVD's video with a generic MPEG-2 decoder. However, a number of folks have written software that can extract the content from any DVD, so DVD can no longer be considered a secure format.

Typically, an MPEG-2 encoder creates an MPEG-2 file that is incorporated into a VOB file by the DVD authoring application. Authoring software usually comes with a Dolby Digital encoder, not with the MPEG-2 encoder itself, and so audio is generally compressed at the authoring stage, not the compression stage.

## Aspect ratio

MPEG-2 for DVD supports two aspect ratios: 4:3 and 16:9. Due to customer demand, pan-and-scanned 4:3 presentations are often optional (two different versions are included), or not available. Modern players do a fine job of converting 16:9 content to 4:3 displays and vice versa (although those with 16:9 sets strongly prefer to have 16:9 content).

Note that 16:9 is still a narrower ratio than most films. Because of this, even 16:9-encoded MPEG-2 for DVDs will have some letterboxing, although much less than the same content would require in 4:3. For example, a typical 1.85:1 movie would need 134 lines of letterboxing in 4:3 while the same movie in 16:9 would only need 20 lines. Overall, 16:9 will have 33 percent more active pixels than 4:3 with wide screen content—which is especially important when displaying on high-definition display devices.

You can gain a slight increase in compression efficiency by aligning the letterbox edges with macroblock or block boundaries. The very sharp edge between the letterbox and the film content can cause noticeable ringing.

## Progressive DVD

Progressive DVD is important to folks on the high end of the home theater market, where customers have high-definition, progressive scan display devices. Some of these also support different frame rates other than 60i, and can display at 3× (72Hz), 4× (96Hz), or my dream NTSC display refresh rate 120Hz, which is 24×5, 30×4, and 60×2 the source frame rate, offering a crisper image and more accurate motion. 120Hz refresh can accurately display all common NTSC frame rates. Because the market life of the DVD format should carry it well into the time HD television sets attain mass-market penetration, the importance of progressive scan DVDs will increase over time.

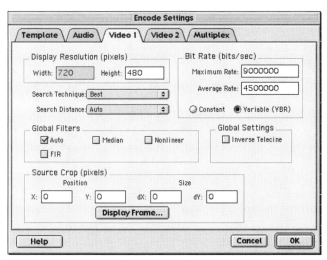

**18.1** Aligned and misaligned letterboxing, showing differences in ringing.

Although MPEG-2 in DVD doesn't support MPEG-2's progressive sequence mode, it does support progressive frames and macroblocks. These are actually several different things that are referred to as "progressive DVD." They all share the same end goal of being able to extract a 480-line progressive 23.976fps video stream for optimal display. On the high-end, DVD players have built-in inverse telecine hardware that is able to take any arbitrary telecined 60i source and convert it back to 24p. This of course can cause glitches where the player can't correctly guess the 3:2 pattern (especially if the source has cadence breaks), and can introduce latency in the video relative to the audio. Good players can delay the audio to match the video, but the use of outboard inverse telecine units can cause sync problems.

A properly encoded disc can send hints to the player on how to decode the file properly. The typical way to handle this is with an MPEG-2 sequence that includes repeat_first_field and top_field_first tags. The repeat_first_field is especially important, because it enables the encoder to not attempt to encode difference data for fields that are identical because they came from the same film frame. This stuff doesn't need to be handled manually—the encoder itself should correctly tag the output. Alas, only a small portion of discs on the market are tagged at all, and a substantial number of those have errors in the hinting.

Because players can't trust in most cases the tags they see, players just ignore the on-disc information and do their own inverse telecine. So, there is generally not much risk in getting the progressive encoding wrong, nor is there an advantage in getting it right. That said, it is my hope that you use tools that enable you to apply inverse telecine tags correctly.

## Multi-angle DVD

One much-hyped feature of DVD is its ability to present multiple camera angles that can be selected and viewed on the fly. If you're encoding this type of content, make sure you are not encoding to Open GOPs, and not doing new GOPs on scene changes. To have seamless mutli-angle switching, the GOPs in each stream must align perfectly.

## Audio for DVD

All DVD players support both PCM (uncompressed audio) and Dolby Digital (AC-3). Players may also support MPEG-2 audio or DTS Digital Audio. DVD audio is always 48kHz.

Due to the licensing cost and complexity of Dolby Digital, it is usually supported only in professional encoders. Consumer DVD authoring tools always use PCM encoding. Because multichannel PCM can take up a large portion of the total bandwidth available on a DVD, PCM is normally kept to stereo, which requires 1.5Mbps.

## All the rest of DVD

DVD is a full-featured, rich, and dauntingly complex format. Compressing MPEG-2 files well is only a small part of the DVD authoring experience.

# MPEG-2 for DVD-ROM

When DVD was first announced, DVD-ROM was heavily promoted. However, only 200 or so consumer DVD-ROM titles are available as of this writing, none of which are particularly popular. DVD-ROM was crippled by two things: OS vendor indifference and the birth of CD-R. Microsoft put some effort into making sure DirectShow could use the video features of DVD-ROM, although the story there is somewhat muddled. Apple supports limited control of DVD Playback via Apple Scripting of the DVD Player application, but so far this implementation has been much less complete and useful.

There are three levels of DVD-ROM. Level 0 is essentially just a big CD-ROM, using the UDF file format. Level 1 uses MPEG-2 files, but not VOB files. Level 2 supports VOB files, and so makes it possible to integrate other applications with normal DVD video information. This is used to make hybrid DVDs, although these are billed as enhanced DVD instead of DVD-ROM. DVD-ROMs are either Level 0 (big discs to store huge amounts of information) or Level 2. The latter is often referred to as "Enhanced DVD."

## Getting enhanced DVD to work

Level 2 DVD-ROM applications need to control the MPEG-2 hardware. Right now, only Windows supports this via DirectShow. There are some older DVD playback PCI cards that only support the older MCI standard for playback control. Many of these legacy cards are buggy, so it's best to target DirectShow-based DVD playback systems, which work with Windows 98 and later with MPEG-2 if there is hardware or software that supports it.

Using DirectShow can be done either through writing a custom application in Visual Basic or C++, or through an application plug-in. For Macromedia Director, the MPEGXtra from Tabulerio Producoes and On StageDVD from Visible Light are both popular options. A lot of DVDs are authored with PCFriendly from InterActual, which offers an HTML/Internet Explorer-based way to author interactive media.

Of course, all these systems are Windows-only.

## MPEG-2 for Digital Cable and Satellite

While DVD gets most of the press, MPEG-2's use in digital satellite and cable systems reach more viewers on a daily basis than view DVDs.

The principal issue in profitability of these delivery systems is how many channels can be pushed down a fixed amount of bandwidth. Because MPEG-2 can squeeze about six channels of decent quality video in the same space as a single analog channel, it was a no-brainer for digital satellite and cable companies to adopt MPEG-2 technology. Those same companies, in their continual quest to increase the number of channels they can offer, have also driven improvements made to real-time encoders. This market has placed a lot of emphasis on statistical multiplexing, which is so cool I need to describe it briefly. In essence, statistical multiplexing is an inter-channel VBR. Because the total bandwidth of these cable and satellite systems is fixed, and any given channel's bandwidth can vary within that overall bandwidth limit, these massive hardware statistical multiplexors simultaneously analyze *all* the channels at once, dynamically adjusting the bandwidth among them to offer the highest average quality possible.

Over the long term, we may see a lot of the existing MPEG-2 market move to MPEG-4, because MPEG-4 offers substantially better video compression, making it theoretically possible to extract yet more money from this market. Broadcasters are especially interested in future H.26L derived MPEG-4 profiles, which could allow two to three times as many channels at the same quality and bandwidth.

### MPEG-2 for Digital Television

The U.S. format for High Definition Digital Television is called ATSC, from the Advanced Television Standards Committee (also known as the Grand Alliance). The Grand Alliance itself was the merger of three different groups of companies, all working on different formats. Like many committees, they had a lot of trouble making a definitive decision, and produced 28 different interlaced/progressive scan, resolution, and frame rate options. These were codified in the infamous

Table 3, back in 1995. Because all receivers of ATSC must support all 28 formats, authoring can get complicated.

Digital, High-Definition Television has been "just a year away" for the better part of a decade now. In Europe, standard definition (SD) digital broadcasts have gained a lot of momentum quickly. And SD digital in the U.S. is popular in cable and satellite networks. But the holy grail—high-resolution digital television—has been much slower in coming. The good news is that HD hardware is available, it works, and many broadcasters are transmitting HD. Gearing up for HD was expensive for the networks and their affiliates, as the broadcasts require all-new video equipment, transmitters, and in some cases even antennas.

The bad news is that it's still expensive (although enormously cheaper than it once was), and not many folks are watching it yet. The primary use of most HD sets is to watch DVDs. However, all the major networks are producing at least some of their content in HD to preserve it for later syndication. They're also broadcasting several hours of HD daily. Many, but far from all, affiliates are broadcasting this HD programming.

However, broader questions remain as to if and how broadcasters will be able to make money from DTV. Other questions remain. Will broadcasters choose to use their extended bandwidth to transmit many channels of SD or a few channels of HD content? How serious is the FCC about its mandate to convert to DTV by 2006? Are satellite and cable operators really going to carry the broadcasts and if so, on what terms? Refer to Table 18.2.

Table 18.2   Compression format constraints.

| Vertical size value | Horizontal size value | Aspect ratio information | Frame rate code | Progressive sequence |
|---|---|---|---|---|
| 1080 | 1920 | 1,3 | 1,2,4,5 | 1 |
|      |      |     | 4,5     | 0 |
| 720  | 1280 | 1,3 | 1,2,4,5,7,8 | 1 |
| 480  | 704  | 2,3 | 1,2,4,5,7,8 | 1 |
|      |      |     | 4,5     | 0 |
|      | 640  | 1,2 | 1,2,4,5,7,8 | 1 |
|      |      |     | 4,5     | 0 |

| Legend for the MPEG-2 coded values in Table 18.2. | |
|---|---|
| aspect_ratio_information | 1 = square samples, 2 = 4:3 display aspect ratio, 3 = 16:9 display aspect ratio |
| frame_rate_code | 1 = 23.976Hz, 2 = 24Hz, 4 = 29.97Hz, 5 = 30Hz, 7 = 59.94Hz, 8 = 60Hz |
| progressive_sequence | 0 = interlaced scan, 1 = progressive scan |

And a more humane presentation of it…

Table 18.3    ATSC options, sanely presented.

| Resolution | Aspect ratio | Interlaced fps (a.k.a. **60i**) | Progressive fps (a.k.a. 24p) |
|---|---|---|---|
| 640×480 | 4:3 | **29.97**, 30 | 23.976, **24**, 29.97, 30, 59.94, **60** |
| 704×480 | 4:3 or 16:9 | **29.97**, 30 | 23.976, **24**, 29.97, 30, 59.94, **60** |
| 1280×720 | 16:9 |  | 23.976, **24**, 29.97, 30, 59.94, **60** |
| 1920×1080 | 16:9 | 29.97, 30 | 23.976, **24**, 29.97, 30 |

I've bolded the ones I think are ever worth using—see the following Note. ATSC uses the consumer 720×480 instead of the professional and DVD 720×480. This is still the same aspect ratio—720 has eight pixels on either side of the frame that aren't included in the 704 broadcasts.

ATSC uses MPEG-2, of course.

### Note

**All the Right ATSC Formats.** Twenty-eight ATSC formats beg the question, "which should you use?". Luckily, the answer is not many.

For standard definition content, I like 640×480 for 4:3 and 704×480 for 16:9, at the native frame rate (typically 24p for film and 29.97i for archival video). Remember the closer you are to square pixel, the more efficient the compression is, so at 4:3, a 640×480 looks better at lower data rates than 704×480 will.

For HD, up-sampled 60i video and new video-style stuff (like news) should use 1280×720 60p. Given the costs of display technologies, it's easier to build in higher frame rates than higher useful resolution, so go for 60p. Trust me—1280×720 is enough pixels. For film content, it's all about 1920×1080, 24fps. However, there are times when it might make sense to use 24p source at 1280×720, as that lower bandwidth would make it possible to include a couple additional SD channels alongside the film channel.

Alas, most broadcasters are adopting a single standard for *all* their stuff, meaning they have to transcode a big chunk of their content from one format to another. I'd much rather they switch the broadcast to the appropriate format. The playback devices can handle it.

## MPEG-2 for CD-ROM

Many were disappointed that MPEG-2 didn't become a format for putting video on CD-ROMs. This isn't because software decoding is that difficult—DVD software decoders can run on any computer. The big problem is simply licensing. There is no way to distribute a free MPEG-2 software decoder, even when it's bundled with free software. This is because there is a minimum per-decoder license fee. This prevents Apple and Microsoft from including MPEG-2 playback inside QuickTime or DirectShow. Apple sells an MPEG-2 software decoder for $19.99, but very few end-users have it installed.

You can buy MPEG-2 decoder boards, and license MPEG-2 software. Either can make a fine solution for playing MPEG-2 in kiosks. The hardware MPEG-2 decoders typically offer video outputs as well, including support for interlaced output. This support for interlaced output is great for projects that mix traditional video with interactivity.

Modern CD-ROM codecs like Sorenson Video 3 and VP3 offer good decode performance, so there is less of a need to have built-in MPEG-2 now. Still, none of these modern codecs offers the quality possible with high bitrate MPEG-2.

## MPEG-2 for Authoring

Unsurprisingly, MPEG-2 is also used as a video production format, both as an intermediate/storage format, and as a real-time editing format. Over the long term, many analysts predict the production and post-production industry will standardize on an MPEG solution for authoring, although MPEG hasn't had that much of an impact in that area yet. The MPEG-4 profiles targeted at this portion of the industry might be the first that are widely adopted.

For high-end nonlinear editing, systems typically use the 422P profiles (also called "Studio Profile"), which offer 4:2:2 chroma support. There are both SD and HD versions. 422P is typically run with only I-frames at 50Mbps (for SD). It's rather analogous to using Motion-JPEG at the same data rate, although it is somewhat more standardized.

## MPEG-2 Encoding Tools

There are zillions of MPEG-2 encoding tools available. Many are hardware-only, but with today's faster computers, software-only encoders run at acceptable speeds. Many DVD authoring packages also include MPEG-2 encoding facilities, although in a number of cases these aren't as good as standalone encoding packages.

One important thing to remember when encoding for DVD or other MPEG-2 formats that are going back to display on a television device instead of a computer: you shouldn't be doing much preprocessing, especially not on clean source. Don't crop out safe area, don't deinterlace, don't correct gamma. If you need to clean up poor quality source, go for it, but if you have professional stuff, you should be able to leave everything alone. Scaling is especially anathema. If you have 486-line input, crop out the extra six lines, don't scale down to 480.

## Canopus ProCoder

ProCoder is my new favorite Windows MPEG-2 encoder. It is extremely fast and good quality in its default mode and of superlative quality and surprisingly swift even in its slow "Mastering Quality" mode. ProCoder also transparently takes care of field order issues and other aspects of transcoding, making it very easy to use for high-end results.

## Cleaner

Out of the box, Discreet Cleaner offers very limited MPEG-2 encoding support. It has two simple presets, one each for NTSC and PAL DVD authoring. You can't change any of the parameters, not even data rate. Its MPEG-2 support is just there as a tease to get you to upgrade to either Charger or SuperCharger.

## Charger

Cleaner MPEG Charger is a software-only upgrade to Cleaner that adds a lot of MPEG-2 functionality. Highlights among these are 1-pass and 2-pass VBR, three encoding quality/speed modes, 4:2:0 and 4:2:2 chroma support, progressive/interlaced modes, and GOP insertion on keyframes. These features go well beyond what DVD is capable of, and include a bunch of modes for which it's hard to find a decoder. However, Cleaner can also read MPEG-2 files, so MPEG-2 can be a good way to store intermediate files for later compression.

Charger can produce very competitive quality, especially in 2-pass VBR mode. Its biggest drawback is that it's glacially slow compared to other solutions, especially on the multiprocessor machines typically used for this application (Cleaner, at least as of v5.1, can't make use of multiple processors on Windows).

## SuperCharger

SuperCharger is a hardware accelerated MPEG-2 board for Cleaner. It's based around the real-time Digital Media Press board (now discontinued as a separate product). The SuperCharger board does MPEG-2 compression in real time, but it still relies on Cleaner to decode the source and do all the preprocessing, and so it's usually much slower than real time. The trick in getting SuperCharger to go fast is to reduce how much processing Cleaner does. Because SuperCharger targets DVD authoring, you don't typically need or want to do any preprocessing, and preprocessing is the slowest part of the process.

SuperCharger doesn't offer the quality of Charger, and it doesn't have 2-pass VBR (but does have 1-pass), nor does it offer 4:2:2 support. Note that, as of Cleaner v5.1, the 1-pass VBR mode didn't work, and Discreet hasn't said whether it will be fixed in a future version.

## Heuris MPEG Power Professional

For most of my MPEG-2 encoding, I use Heuris MPEG Power Professional. Its interface is clunky, and it isn't as flexible as Cleaner, but it reliably produces excellent quality, and it is reasonably speedy. As a bonus, MPP can directly generate a VOB file, which is nice for DVD authoring.

## Apple iDVD

Apple's iDVD is a consumer-class DVD authoring package that runs under Mac OS X. It comes free with Apple computers with built-in DVD-R drives. iDVD only supports PCM audio, and has a fixed 6Mbps data rate. However, it takes incredible advantage of the Altivec Velocity Engine on Apple's G4 machines, yielding real-time software transcoding on 800MHz dual G4 machines. As future machines get faster, iDVD will presumably be able to offer higher quality compression.

You can't extract the MPEG file from this tool—the encoder only works in optimized mode.

## Apple DVD Studio Pro

DVD Studio Pro is Apple's professional DVD authoring package. It was originally produced by Astarte under the name DVDirector. One bonus is that it provides its MPEG-2 encoder as a QuickTime component, so you can directly render to MPEG-2 from any application that supports QuickTime export. The encoder is a single-pass VBR, with no non-DVD options. As a software-only encoder, it's very fast on a G4, although it's tuned more for quality than for speed compared to the encoder in iDVD.

DVD Studio Pro also includes a Dolby Digital compression app called A.Pack. It's a pretty full-featured encoder.

# For Further Reading

*DVD Authoring and Production; An Authoritative Guide to DVD-Video, DVD-ROM, and WebDVD* by Ralph LaBarge. 2001. CMP Books.

# Chapter 19

# MPEG-4

## Introduction[1]

The Moving Picture Experts Group (MPEG) has long been a leader in defining open standards for digital video. Established in 1988 as an International Organization for Standardization (ISO—and yes, the initials don't match; it's a linguistic compromise thing) workgroup, they brought us MPEG-1 in 1992 and the enormously popular MPEG-2 in 1994. They've focused on MPEG-4 since 1993. Version 1 of the specification was approved in 1998, v2 followed in 2000. Some industry pundits awarded MPEG-4 Next Big Thing status while others declared it dead on arrival. I suspect its level of success will vary wildly across the plethora of industry sectors that might use it.

In general terms, MPEG-4 is a standard way to define, encode, and play back time-based media. In practice, MPEG-4 can be used in radically different ways in radically different situations. MPEG-4 is designed to be used for all sorts of things, including delivering 2D still images, controlling animated 3D models, handling two-way video conferences, and streaming video on the Web through set-top boxes and even on wireless devices.

## MPEG-4 Architecture

MPEG-4's official designation as a standard is ISO/IEC-14496. MPEG-4 picks up where MPEG-1 and MPEG-2 left off, in that it defines methods for encoding, storing, transporting, and decoding multimedia objects on a variety of playback devices. Where MPEG-1 and -2 were all about compressing and decompressing video and audio, MPEG-4 is about defining "audio-visual or media objects" that can have very complex behavior.

---

1. Thanks to Rob Koenen, President of the MPEG-4 Industry Forum, for his invaluable help with this chapter.

There is no such thing as a generic MPEG-4 player or encoder. Because the specification is so huge and flexible, subsets of it have been defined for various industries. A file intended for delivery to a digital film projector isn't meant to work on a cell phone! To differentiate among its many possible applications, MPEG-4 breaks things down into Profiles and Levels.

## MPEG-4 Profiles and Levels

When someone asks if a piece of technology is "MPEG-4 compatible" the real question is, "Which MPEG-4 Profiles and Levels does it support?". A Profile is a subset of the MPEG-4 standard that defines a set of features needed to support a particular delivery medium, like particular codec features. A Level specifies particular constraints within a profile, such as maximum resolution or bitrate. Combined, these are described as Profile@Level, e.g. Advanced Simple@Level 3. Compliance with profiles is an all-or-nothing proposition—if you have a device that can handle the Advanced Simple Visual profile Level 3, it has to be able to play any file that meets that specification. This is in sharp contrast to traditional playback of files on personal computers, an indeterminate situation in which you don't know in advance which features will be supported by a user's hardware or operating system or installed player. With MPEG-4, you never have to worry about whether the user is willing to click "okay" to download a new codec or player, as long as you stick to your target profile.

There are a ton of profiles today, although only a small subset of these are likely to be used in mass-market products, or in any products at all. As new markets are identified, as codecs are enhanced, and as CPUs get faster, there may be more profiles added. This is a good thing. Each profile is forever fixed in time, so there's no guesswork in determining what you'll need to do in order to support older devices.

It's important to understand that MPEG-4 specifies what is a compatible bitstream. Individual encoders are free to use whatever techniques they can to make a compatible bitstream, and decoders are free to decode that bitstream in different ways.

Profiles define media objects. So what's an "object"? Anything from a standard audio or video stream to a still image, text, synthesized speech, 3D model, and so on. Levels can be huge collections of parameters that need to be supported on decode. I'm going to only specify those that are most relevant to compression, especially resolution and frame rate.

A whole book could be devoted to describing the plethora of MPEG-4 profiles. There are profiles for such cool things as controlling facial and body animation on 3D characters and for doing synthesized speech and music. There's even a profile for controlling MPEG-4 through Java. Descriptions of the video and audio profiles relevant to compression can be found later in this chapter.

## MPEG-4 File Format

The MPEG-4 file format, called MP4, is based on Apple QuickTime. Anyone who has ever done QuickTime engineering will find the file format very familiar down to the use of a "moov" atom, track structure, and so on. MPEG-4 also borrows hint tracks from QuickTime, abstracting information about how content is to be streamed from the actual content. MPEG-4 expands this to support multiple hint tracks, so a single file can define how it should be streamed in a variety of different environments.

Although the file format comes from QuickTime, MPEG-4 solves many important problems that have yet to be addressed in QuickTime. For example, QuickTime is unable to dynamically adjust to a user's connection speed, much to the frustration of many content providers. MPEG-4's scalable profiles contain a set of tools for handling this.

MPEG-4's support for interactivity and rich media is also very different than QuickTime's wired sprites. Instead, it uses the VRML derived BIFS, or BInary Format for Scenes. Companies, especially Envivio and iVast, are building authoring and playback tools for BIFS. It's important to note that the initial ISMA Profiles don't include any support for interactivity, though.

## MPEG-4 Servers

MPEG-4 has a well-developed server spec, which includes a robust transport stream for use in lossy environments, including over-the-air broadcast. Unlike QuickTime, MPEG-4 provides robust playback for streaming all media types, not just audio and video, but synthetic content as well. For example, the spec supports a text-to-speech engine, face and body animation parameter sets, streaming 3D textures, and more. MPEG-4 also describes IPMP (Intellectual Property Management and Protection) support suitable for use in secure communications and pay-per-view applications, and MPEG-4 includes systems for client- and server-based interaction such as a general event model for triggering events or routing user actions to objects in a scene and three modes for quality of service monitoring.

Many of the initial generation of MPEG-4 servers are based on Apple's open source Darwin Streaming Server, which in practice appear very similar. One non-Darwin based server is PacketVideo's pvServer. pvServer is the first to support the Simple Scalable profile, and it can automatically reduce the bandwidth as appropriate when delivering to wireless devices. Philips offers the Windows-based WebCine Server and the Solaris-based WebCine Enterprise Server. RealNetworks' Helix Universal Server also supports MPEG-4.

## MPEG-4 Players

While there is a reference player, MPEG-4 player implementation is left open to interpretation by the many different vendors. Different players can support different profiles. They can also support different levels of decompression quality within a single profile, which can result in lower-quality playback on lower-speed devices. For example, postprocessing (deblocking and deringing) may not be applied—these tools are not required in the standard.

This implementation-dependence is sure to cause some confusion among end users and developers alike in that it's quite possible that content that will play on one vendor's implementation but won't play on another's, if they support different profiles or levels. Initially, the most popular players are QuickTime 6 and Envivio's plug-in for RealOne and QuickTime for Windows. Apple's player only supports the ISMA Profiles, while Envivo's also does Advanced 2D, enabling interactivity.

## MPEG-4 Video Codecs

One wonderful thing about MPEG-4 is that it is a fully documented modern codec. Most other codecs keep their internals extremely proprietary, making it impossible to teach compression principles using specific examples. I expect MPEG-4 to play a primary role in codec education of the future.

I'm going to resist the temptation to describe the MPEG-4 video codec in loving detail, as that would fill an entire book in itself and would put most of you to sleep. But I'll give you a quick, high-level overview. Given its MPEG heritage, MPEG-4 shares many features with MPEG-1 and -2, such as DCT (Discrete Cosine Transformation) encoding and I (keyframe), P (delta), and B (bi-directional) frames, all tied together inside of GOPs (Group of Pictures). It also has a lot of enhancements, especially for low data rate use. These include better motion estimation, a deblocking filter, and error resilience tools. Its quality at Web data rates (20–1000Kbps) are enormously better than MPEG-1, and ballpark competitive with other Web video solutions.

Unlike most Web codecs, MPEG-4 also has full support for interlaced content, support in the specification for resolutions up to 4096×4096, and an astounding data rate range from 5Kbps to 10Mbps in v1. MPEG-4's theoretical sweet spot goes all the way from extremely low bandwidth mobile devices up to HDTV. Of course, most MPEG-4 profiles and levels don't support playback of the full range the video spec alone defines.

The MPEG-4 video codec natively supports alpha channels in the Core and Main profiles, so video can be composited in real time over a background, resulting in a more flexible and higher quality image. This can be used for segmentation or "shape coding," in which it is able to internally separate out background and foreground elements on a scene-by-scene basis. Ideally, segmentation information can be provided via an alpha channel in the source, but it can also be dynamically generated if the encoding software supports this feature.

**Note**

It's a common misconception that MPEG-4 includes a magic encoder that somehow extracts objects from the background. It doesn't, but it does provide the ability to deliver a stream that's made up of multiple layers of video; vendors who encode still have to figure out how to make those layers.

This segmentation improves quality quite a lot at a given bitrate, but is obviously somewhat more complex on playback, and can be much more complex to encode. It's unlikely that many MPEG-4 video encoders will support segmentation, but the now-defunct ObjectVideo demonstrated a promising low-bitrate MPEG-4–based tool that utilized MPEG-4's support for object segmentation. Other vendors are working on similar technology.

To understand the value of segmentation, imagine a video of a presenter lecturing in the front of a room. With a conventional codec, every time the presenter moves away from the background, and then back again, the background must be retransmitted. By segmenting the background away from the foreground objects, the codec can remember what was there before, and doesn't have to resend that data every time the background is revealed again.

While the MPEG-4 video codec's decoder standard for any Profile@Level is fixed, there is still a lot of room for innovation in authoring. The pace of innovation in MPEG-4 compression won't be as fast as for codecs from vendors who can change their decoder specs at will, but we can expect to see incremental improvement over the years, similar to what's happened with MPEG-1 and MPEG-2. The first VideoCDs and DVDs often offered poor quality compared to that of modern discs created with the same spec, but with improved compression tools. The continuing speed gains of Moore's Law help enormously here, as there is continually more CPU power to throw at motion searching, preprocessing, and other techniques to improve quality.

Many different vendors are going to be providing MPEG-4 codecs. The following figures show interface screen shots from Apple's QuickTime v6 MPEG-4 exporter (Figure 19.1) and PacketVideo's PVAuthor tools (Figure 19.2). The QuickTime exporter supports Simple and PVAuthor Simple Scalable.

## MPEG-4 Audio Codecs

MPEG-4 has a rich set of audio features. Like most platforms, it provides separate codecs for low bitrate speech and general-purpose audio. Given how MP3 (a.k.a. MPEG-1, Layer III audio) was born of MPEG, it seems quite possible that MPEG-4 audio may wind up becoming very important as a consumer music file format.

Most, but not all, MPEG-4 audio codecs come in error resilient (ER) versions. These are for lossy environments such as RTSP streaming. Error resiliency requires slightly higher bitrates than the

## Rectangular Visual Profiles

The visual profiles garner the lion's share of attention, although audio and systems are also extremely important. Also, video tends to take up more CPU power than audio, so visual profiles tend to be more creator-constrained for smaller devices. The most basic kind of video profile is rectangular, as opposed to video that can fit into an arbitrary shape, supported in the more complex MPEG-4 codecs. Some of the more advanced profiles can use rectangular video for mapping onto other kinds of surfaces and objects.

There is a new codec in development, code named H.26L, that is a continuation of H.261 and H.263. It looks to offer substantial improvements in compression efficiency. H.26L's development is pursued by MPEG and ITU-T together in the "Joint Video Team", and the codec will be included in MPEG-4 as MPEG-4 part 10 (Advanced Video Coding).

Resolution is an important aspect of visual profiles. Resolution is defined with such terms as QCIF, CIF, 2CIF, and 4CIF, derived from the MPEG "Common Intermediate (meaning between PAL and NTSC) Format". Each has a canonical resolution. However, the actual limitation in the spec is the maximum number of 16×16 macroblocks allowed. So, you can redistribute the blocks into any shape you like, and any profile-compliant player must be able to decode them, although they may display strangely if the output resolution on either axis is greater than what the device supports. So, for QCIF, instead of 176×144, you could deliver 256×969 (for streaming *Ben-Hur*, perhaps). The one exception to this redistribution rule is Simple Level 0, which is limited to 176×144 actual pixels.

| Name | Canonical resolutions | 16×16 blocks |
|---|---|---|
| QCIF | 176×144 | 99 |
| CIF | 352×288 | 396 |
| 2CIF | 352×576 | 792 |
| 4CIF | 720×576 | 1620 |

## MPEG-4 Audio Codecs    305

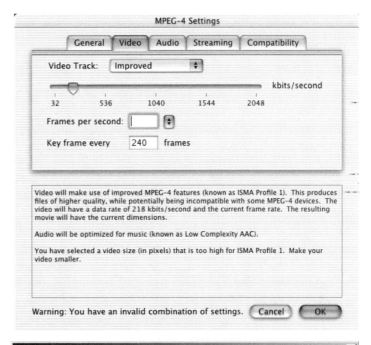

**19.1** The MPEG-4 exporter from Apple's QuickTime v6. It is the first widely available MPEG-4 encoder. I don't like the sliders for data rate control, though.

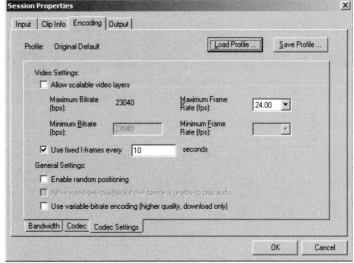

**19.2** PacketVideo's PVAuthor v3. PVAuthor is the first tool for the Simple Scalable profile. It targets mobile devices, and doesn't support resolutions above 384×288.

non-ER versions, and so shouldn't be used for file-based applications such as progressive download.

Audio scalability in MPEG-4, unlike that in most other formats, is based around enhancement layers. Instead of having to store multiple versions of the same audio content at different bitrates, the audio is split into a base layer and multiple enhancement layers. A user must receive the base layer to hear anything. After that, a user can receive from none to all of the enhancement layers.

### Simple

The Simple Profile is the only one most people have seen so far. It's the easiest to implement and decode, even on mobile devices. It may also be used for recording video with low-power devices. Simple is the video profile for ISMA Profile 0. Level 0 also has a limit of 15fps.

| Level | Max size | Max objects | Max data rate |
|---|---|---|---|
| 0 | QCIF (fixed to 176x144) | 1 | 64 |
| 1 | Equivalent to QCIF | 4 | 64 |
| 2 | Equivalent to CIF | 4 | 128 |
| 3 | Equivalent to CIF | 4 | 384 |

### Advanced Simple

Advanced Simple is a superset of Simple, so it can play all Simple content. This profile adds a number of enhancements to support better visual quality. Interlaced content is supported in levels 4–5. Advanced Simple Level 3b is used in ISMA Profile 1, and support for higher levels are planned for later ISMA Profiles.

| Level | Max size | Max objects | Max data rate (Kbps) | Interlace |
|---|---|---|---|---|
| 0 | QCIF | 1 | 128 | No |
| 1 | QCIF | 4 | 128 | No |
| 2 | CIF | 4 | 384 | No |
| 3 | CIF | 4 | 768 | No |
| 3b | 2CIF | 4 | 3000 | No |
| 4 | 2CIF | 4 | 3000 | Yes |
| 5 | 4CIF | 4 | 8000 | Yes |

### Simple Scalable

Simple Scalable adds scalability to Simple through a single enhancement layer. Scalability enables the server to dynamically reduce bitrate through lowered **image** quality, frame rate, and/or resolution.

| Level | Max size | Max objects | Max base layer data rate |
|---|---|---|---|
| 1 | CIF | 4 | 128 |
| 2 | CIF | 4 | 256 |

These enhancements can include better overall audio quality, additional channels, and increased sample rate.

The first generation of MPEG-4 encoding tools just support AAC-LC and maybe the CELP (Code Excited Linear Prediction) for voice coding.

Following are the audio object types currently in MPEG-4. Note that many of the object types don't exist in any of the profiles, and so might never be used in any real products.

## Advanced Audio Coding (AAC)

AAC was originally defined as a part of MPEG-2, although AAC wasn't widely used in MPEG-2 applications. AAC has a host of variations.

*AAC Main*    AAC Main adds perceptual noise shaping (PNS) to the MPEG-2 AAC, improving quality at lower bitrates. AAC Main can handle up to five channels plus one subwoofer/LFE channel in a single audio object. This configuration is commonly referred to as 5.1 (pronounced *five-dot-one*). Note that an MPEG-4 implementation may use more than one AAC 'object.'

*AAC Low Complexity (AAC LC)*    AAC LC is a low complexity implementation of AAC. This means it requires much less CPU power to encode and decode, but sacrifices some compression efficiency. In real-world listening tests with optimized encoders, AAC LC quality at 128Kbps has been almost as good as AAC Main. AAC-LC is used in the ISMA profiles.

*Error Resilient AAC Low Complexity (ER ACC LC)*    The error resilient version of AAC LC.

*AAC Scalable Sampling Rate*    The same as the MPEG-2 Scalable Sample Rate (which isn't used in any actual MPEG-2 implementations I know of), but with PNS (perpetual noise shaping).

*AAC Long Term Predictor (AAC LTP)*    This replaces the MPEG-2 AAC predictor (the part of the audio codec that predicts what the audio will sound like based on the previous audio), with one that provides higher quality.

*Error Resilient AAC Long Term Predictor (ER AAC LTP)*    The error resilient version of AAC LTP.

*Error Resilient AAC Low Delay (ER AAC LD)*    This AAC codec is targeted for extremely low latency, so it can be used in two-way, real-time communication. It uses LTP and PNS. There is no non-ER version. The codec was introduced in MPEG-4 version 2, where all audio codecs were resilient.

### Fine Granularity Scalable (FGS)

FGS adds scalability to existing Simple and Advanced Simple profiles. Up to 11 scalable layers are supported.

| Level | Max size | Max objects | Max base layer data rate |
|---|---|---|---|
| 0 | QCIF | 1 | 128 |
| 1 | QCIF | 4 | 128 |
| 2 | CIF | 4 | 384 |
| 3 | CIF | 4 | 768 |
| 4 | 2CIF | 4 | 3000 |
| 5 | 4CIF | 4 | 8000 |

### Advanced Real Time Simple (ARTS)

ARTS is for video conferencing and other applications requiring low-latency real-time performance. It's a superset of Simple. It adjusts encoding in real time based on feedback from the decoder, so it's only useful in real-time encoding situations.

| Level | Max size | Max objects | Max data rate |
|---|---|---|---|
| 1 | QCIF | 4 | 64 |
| 2 | CIF | 4 | 128 |
| 3 | CIF | 4 | 384 |
| 4 | CIF | 16 | 2000 |

## Arbitrarily Shaped Video

The Arbitrarily Shaped Video profiles are for complex playback of video in more than simple rectangles, which includes interactive video, sprites, and other rich media.

### Core

Core is a superset of Simple, adding B-frames and 1-bit shape coding. This shape coding precisely defines the shape of the edge of the object, unlike traditional alpha channel methods. The transparency

## TwinVQ

TwinVQ offers better quality at lower bitrates than AAC. It's limited to mono or stereo. ER TwinVQ is the error resilient version.

## Error Resilient Bit Sliced Arithmetic Coding (ER BSAC)

ER BSAC's big advantage is that it supports very fine-grain scalability capable of varying bandwidth by as little as 1Kbps per channel. This would be useful for real-time audio streaming over networks with variable bandwidth.

## Error Resilient for Harmonic and Individual Lines plus Noise (ER HILN)

ER HILN, a real mouthful of a general purpose audio codec, is for bitrates of 2k to 16Kbps, lower than the other general audio codecs in MPEG-4. It also has bitrate scalability, and supports speed and pitch changes during playback.

## Harmonic Vector eXcitation Coding (HVXC)

HVXC is targeted for very low bitrate speech at 8kHz mono. In CBR mode, it requires 2k to 4Kbps. In VBR, it can drop below 2Kbps. Its quality isn't great, but no codec before it has been able to delivery intelligible audio at these bitrates. There is an error resilient version of HVXC.

## Code-Excited Linear Prediction (CELP)

This is another implementation of the CELP low bitrate speech technology, like the ACELP.net codec in Windows Media. Code Excited Linear Predictive (CELP) is used at 4k to 24Kbps and 8kHz or 16kHz mono. CELP provides better quality than HVXC at higher bitrates, but of course, it has to have higher bitrates. The error resilient version is called ER CELP.

# ISMA Standards

The International Streaming Media Alliance (ISMA) is a consortium of companies pushing MPEG-4 as a streaming solution. The core members are Apple, Cisco, IBM, Kasenna, Philips, and Sun. Where the M4IF (MPEG-4 Industry Forum) is an organization focused on more high-level issues, the ISMA has focused on the nitty gritty of making MPEG-4 a real-world Internet video solution. M4IF is horizontal, and seeks adoption of MPEG-4 in all markets. It does not address issues that are not covered by MPEG-4. ISMA is vertical, and seeks to specify an end-to-end solution for real-world Internet applications.

The ISMA has released its first standard, which includes complete documentation on how to stream MPEG-4 over RTSP, and two "profiles." These are a different kind of profile than itself is the same over the entire image. The levels of the profile are set up with enough extra CPU power to handle multiple overlapping objects and video on the screen at the same time.

| Level | Max size | Max objects | Max data rate |
|---|---|---|---|
| 1 | QCIF | 4 | 384 |
| 2 | CIF | 16 | 2000 |

### Advanced Core

Adds Advanced Scalable Texture (an improved still image format) objects to Core.

| Level | Max size | Max objects | Max data rate |
|---|---|---|---|
| 1 | QCIF | 4 | 384 |
| 2 | CIF | 16 | 2000 |

### Core Scalable

Core Scalable adds scalability via frame rate and resolution reduction. Both the video and other objects can have lower resolution. This is for MBR bandwidth reduction.

| Level | Max size | Max objects | Max data rate |
|---|---|---|---|
| 1 | QCIF | 4 | 768 |
| 2 | CIF | 8 | 1500 |
| 3 | 4CIF | 16 | 4000 |

### Main

This profile is targeted at interactive broadcasting, and so supports interlaced video. When broadcasters talk about "the interactive future with MPEG-4," the Main profile is what they're referring to. At its maximum level, this profile could replace the current DTV system for HDTV delivery. The bitrates of Simple, Core, and Main are the same at the same level, thus there isn't a Level 1 for Main.

| Level | Max size | Max objects | Max data rate |
|---|---|---|---|
| 2 | CIF | 16 | 768 |
| 3 | 2CIF | 32 | 1500 |
| 4 | 1920×1088 | 32 | 4000 |

MPEG-4 profiles, in that they specify specific MPEG-4 profiles for use—somewhat confusing. Profile 0 targets embedded and mobile devices like cell phones. Profile 1 targets more powerful devices and desktop computers, and will probably become the initial default for MPEG-4 files viewed on computers. Both profiles are limited to one video and one audio object.

| Profile | Visual Profile@Level | Audio Profile@Level |
|---|---|---|
| ISMA 0 | Simple @ Level 1<br>Typical resolution: 176×144<br>64Kbps max | High Quality Audio @ Level 2<br>Up 2 2 channels<br>Up to 48kHz<br>CELP and AAC LC |
| ISMA 1 | Advanced Simple @ Level 4 w/o interlace (may become 3b)<br>Typical resolution: 352×288<br>Total data rate max 1500 | |

## Wireless Standards

The MPEG-4 wireless standards come from 3GPP (Third Generation Partnership Project). These formats target mobile phones and other low-power devices with small screens, similar but not the same as ISMA 0. Note that 3GPP2 isn't a more advanced version of 3GPP, it's just the format that targets CDMA2000 (that's Code Division Multiple Access 2000) networks instead of UMTS networks.

| Standard | Network Type | Visual Profile@Level | Audio Profile@Level |
|---|---|---|---|
| 3GPP | UMTS | Simple @ Level 0<br>Max resolution: 176×144<br>64Kbps max | MPEG-4 AAC LC<br>GSM-ASR (speech) |
| 3GPP2 | CDMA2000 | Simple @ Level 0<br>Max resolution: 176×144<br>64Kbps max | MPEG-4 AAC LC<br>EVRC (speech) |

## DRM in MPEG-4

Digital Rights Management is a critical feature of any media architecture. The ISMA has been hard at work at developing a robust, secure DRM infrastructure for MPEG-4. However, it hasn't been released, and so DRM isn't included in the initial profiles.

### Advanced Coding Efficiency (ACE)

ACE adds a number of codec improvements to Main, but drops sprites. ACE typically provides similar quality to that of Main at a substantially lower data rate. I wish I could watch TV at ACE Level 4.

| Level | Max size  | Max objects | Max data rate |
|-------|-----------|-------------|---------------|
| 1     | CIF       | 16          | 384           |
| 2     | CIF       | 32          | 2,000         |
| 3     | 4CIF      | 32          | 15,000        |
| 4     | 1920×1088 | 32          | 38,400        |

### N-bit

N-bit is aimed at surveillance and medical applications, and supports from 4 to 12 bits per luma and/or chroma channel. N-bit is not likely to be used in consumer applications. The only Level is L2: 2Mbps with 16 objects.

| Level | Max size | Max objects | Max data rate |
|-------|----------|-------------|---------------|
| 2     | CIF      | 16          | 2,000         |

### Simple Studio

Simple Studio is for video editing and similar high-quality, low-compression applications. Its use is analogous to that of modern Motion-JPEG or I-frame only MPEG-2 in editing systems, although Simple Studio offers more efficient compression. It's for authoring, not delivery. It has multiple alpha channels, and thus supports arbitrary-shape coding.

### Core Studio

Core Studio is a superset of Simple Studio that adds support for P-frames, improving codec efficiency enormously. It is analogous to MPEG-2 in editing systems that don't have to be I-frame only. Core Studio is an authoring format, not a delivery format.

### Still Visual Profiles

The Still Visual profiles don't use video, just stills. Other profiles often support Still Visual profile objects. They both use wavelet compression, which means they can draw a lower quality version of the image for a partially transmitted image.

# Applications

The importance of MPEG-4 can best be seen in the breadth of applications that have been announced and proposed for it. The cable industry has already announced their intent to replace their current MPEG-2 implementations with MPEG-4 in the future. This is an obvious decision, as MPEG-4 will enable them to double or triple the number of channels available on their existing bandwidth, and allow them to take advantage of MPEG-4's interactive capabilities for delivering ITV and video on demand (VOD) to set-top boxes.

Another cool feature of MPEG-4 is MPEG-J, a Java library for controlling MPEG-4. Combining Java and MPEG-4 makes it possible to create applications far beyond Java in media complexity or MPEG-4 in application complexity, while making applets that run in the Java sandbox, removing security and stability concerns. And even more exciting, such Java applets can be embedded in the MPEG-4 stream. This would enable a one-way communication system, such as cable, that could deliver full Java applets inside set-top boxes. Applications might include much-improved navigation systems, interactive advertisements, home shopping, dynamic sports score information, and innumerable other possibilities. Cool, huh?

On the other end of the video compression industry, future versions of digital film systems designed to replace the projectors in movie theaters may be built around MPEG-4. These involve extremely high data rates and correspondingly brilliant quality.

Other proposals include replacing AM radio broadcasting with higher quality, data-enabled MPEG-4 audio streams, streaming video to mobile devices, video conferencing, surveillance, and in-studio video transport. It's hard to predict which of any of these will come to market, but the advantages of MPEG-4 make it likely it will be the basis of many future media technologies. The satellite radio broadcaster XM Radio uses a version of AAC.

## MPEG-4 still images

MPEG-4 includes a wavelet-based encoder for still images that can be used for backgrounds, texture maps for 2D and 3D objects, and so on. Compared to JPEG, wavelet compression offers typical Web quality at about 25 percent the data rate of JPEG. Wavelets allow the server to dynamically reduce the file size (and quality) of the bitmap for lower bandwidths, reducing the need to author multiple versions of a presentation for different connection speeds.

This facet of MPEG-4 has value all its own. E-Vue's line of image tools and Web plug-ins are based on MPEG-4's wavelet compression. While the wavelet based JPEG 2000 is only now rolling out, and other Web-based wavelet formats have been released before, none got past the chicken-and-egg problem of producing sufficient content to drive downloads. If support for MPEG-4 becomes ubiquitous, MPEG-4 images might become popular as a replacement for GIF and JPEG in Web pages and elsewhere.

### Scalable Texture

These can be used for digital photography and come in three levels, 400K pixels to 6M pixels.

### Advanced Scalable Texture

Advanced Scalable Texture improves the error resilience and other features of Scalable Texture. Three levels. New object type.

## Audio Codecs and Profiles

Unlike Video, where the all profiles share one basic codec with some enhancements in each, the MPEG-4 Audio has a number of very different codecs, several of which are usually supported in a given profile. The codecs themselves are described here.

### Scalable Audio

Scalable Audio is the basic audio profile for most uses. It includes the AAC LC, AAC LTP, AAC Scalable, and HVXC. It has four Levels.

### High Quality Audio

High Quality Audio has CELP, AAC LC, AAC and their error resilient modes. It's used in ISMA Profiles 0 and 1.

| Level | Max. channels/object | Max. sampling rate [kHz] |
|---|---|---|
| 1 | 2 | 22.05 |
| 2 | 2 | 48 |
| 3 | 5.1 | 48 |
| 4 | 5.1 | 48 |
| 5 | 2 | 22.05 |
| 6 | 2 | 48 |
| 7 | 5.1 | 48 |
| 8 | 5.1 | 48 |

## Rich media

MPEG-4 gains enormous power in rich media from, believe it or not, its VRML heritage. Instead of defining the content as just a stream of audio samples and video frames, everything is constructed out of media objects, which can be a standard audio or video stream, or a still image, text, synthesized speech, 3D model, or so on. Rather than simply display video in a rectangle, these media objects allow any type of supported media to be mapped to objects in a scene. So, for example, audio can be mapped to an object and thus be given attributes that would position that audio in 3D space. These media objects can be interactive as well. More broadly, MPEG-4 is capable of mixing the best of Shockwave, Flash, VRML, and traditional digital video into a single file format, server, and player. This enables the creation of graphically rich, complex presentations that can be deployed over slow connections.

One specific type of rich media in MPEG-4 is the protocol for controlling facial animation, where a 3D model of a face can be animated in real time. When combined with either audio samples or synthesized speech via the text-to-speech converter, you'll be able to achieve precise lip sync.

MPEG-4 is so deep that building a set of authoring tools that support more than a fraction of its potential as a rich media authoring and delivery environment is going to be a major challenge. I hope broad acceptance will create a robust market for these sorts of tools.

## Where and How MPEG-4 Matters

MPEG-4 is going to be hugely important, except in the place that most folks seem to think it will matter most. MPEG-4's big advantage, and big limitation, is that it is an international standard.

MPEG-4 will matter because it, like all standards, provides stability and interoperability. As long as a device can meet the tests for a given profile@level, any content targeting that profile@level should work on that device. It also provides a basis for compatibility between products from different vendors. One company's cell phone can work with another's video encoder without having to do specific compatibility development. As a counterexample, take the mobile audio players that supported the RealAudio G2 format. Because the RealAudio format was updated with the RA8 codecs, the older mobile devices don't work with the latest and greatest files created for desktop playback unless they received a firmware update.

However, the slowness of the specification process is also MPEG-4's biggest weakness in the main environment for Web video today: desktop computers running mainstream operating systems. QuickTime, Windows Media, and Real all provide video codecs better than that of MPEG-4. This isn't surprising—they've had a couple years to digest the drafts, apply the good ideas, and add proprietary enhancements that improve quality, but break compatibility. And that's a good thing—you can update software a lot easier than silicon. And because they all provide dynamic download of new codecs, we can expect the quality gap between MPEG-4 implementations and

### Natural Audio

Natural Audio includes all the audio codecs of MPEG-4, including error resiliency modes for those that have them.

| Level | Max. sampling rate [kHz] |
|---|---|
| 1 | 48 |
| 2 | 96 |
| 3 | 48 |
| 4 | 96 |

### Mobile Audio Internetworking (MAUI)

| Level | Max. channels | Max. sampling rate [kHz] |
|---|---|---|
| 1 | 1 | 24 |
| 2 | 2 | 48 |
| 3 | 5.1 | 48 |
| 4 | 1 | 24 |
| 5 | 2 | 48 |
| 6 | 5.1 | 48 |

the Big Three to continue. However, compression efficiency isn't always going to be the critical factor for video streaming—in many cases, interoperability trumps it. It will be useful for desktop playback to be able to view MPEG-4 content, especially for content providers who only want to create content for a single format. But the big story with MPEG-4 will not be playback on personal computers.

A good analogy is MP3. Although other audio formats offer much better compression efficiency, MP3 is "good enough" and ubiquitous, and shows no sign of being replaced by other formats.

It remains to be seen how MPEG-4 support is going to evolve. The Internet Streaming Media Alliance, which includes Apple, Sun, Phillips, and Cisco, among others is focused on MPEG-4. Notably absent from that group are RealNetworks (although they do include MPEG-4 playback in the current player) and Microsoft. QuickTime is poised to take a huge lead as Apple pursues MPEG-4 compatibility. QuickTime has been extended to both play back ISMA MPEG-4 content and provide a development environment for it, leveraging the broad number of existing QuickTime authoring tools. While Windows Media and RealVideo can easily provide simple playback of audio and video streams, they would have to replicate much of the full functionality of QuickTime to handle authoring of the more complex MPEG-4 profiles—a massive engineering project.

One big issue that had delayed the release of MPEG-4 solutions was licensing. The MPEG Licensing Authority (MPEG-LA) controls the pool of patents behind MPEG-4. Their initial proposed licensing terms were roundly criticized for being too expensive and too difficult to implement. However, the final terms are much clearer and won't require most content creators to pay anything at all. There are a variety of fees to pay in different industries.

For those putting content on the Web with a subscription, there is a $0.02 per hour fee for content viewed or a flat $0.25 fee per subscriber. However, those fees only apply after the first 50,000 subscribers—which means that most subscription sites won't have to pay. The rates for advertising sites haven't been announced yet, but are assumed to be based on some sort of revenue share. Non-commercial and not-for-profit sites don't have to pay anything at all.

# MPEG-4 Authoring Tools

While there are dozens of codec vendors working on licensable MPEG-4 technology, there are few shipping products as of this writing. More will come now that the licensing issues are worked out.

## Philips WebCine

The Philips WebCine was the first commercial MPEG-4 authoring tool (see Figure 19.3). At a cool $27,000, it provides (DirectShow) file based and live compression. It's an integrated dual Pentium computer, with a Matrox Digisuite LE hardware card, which includes a full range of

19.3 Philips WebCine is a high-end encoding solution for real-time broadcasting and on-demand compression of MPEG-4. It was the first commerically-available MPEG-4 encoder.

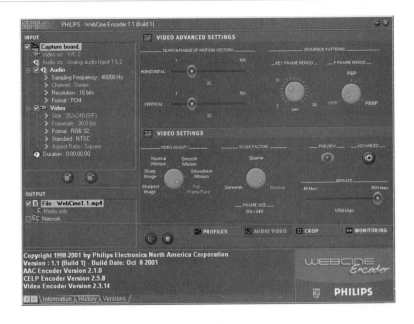

professional inputs, including SDI and component analog. The current v1.1 only supports Simple and B-frames from Advanced Simple. WebCine is limited to the 352×288 CIF resolution.

## PacketVideo PVAuthor v3

PacketVideo's PVAuthor ($99) is a DirectShow-based MPEG-4 compression tool. PacketVideo is focused on wireless MPEG-4 applications, so PVAuthor is the first commercial tool that supports the Simple Scalable profile. PVAuthor is also limited to 352×288 resolution.

## QuickTime 6

Apple's QuickTime 6 includes an MPEG-4 exporter with Simple Visual and AAC-LC support. While it is only 1-pass, it is very fast and provides good baseline support. Because it is implemented as a QuickTime component, this means other apps that use QuickTime can quickly adopt MPEG-4 export.

## Things Called "MPEG-4" That Aren't MPEG-4

### MS MPEG-4v1-v3

Microsoft thought that their Advanced Streaming Format (ASF) was going to be the file format for MPEG-4, and so built their implementation of the draft spec codec inside Windows Media Player, in an ASF file. Had ASF been the file format for MPEG-4, this would have been an interesting beta pre-release. However, because the standard actually started with QuickTime as the basis for their file format, MS ASF files aren't MPEG-4 files. After MS MPEG-4 v3, the next version of the codec was renamed Windows Media Video v7, to avoid confusion with the real MPEG-4, and to bring the numbering in sync with the rest of the Windows Media architecture.

Microsoft's Windows Media v7 architecture has what is labeled as an ISO-standard MPEG-4 encoder and decoder, but it still uses the ASF file format. However, this may enable the Windows Media server and player to create and play back ISO MPEG-4 streams, once Windows Media includes support for the MPEG-4 audio codecs and file or streaming format.

### DivX

The initial versions of DivX were a hack to enable the use of the MS MPEG-4 codecs in AVI files. Hence they were no more MPEG-4 compatible than those were. Since v4.0, DivX has been based on their own code for using MPEG-4 video. However, as the content is in an AVI file, there's no way to run compliance tests on it. Thus, ".divx" files won't interoperate with any other MPEG-4 solutions.

However, the Divx v5 codec and encoding tools now include support for exporting to video-only .mp4 files—including support for Advanced Simple.

### Flask MPEG

The Flask MPEG encoder isn't even really an encoder. It's a preprocessing/DVD ripping tool that sits on top of either Windows Media/DirectShow codecs or Adobe Premiere plug-ins. It's often erroneously described as an MPEG-4 encoder, as it is generally used with the old MS MPEG-4 v3 codec. As an MPEG to AVI converter, it is a pretty slick tool, though.

# Chapter 20

# AVI

Microsoft's Audio Video Interleave (AVI) was one of the early formats that allowed audio and video to be combined within a single file that could play off a CD-ROM. AVI's architecture is simple and unambitious when compared with QuickTime's architecture. That simplicity is AVI's strength and its weakness. It's a weakness in that AVI files hold only very simple kinds of data. It's a strength because that simplicity allowed a large number of vendors to provide complete support for authoring and playback of AVI files with and without using Microsoft's code.

While complex QuickTime files will only play through the Apple QuickTime API, AVIs will play on an incredible number of platforms, including applications like XAnim running under Linux. AVI also forms the basis for authoring systems from such companies as Matrox and Canopus. While the AVI format doesn't offer reference movies and other handy features found in Quick-Time, AVI works adequately for end-to-end authoring systems with integrated hardware and software.

## The AVI Format

AVI is a very simple format. It stores a single track of audio and a single track of video, each of which use the same codec throughout. The AVI video codec itself has to have a fixed frame rate throughout.

AVI is a derivative of the IBM/Microsoft Resource Interchange File Format (RIFF), which came from Electronic Arts' Deluxe Paint program for the Amiga. The original AVI format was limited to 1GB in total file size, and used Microsoft's Video for Windows (VfW) API, included with Windows 3.0 and beyond. To address the file size limit and other issues, the Open Digital Media Consortium (OpenDML) created extensions to the AVI format that most (but not all) authoring tools support, including current versions of Microsoft's DirectShow API, included with Windows 95 and beyond. The OpenDML extensions increased AVI's maximum file size to terabytes (though they still require a file system that handles files that large). The extensions also added support for fields.

Microsoft has been backing away from AVI for many years, trying to push Windows Media Video (WMV) as the delivery format and Advanced Authoring Format (AAF) as the video editing format. WMV is succeeding for Web use, although it's lagging behind for CD-ROM use. AAF may be DOA—no products have shipped that use it.

## AVI Interleave

One tweaky feature of AVI is its ability to specify the audio/video interleave. This was a critically important issue back in the 2x CD-ROM era. In essence, interleave defines how large audio samples should be relative to video samples. The smaller the chunks, the more accurate the audio sync can be, but the bigger the file. This is because blocks are in 2K chunks, rounding up to the next 2K mark, with an average of 1K wasted per chunk. Typically, to achieve sync as good as possible, you would use the 1 frame sync option, which stores each video frame's audio next to it in the file. However, this would raise the data rate by an average of 1Kbps (there being 8 Kilobits per K) per fps of the file. This was the reason "Why are my QuickTime files so much smaller than my AVIs?" became the most common compression question of that era. Today, it's fine to stick with the default of half-second interleave when targeting modern computers.

## Delivering Files in AVI

### AVI for CD-ROM

For most of their respective lives, AVI has been nearly as popular as QuickTime for CD-ROM. They offer similar codecs and broadly similar features.

Because AVI was designed for CD-ROM use, there really aren't any specific tricks you'll need for improving compression quality. Just make sure the user has the correct codec installed on their machine. The standard set of codecs in AVI is ancient and not very good, as it ignores modern codecs like VP3 and DivX. For best results, you'll need to have a codec installer.

### AVI for the Web

AVI is not a format well suited to Web playback. AVI files themselves can be played with a number of media players, such as RealVideo, Windows Media, and QuickTime (if QuickTime has the right codecs installed). However, there is no "native" AVI player. Thus, most AVI files on the Internet are intended to be downloaded and played off the local computer.

# AVI Architectures

Because AVI is a relatively simple, open format, a number of different libraries have been created that applications can use to play the files. Also, Microsoft has released multiple generations of the libraries, so applications might support significantly different feature sets depending on which libraries were targeted.

*Video for Windows*   Video for Windows (VfW) was the original API used for AVI files. It was introduced back in the days of Windows 3.x. A 32-bit version of VfW was introduced for Windows 95, although Microsoft swiftly made ActiveMovie the API of choice for working with AVI.

*ActiveMovie*   ActiveMovie was a 32-bit only reengineered API for video playback, replacing Video for Windows.

*DirectShow*   DirectShow was more ActiveMovie v2.0 than a new architecture; applications written for ActiveMovie can work with the DirectShow libraries. DirectShow is now distributed as part of the DirectX multimedia and gaming API software package.

*QuickTime*   QuickTime also sports the ability to author and play back AVI files that use the QuickTime codecs. This includes the usual suspects like Cinepak and Indeo. QuickTime can also read DV .avi files, which is great for interoperability.

# AVI Authoring Tools

*Cleaner 5*   Cleaner came pretty late to the AVI party. Early versions relied on QuickTime's AVI export ability, and were limited in their codec options. This is still true for Cleaner on Mac, but the Windows version can now use the full suite of installed AVI codecs. However, Cleaner is still inexplicably limited to a maximum 4GB file size, which limits Cleaner's ability to make intermediate files for compression with other AVI-only apps. Cleaner's built-in MP3 codec doesn't work for AVI.

*Adobe Premiere*   Before Movie Cleaner Pro, Adobe Premiere was the main tool used for CD-ROM compression. Alas, Premiere's AVI authoring tools haven't been significantly updated since then. Overall, the UI closely matches that of QuickTime.

A couple of handy functions are offered in Premiere: Frames Only at Markers makes a slide show by creating slides of frames at which you place markers. Just pretend it's 1982 and you're watching a filmstrip, but without those annoying beeps.

Add Keyframes At Markers and At Edits let you do just that. If you're having trouble getting natural keyframes during transitions, At Edits will automatically make the first frame after a cut a keyframe. At Markers lets you specify exactly where you want keyframes to appear.

***Adobe After Effects*** Adobe After Effects was frequently used for video compression back in ancient times. The application itself supports excellent preprocessing. However, the Output module in After Effects is intended for interoperation with other video apps, not to make final deliverable files.

Encoding to AVI from within After Effects uses the same dialog as for QuickTime compression. It supports all installed codecs for video, but only uncompressed audio.

***VirtualDub*** VirtualDub is a neat, free AVI processing and encoding utility from programmer Avery Lee. It's Free Software, under the Free Software Foundation's Gnu Public License (GPL), so programmers can download the source code and modify it, as long as they also distribute their modified source code along with any compiled versions. And like most free software, it seems to be perennially locked in a pre-1.0 version (it's at v0.6 beta as I write this). It's filled with a lot of useful and useless features that weren't built with usability in mind.

That said, if you need to do something that VirtualDub can do, you can almost certainly do it faster there any anywhere else. One very welcome feature is that it can do Y'CbCr-native processing if you're doing straight recompression in "Fast Recompression" mode, which is extremely high quality and, well, fast. VirtualDub also comes with a long list of interesting filters of highly varying quality and utility, as well as a striking lack of documentation. VirtualDub's filter model supports plug-ins, and there's a healthy culture of folks making new useful and/or neat filters. Some of the noise reduction options available beat what's in Cleaner 5.1.

Another very neat, unique feature is its frameserver support. This hands off a virtual AVI file, generated in real time, to another app. The virtual file that other applications see as a normal AVI file, even though it is being generated in RAM on the fly, saves the drive space an intermediate file might otherwise take up. VirtualDub also supports integrated AVI capture, although it doesn't have device control.

For doing unusual or one-off AVI based projects, VirtualDub is as useful a tool in my toolbox as After Effects. VirtualDub is very, very useful for is making movies from unencrypted .VOB files from DVD. One awesome VirtualDub plug-in is the flaXen VHS Filter, which is one of the best noise reduction filters I've ever used, especially with the chroma noise typical of VHS.

***AviSynth*** AviSynth, from Ben Rudiak-Gould (who also wrote the invaluable Huffyuv codec) is another frameserver app. It's more flexible than the one integrated into VirtualDub, and includes filters and a scripting language.

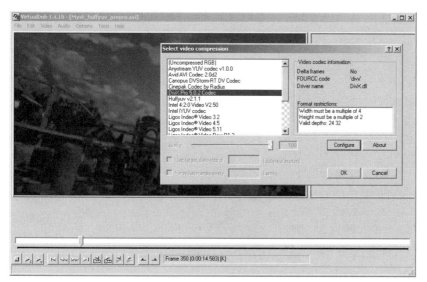

**20.1** VirtualDub's interface was obviously done by an engineer, but it has lots of useful features, like this dialog showing the restrictions of the selected codec.

As I write this, v1.0 is in beta. It doesn't support sound yet, but otherwise looks extremely promising for those who need something like this.

*Macromedia Director*   Macromedia Director has been the CD-ROM authoring tool of choice since before CD-ROMs were standard equipment on computers. Director has had good support for AVI on Windows for years. The Mac version of Director uses QuickTime's AVI playback capabilities, which are generally quite good. Out of the box, Director supports a lot more control for QuickTime movies. If you need better support for .avi playback under Windows, Tabuleiro's DirectMedia Xtra adds a lot of useful features, including dynamic volume and speed control.

*Tip*
Director likes the resolution of all AVI files to be divisible by eight for both height and width.

## AVI Delivery Codecs

There is an enormous number of AVI codecs available, many of them ancient or useless. Here are the usable, modern ones:

*VP3*   VP3, from On2 (formerly the Duck Corporation), has garnered a lot of positive attention. The codec is open source, so you can download it free from http://www.vp3.com. However, you have to buy the commercial version for $395 to get support. Beyond support, there aren't any quality or other differences between the commercial and free versions. VP3 isn't preinstalled, and there's no open downloadable component method for adding AVI codecs, so you'll need to verify that the viewers have it installed. This is no big deal for CD-ROM apps, but it can be difficult for Web video. VP3 .mov files are playable within DirectShow players if the VP3 DirectShow codec is installed.

Compared to Indeo, VP3 offers better quality with action footage and highly saturated colors, nearly as fast a decoder, and much, much lower data rates. It lacks, however, the various cool features of Indeo 5. VP3 has a bad habit of substantially exceeding the target data rate if the content is more difficult than the data rate allowed. VP3 has a deblocking filter that improves quality when running on G4, PIII, or better machines.

VP3 is a fairly standard codec enhanced by a few special options. It defaults to the Quick Compress mode, which produces less of a quality hit than most other Quick Compress modes and speeds up encoding quite a bit. You can specify whether or not the codec drops frames to maintain image quality and data rate. It also features a fairly powerful Auto-keyframe mode. For most content, the default options are good, although I tend to change the Maximum Keyframe Distance to 10 times the frame rate. If you use this function—and I recommend you do—you should set the keyframe rate in the standard AVI dialog to 9,999 (or off, if your application supports it) so it won't try to insert keyframes.

**20.2** The VP3 codec UI. Note the current AVI version has features not included in the QuickTime implementation.

The Lowest Allowed Quality value sets a minimum threshold for quality. The higher this number (up to 63), the more flexibility the codec has to hit its target data rates. The sharpness control affects how much smoothing is applied during compression. The High Detail setting can look great at higher data rates, but Normal and especially Smooth are better as data rates go lower.

The VP3 encoder for DirectShow currently has some features not seen in the QuickTime encoder. A file encoded with these features will still play under QuickTime. However, VP3 in DirectShow needs resolution divisible by 16 on both axes—not a limitation in the QuickTime version.

*Indeo 5* The various Indeo codecs were originally produced by Intel Architecture Labs, and are now owned by Ligos. Intel's motive was to make a superior technology that would encourage users to buy new Intel processors. Early versions of Indeo were intended as enabling technology to grow the multimedia industry (and thus sell more processors). In an era where Macintosh computers dominated digital video production, developing Indeo was a means to encourage users to use Intel video capture and video playback cards in Wintel machines. In the early days of CD-ROM-based digital video, the Indeo 3.2 codec was (a distant) second only to Cinepak in market penetration and had strong advantages for talking-head content.

When Pentium machines came out, Intel wanted a new codec to demonstrate the Pentium's power while still being usable on older machines. They redeveloped Indeo from scratch and renamed it Indeo Video Interactive (IVI) 4.0, a scalable wavelet-based compression technology. The current version, IVI5, is a further enhancement of that technology, originally optimized for the Pentium II processor.

Wavelet's promise is smaller files, better quality, less digital-looking artifacts, and the ability to drop quality instead of frames on slower processors. Like Sorenson Video 2, the Indeo codecs were hampered by their reliance on YUV-9 color subsampling. To help alleviate the YUV-9 problem, Indeo 5 running on MMX systems will interpolate between the 4×4 color blocks, reducing (but not eliminating) the YUV-9 problem.

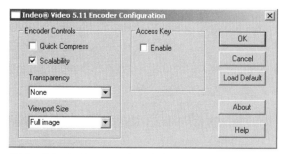

**20.3** The Indeo 5 codec dialog with its myriad of controls.

Indeo AVI files still have something to offer CD-ROM producers and game developers. IVI supports a one-bit alpha channel, and can do good real-time compositing during playback, either through an API or via a Director Xtra. Real-time compositing at playback is great for achieving a number of game effects, such as compositing an actor over a dynamically generated background.

The Indeo API can also do local decoding of a region of a movie. For instance, imagine a submarine game in which the user looks through a periscope. One could render a 1,024×192 Indeo movie, representing a panoramic image around the sub. Programmatically, the Indeo codec could then be told to decode only a specific 192×192 region that the periscope is looking at, dramatically saving processing time.

In its early days of development, one knock against Indeo came from its processor requirements for playback, which were steep in those days. Today, requiring a Pentium II CPU for 640×480 playback is nothing.

The Indeo codecs for DirectShow are available free from Ligos. Some Linux/UNIX apps such as XAnim also include binary support for Indeo 5.

### Scalability

Because Indeo is wavelet-based, it is built around successive bands of increasing quality that are decoded in sequence, creating the final image (see Figure 3.16 on page 47). With scalability turned on, under high processor loads, the codec skips decoding higher quality bands instead of dropping frames. This preserves motion at the expense of some image quality—generally a good trade to make. In the real world, however, this is less useful than it sounds. The lowest quality decode is about 80 percent of the decode complexity as the high mode. Still, every little bit can help, and I recommend leaving this on in most cases.

### Transparency

Indeo was the first delivery codec with built-in support for transparency effects. The "None" mode has no transparency. The "First Frame" uses a key color in the background of the first frame to define the alpha channel for the rest of the frames. Alpha channels let you attach an existing 1-bit alpha channel to the movie. Note there is no edge feathering in a 1-bit alpha channel, so every pixel is either entirely in the image or entirely out of it, which can cause some ugly effects around the edge. If you know in advance which background is going to be attached to the image, you can premultiply the background into edge pixels.

*Tip*
- Never use the "Quick Compress" mode for final rendering—the quality isn't as high. But it's great for rapid prototyping.
- Indeo does well in quality-limited mode, using the Quality slider instead of data rate controls. Setting a Q=85 instead of limiting data rate can often yield better results and a smaller total file size. Unfortunately, Indeo doesn't support explicit VBR, so the final data rate can be difficult to predict.

**DivX**  DivX is a strange beast, and has been a source of a lot of controversy in our industry. First off, it is not related to DIVX, the failed DVD format from Circuit City.

Early versions of Windows Media and DirectShow shared the same codecs, so using Windows Media video and audio codecs in AVI files worked fine. Many folks—especially in the DVD ripping community—preferred to make files in AVI instead of ASF (the old extension for Windows Media's Advanced Streaming Format) because they could use a wider variety of players with them and could use an AVI file as a source for further compression, which wasn't possible with ASF. However, with MS MPEG-4 v3, Microsoft no longer allowed the Windows Media codecs to be available from within DirectShow. So, some intrepid engineers came up with a hack to make this work, and claimed they had created a new codec. This put a bad taste in many folks' mouths, mine included. They dubbed this hack the "DivX ;-> codec," implying they had actually created a compression technology instead of just hijacking Microsoft's.

Microsoft fixed whatever loophole made this possible, so the Window Media Video codecs were not hacked in similar fashion. This meant the gap between what DivX could do (which they couldn't improve because they didn't write the code) and what other codecs were capable of kept growing.

Eventually, the private company that owned DivX, DivXNetworks, and an open-source based sibling, Project Mayo, made a new codec. It was based on the MPEG-4 Simple reference code from MoMuSys, extensively tweaked up. This became DivX 4, which wasn't at all based on Microsoft code.

Next came DivX 5, which is also based on MPEG-4, but not on the code from either MoMuSys, DivX 4, or OpenDivX. DivX 5 comes in a free, feature limited version, a free, full featured version that pops up advertising, and a Pro full-featured, no ad version for $29.95.

DivX 5 Pro has a lot of options for encoding. Most are safely ignored, but there are a few I want to highlight. DivX 5 added support for tools from MPEG-4 Advanced Simple: Quarter Pixel motion estimation, GMC, and Bidirectional Encoding. All can improve compression efficiency, and all increase decode requirements somewhat. Quarter Pixel and Bi-directional Encoding both give a good efficiency boost without detracting from performance too much. GMC doesn't help a whole lot and is fairly expensive for decode, so it should only be used for low data rate files. Note some third-party MPEG-4 decoders like QuickTime can't handle bi-directional encoded frames, so it's best to turn that off if you don't know what players are likely to be used to view the file.

DivX is the only AVI codec that supports a 2-pass VBR mode. This is handled in a unique way—you encode the file twice. The first time, you set the file to the first pass mode, which does the analysis and writes out information about the relative complexity of each video frame. Then you set the codec to the second pass mode, encode the file again, doing the actual compression. It sounds kludgy, but isn't actually that hard. I hope future encoding tools will automate this process.

Psychovisual enhancements redistribute bits between macroblocks in a given frame according to a software model of those portions of the image that are going to be most quality-critical to a viewer. This generally improves quality somewhat. The Light setting is subtle, and safe to always leave on. Normal is fine for most content, but in some cases Extreme can cause some strange side effects. QA your output carefully if you encode with Extreme.

DivX can encode both progressive and interlaced source, making DivX a decent choice to store interlaced files at high data rates.

In the Advanced Parameters tab, you'll definitely want to have Performance/Quality set to Slowest, but stay away from the other options until you have a day or two to experiment with them. The defaults are generally just fine.

DivX also includes some built-in preprocessing filters. While these work okay, they're really meant for doing a live capture. It's best to preprocess prior to applying the codec.

The current DivX decoder can still decode all the previous flavors, but other MPEG-4 codecs only handle DivX 5. DivXNetworks encourages the use of files with the extension .divx, which use the DivX video codec and MP3 audio. Players that can handle this are available on a wide variety of platforms.

Given the complexity of DivX settings, I appreciate that they provide a way to save and load settings configurations within the codec itself.

With installed codecs, DivX lays back inside any DirectShow or Windows Media app, QuickTime, and a special player called "The Playa." You can also define a default postprocessing level, although this can be overridden by the user and will automatically be reduced on slower machines.

**20.4** The many settings tabs of DivX 5.02 Pro. Most of these don't need to be tweaked, fortunately. It's great they allow you to save and load particular settings within the codec itself.

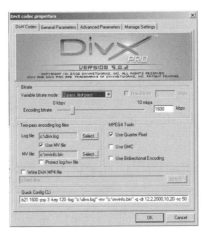
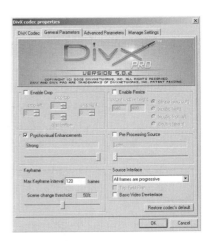

# Legacy Delivery Codecs

*Indeo 4.4*   Indeo 4.4 was an earlier, lesser incarnation of Indeo 5. Its features were overall very similar to those of IVI5. The codecs are still available from Apple, mainly for playback of legacy content. If you need to make new files with Indeo, use 5. Indeo 4.4 can play back files authored with Indeo 4.3 but not 4.2 or earlier.

*Cinepak*   Cinepak was the reigning CD-ROM video codec for many years. Originally created as Compact Video by SuperMac and brought to Radius when the two companies merged, the codec underwent a last-minute trademark infringement-induced name change to Cinepak. It was licensed for inclusion in everything from QuickTime to Video for Windows to the Atari Jaguar and the Java Media Layer. In 1998, some former Radius employees founded Compression Technologies and released an updated version. Cinepak is easily the most used, most ubiquitous codec ever.

Free to end users, its combination of good overall performance and great price made it the dominant codec in multimedia until last year. Virtually every personal computer shipped since 1994 has Cinepak preinstalled. Alas, progress has finally passed it by. There is little reason to use Cinepak anymore, unless you need to target 200 MHz or lesser computers. MPEG-1 is similarly ubiquitous, and any number of codecs provide better quality at lower data rates.

If you're targeting older machines, especially in the educational market, Cinepak can be a good choice. It's also useful for making QuickTime fallback movies for compatibility with Windows Media Player, RealPlayer, UNI, QuickTime readers such as Xanim, and older versions of QuickTime.

The Cinepak Pro codec available for QuickTime isn't available for authoring AVI (although you can losslessly convert the QuickTime-compressed files to AVI). Compression Technologies does, however, make available an updated VfW component for AVI applications that includes some bug fixes and slight quality and performance enhancements.

## *Tip*

- A favorite trick with Cinepak is to have low-rate keyframes. If a video requires no random access, and has a mostly static background, Cinepak can use a single keyframe followed by literally hundreds of interframes. A keyframe will be automatically inserted if a frame is more than 85 percent different than the previous frame, so you'll generally get a keyframe with every camera cut.
- For black-and-white content, the Cinepak 256 Grays mode can provide beautiful results at reasonable data rates. Back before PCI video cards, this mode could actually play back 640x480 15fps full-screen.
- The 256-color Cinepak mode is rarely used, but can occasionally work well for animated content containing too much motion for other codecs to handle.

***Indeo 3.2*** Indeo 3.2 was Cinepak's biggest competitor in the mid-90s video codec industry. It was YUV-9, like all Indeo codecs. Version 3.2 was freely available, but required an install for QuickTime on both Mac and Windows, as well as in Video for Windows. Compared to Cinepak, Indeo 3.2 offered somewhat lower data rates and higher quality for low-motion talking head-style content. It encoded somewhat faster than Cinepak, but had higher decode requirements. Indeo 3.2 is still seen on some legacy CD-ROMs, but there is no reason to create content in it anymore.

*Tip*

- The resolution of Indeo files must be divisible by four on both axes, or playback may be unstable on some systems.
- The width of an Indeo file must be greater than its height, or playback will be unstable on some systems.

***Microsoft Video 1*** This was Microsoft's pre-Cinepak codec. Like Apple Video, it was also quite lousy with terrible compression efficiency, no data rate control (quality limited), and only provided 16-bit and 8-bit playback. Video 1 hasn't been updated since 1992, and there is no reason to use it anymore.

***ClearVideo*** Iterated's ClearVideo codec was a slightly successful entrée to the Web video market. It was available as a Windows AVI and QuickTime codec, as well as RealVideo (Fractal). It's the only commercially released fractal codec. Intended as an Internet broadcast codec, its use in QuickTime and AVI was hampered by customers' needs to download and install it to view video. It was very slow to compress and required a substantial computer to play back movies at even moderate data rates. When Microsoft abandoned AVI and made Windows Media their Web video codec of choice, ClearVideo's end as an AVI solution had come.

Iterated has been reborn as a vendor of asset management products called MediaBin. They've sold ClearVideo to Enxnet, who are making the Windows decoders available for download. They also claim to be working on an improved version of the technology, but haven't made anything public as of the writing of this book.

***TrueMotion*** The Duck Corporation's TrueMotion family of codecs were focused on the games market, and enjoyed some success there. The TrueMotion family included three main codecs: TrueMotion-S, TrueMotion 2.0, and TrueMotion 2X. A fourth member of the family, TrueMotion 3, became the On2 VP3 codec. TrueMotion 2X is still being sold, as its better decode performance makes it better for some very high-resolution applications (although its data rate requirements are massively higher). TrueMotion 2X targets 640×480 30fps on 200MHz machines. It's $250.

TrueMotion-S was an intraframe-only codec with limited quality but excellent playback performance. Its main selling point was the ability to play back reasonable-quality full-screen video on a variety of platforms. While it was used in a number of games, the sales model didn't prove as successful as Duck hoped. Duck fired Horizon as their distributor, and Horizon began selling a TM-S derived codec called Power!Video (and they'd insist you call it "Power-bang-video"), which quickly vanished from the marketplace.

Duck decided to produce a new interframe codec that would maintain the good playback of TrueMotion-S, while providing higher quality at lower data rates. This product was TrueMotion 2.0, of which TrueMotion 2X was enhancement.

## AVI Authoring Codecs

While AVI is commonly used as an authoring codec, it doesn't ship by default with a broad range of authoring codecs. It doesn't even include a Motion-JPEG or Photo-JPEG codec. Typically, an authoring system will come with a custom codec for its own file format that won't interoperate with hardware from other vendors. However, authoring systems generally offer a software codec, so you can transcode.

*DV*   AVI doesn't support a native DV bitstream in the same way that QuickTime does. Instead, you have an AVI file that contains the binary data from the DV format source, with an .avi extension.

The biggest limitation of DV comes from its fixed data rate. While it's generally adequate for content originally produced in DV, and hence cleanly captured and compressed in-camera, it often doesn't have enough bandwidth to provide adequate quality with noisy source, such as VHS dub. For this reason, I don't usually recommend transcoding existing analog source to DV before capture—DV's 25Mbps is often not enough bandwidth to provide sufficient quality.

DV is the only real authoring codec besides compressed that ships with Windows by default.

*Motion-JPEG*   Motion-JPEG (M-JPEG) is the grand-daddy authoring codec used by most compression hardware. Essentially, it takes JPEG, makes it 4:2:2 instead of 4:2:0, and allows storing each field as its own bitmap (although you can do progressive M-JPEG as well).

Different hardware vendors usually supply their own, non-compatible M-JPEG codecs. There are also some software JPEG codecs that can be used. Windows applications that can read .mov files with DirectShow codecs can generally use these to read QuickTime JPEG files. I use the ones from PICVideo, and have found them trouble free.

*Huffyuv*   The most useful authoring codec for me is Ben Rudiak-Gould's Huffyuv. This is a lossless Huffman encoding codec. It's extremely fast for both encode and decode, and can easily do a real-time capture of 640×480 30fps video on a modern computer with a fast hard drive. With Huffyuv, typical content is about 50 percent the size of uncompressed video. Huffyuv can work in both Y'CbCr 4:2:0 and RGB color space, so you can pick the native color space of your tools. It's not included anywhere, but the codec is free.

Any programmers who wish to earn my eternal gratitude should port Huffyuv to QuickTime. It would be useful in its own right, and we could also use it to save preprocessed Huffyuv AVI files on a Mac for final compression on Windows.

*Indeo 4:2:0 Video/Indeo Raw* This one is a strange bird. Many moons ago, Intel made the Intel Smart Video recorder line, a series of budget-priced capture cards. They were all YUV-9-native, which is why the Indeo codecs were similarly YUV-9. This codec (which goes under these two names) puts out an uncompressed YUV-9 bitstream. Great if you need to store uncompressed YUV-9. Otherwise, it's pretty useless.

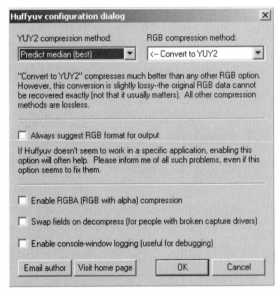

**20.5** The codec dialog for Huffyuv. Note you can compress source from RGB apps as either RGB or Y'CbCr.

*None* My second least favorite authoring codec in AVI is None, which is an RGB uncompressed option. I really wish the default DirectShow included a decent compressed intermediate format.

*Microsoft RLE* My least favorite authoring codec in AVI is Microsoft RLE. This is an 8-bit only RLE (Run-length encoded, as described in Chapter 3) codec. If you already have 8-bit content, it's an okay way to transport it around, but otherwise RLE is a dog.

## AVI Audio Codecs

Alas, AVI doesn't have a great suite of audio codecs. Microsoft has made most of its recent efforts proprietary to Windows Media instead of updating the AVI audio codecs.

*MP3* AVI supports MP3 as an audio codec for playback, although DirectShow in Windows XP doesn't include an MP3 compressor directly. Older versions of Windows shipped with a very limited MP3 encoder with a maximum data rate of 56Kbps. However, you can plug in other MP3 DirectShow codecs that enable general AVI apps to encode to MP3.

While the Fraunhoffer codec is a good one, its VBR mode make it impossible to predict its final file size, and the CBR mode isn't as efficient as it could be. Finally, it's limited in its sample rate and channel combinations; for example, 96Kbps is the minimum data rate that can do 44.1kHz stereo. If absolute maximum quality VBR is required, I recommend using the open-source LAME (LAME Ain't an MP3 Encoder) MP3 encoder instead, which lets you specify a target data rate for the VBR file. It also allows a broader range of sample and data rate combinations.

AVI in general allows the use of VBR MP3 audio. However, some non-DirectShow players may not support VBR MP3.

*IMA ADPCM*   The IMA ADPCM (Interactive Multimedia Association—Adaptive Pulse Code Modulation) audio codec was the dominant CD-ROM codec of the Cinepak era. It offered straight 4:1 compression over uncompressed, so where a 44.1 mono track would be 705.6Kbps, an IMA version would be 176.4Kbps. Stereo required twice the data rate, so a full 44.1kHz stereo in IMA would take 352.8Kbps. Audio quality was quite good at 44.1kHz, but a loss of frequencies in the higher ranges of the sample rate quickly degraded quality at lower data rates.

Today, there are two big reasons to use IMA: performance and compatibility. IMA is dead simple to decode, and so takes up a negligible amount of the available CPU time even on moderately powerful machines. This can make IMA a good choice for local playback when low performance requirements are more important than compression efficiency. IMA is also a published standard, so virtually anything that can play an AVI file can play one with IMA.

*Ogg Vorbis*   Ogg Vorbis is an open-source effort to make an MP3-like audio codec. It has been publicly available for quite a while and just had its 1.0 release. There is a DirectShow component for playing back Ogg Vorbis audio.

*None*   As with the other formats, AVI's None mode is uncompressed PCM. As such, it's mainly used for authoring. However, a lot of older files that predate IMA (before QuickTime 2.1) will use the None codec, either in 16-bit or (heavens forbid) 8-bit.

# Legacy Audio Codecs

*μ-Law*   μ-Law, (that's the Greek letter which is pronounced "mu" and rhymes with "you") is an old-school telephony codec, offering 2:1 compression, and hence 16-bit quality over 8-bit connections. It sounds okay for speech, but its poor compression efficiency makes it an unusual choice for any modern applications. It is also called "U-law" (for those who can't easily type the μ character), and G.711. It has the distinction of being the best compressed audio codec supported in older versions of Sun's Java Media Layer. The current versions support IMA.

*A-Law*  A-Law is quite similar to μ-Law, being a 2:1, 16-bit telephony codec. It is less commonly used and offers no significant advantages.

## For Further Reading

| | |
|---|---|
| http://www.microsoft.com/hwdev/archive/devdes/fourcc.asp | Microsoft's list of AVI codecs and their four-letter codes. Very handy for finding out what codec that AVI you have is in. |
| http://www.jmcgowan.com/avi.html | John McGowan's AVI Overview: an excellent, detailed resource of the history and the use of the AVI format. |

# Chapter 21

# Flash MX

## Introduction

Macromedia Flash MX is the latest video format, and may become quite important. Flash MX replaces Flash 5. Flash was born as an animated vector graphic format, which has been progressively enhanced to add rich and interactive features. While Flash may be best known for those annoying home page animations we all skip past, the format itself is quite powerful, and can deliver a much more media-rich experience than traditional HTML-based sites.

While there have been various third-party video-in-Flash products over the years, Flash MX adds a built-in video codec, Spark, licensed from Sorenson Media. Spark is progressive download only, and designed more for efficient playback than maximum compression, but can still offer darn good quality.

So, why is Flash MX an important video format? Because of the sheer ubiquity of Flash. Macromedia estimates that 98 percent of desktop computers have the Flash Player installed—a number that's well beyond the market penetration of any other software. Flash is commonly available on alternate OSs, mobile devices, set-top boxes, and so on. Macromedia's goal is to have Flash installed on anything that's connected to the Web and that has a screen, and they're doing a fine job of achieving that goal. The Flash format is open, so other companies can create implementations for their own platforms without assistance from Macromedia, which helps Flash burrow into the many nooks and crannies of the computing world.

For folks who want to integrate Flash and video, Spark is likely to become the standard way to do it. Embedding a Flash track in QuickTime will remain a superior option for many projects, especially those that are more video-oriented than vector animation-oriented. QuickTime and its codecs provide real-time streaming and better compression efficiency than Spark.

## Flash MX for CD-ROM

Flash can work just fine for CD-ROM projects. You can also embed Flash within a Director presentation, but other formats could be used with Director. Flash MX would have the advantages of not requiring an installation. The Flash MX application lets you build stand-alone applications for Windows and Mac (a Carbonized Classic/X). Unlike Director, you can build projectors for both platforms from either platform.

## Flash MX for progressive download

Flash MX is natively progressive download, and does an excellent job of it. Make an .swf file that just contains the Spark video and audio, and you'll get a standard progressive download experience. You can also use Flash to control user download experience.

## Flash MX for real-time streaming

Flash MX can be streamed in real time via Macromedia's new Flash Communication Server MX. The Communication Server has a rich set of other features for deploying integrated rich media presentations with just Flash Player 6 as the client. Among other things, this includes videoconferencing—taking advantage of Spark's H.263 heritage. Videoconferencing and video capture can work from any Flash Player 6 client on MacOS or Windows with an attached capture device.

The Communication Server only does real-time streaming of Flash (through its proprietary RTMP protocol), not progressive download. You'll need to encode files a little differently if they're going to be served by Communication Server.

## Flash MX for interactive media

Flash MX is an interactive and rich media format that includes video. Flash itself is a topic for many, many books, and most people have seen plenty of Flash movies, so I won't belabor the topic here. Simply stated, Flash MX can do pretty much anything you can image on the interactive side, and is much easier to author for than other formats. If you choose not to use Flash MX for a certain task, it would likely be because of its video limitations, not its interactive limitations.

# Authoring Flash MX

Macromedia licensed Spark from Sorenson, and Sorenson has reserved the full Spark Pro encoder feature set for themselves. So far, there hasn't been any indication that Sorenson will license the technology to other vendors, so for now the only game in town for professional Spark encoding is Sorenson Squeeze. Personally, I'd love to see them release a QuickTime

**21.1** The Sorenson Squeeze UI for Spark Pro. It includes support for 2-pass VBR and other professional options.

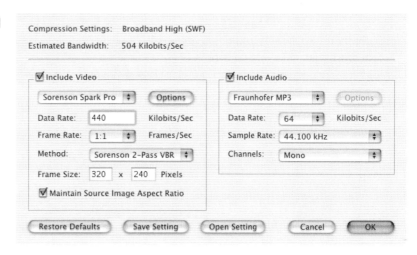

Export Component for Spark Pro, so other QuickTime compression tools like Cleaner could export to Spark Pro.

## Sorenson Squeeze

Sorenson Squeeze is the first and so far only tool that provides the Spark Pro features, so it's what you'll want to be using to produce high quality Flash MX video content. Squeeze can output Spark to both .swf and .flv files. The .swf files are complete Shockwave files that you can play as is. The .flv files are used as source within the Flash MX authoring tool itself, so you can integrate the compressed video with the rest of the presentation.

Note Squeeze only supports MP3 audio in .swf files. If you want to use either of the other Flash MX audio codecs (described later), you'll need to make an .flv with uncompressed audio, and do the final compression within Flash MX. Squeeze supports both 1-pass and 2-pass encoding modes. For progressive download, files should be with 2-pass VBR for maximum compression efficiency. For files to be streamed from the Flash Communications Server, encode with CBR to get a flat bitrate.

## Squeeze and Flash MX

Flash MX includes a subset of the Spark Pro features of Squeeze, and lacks any preprocessing besides scaling. It doesn't even have data rate control (see Figure 21.2); it simply provides a quality-limited 1-pass encode. While this may be adequate for projects that have a little bit of video, any video-centric project should be encoded with Squeeze.

However, only the Flash MX tool has access to audio codecs other than MP3.

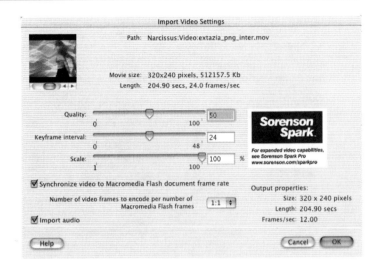

**21.2** The Flash MX application's Spark dialog. It lacks critical features like data rate control.

## Spark Video Codec

The Spark video codec is based on (but not compatible with) the standard H.263 videoconferencing codec. Support for the baseline version of H.263 is required in MPEG-4. Sorenson had already implemented the baseline version of H.263 as part of their MPEG-4 codec, and this was used as the basis of Spark. The principal design goal of Spark is that it can easily be ported to and decoded on a wide variety of devices. Hence decoder speed and size (it takes less than 100K) were more important than compression efficiency.

Given all of the above, Spark offers decent compression efficiency, especially when encoded in 2-pass VBR within Squeeze. The encoding implementation in Flash MX, as was previously mentioned, doesn't really offer any meaningful options (not even data rate), but Spark Pro in Squeeze has a number of options reminiscent of other Sorenson codecs.

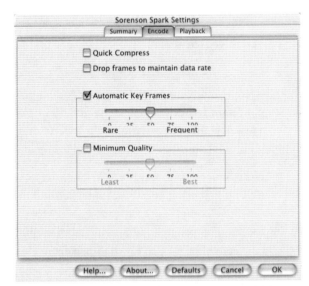

**21.3** The Spark Pro Encode options, from within Sorenson Squeeze for Flash MX.

### Quick Compress

Quick Compress makes the file encode slightly faster, with a small quality hit. Ignore Quick Compress, and encode all your files in full-quality mode.

## Drop Frames

If this option is on, and you're encoding in CBR mode, the frame rate will drop to maintain data rate. This option doesn't do anything in 2-pass VBR encoding.

## Automatic Keyframes

This control lets the codec insert keyframes with scene changes and high motion. It should always be on. The slider controls how sensitive the codec is to keyframes. The default of 50 is good for most content.

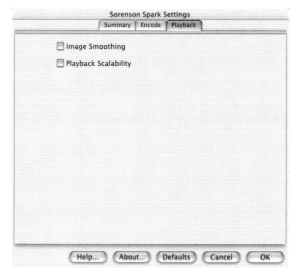

**21.4** Spark Pro Playback tab. Both options should only be used with care.

## Minimum Quality

Minimum Quality sets a minimum visual quality for each frame below which the codec won't drop. Instead, it'll raise the quality of each frame until it meets the minimum quality target. With Drop Frames turned off, the Minimum Quality feature can cause the data rate to overshoot the target substantially. Turn drop frames on if you must use Minimum Quality and don't want to overshoot the target. Minimum Quality should generally be off unless legibility of elements on the screen is critically important.

## Image smoothing

Image smoothing turns on a deblocking filter in the codec. This helps improve quality, especially with aggressive compression where the video would be very blocky. It doesn't help quality much if at all with video that otherwise wouldn't be blocky. However, it does increase decode complexity, so it shouldn't be used when targeting slower devices at higher data rates and resolutions.

## Playback scalability

With the playback scalability option, the video can drop the frame rate in half instead of having uneven stuttering on slower machines. Playback scalability uses the Sorenson Video 2 scheme, where the even and odd video frames are encoded in two separate streams, and scalability is achieved by just not playing one of the tracks. This reduces compression efficiency because it doubles the temporal distance between frames, making delta frames less efficient. It also makes decoding both streams somewhat slower than just doing a single stream. Because of these limitations, you're best off only using playback scalability with progressive download in cases where it's explicitly required. A good alternative would be to create multiple versions of the file, and have the Flash presentation switch among them as needed. For streaming from the Communication Server, Playback Scalability provides bandwidth scalability. For files to be streamed over the public internet, it is a good idea to use it.

## Audio Codecs in Flash MX

If you're making an .swf file straight from Squeeze, you'll just use the MP3 audio codec. If you're making a .flv file in Squeeze for incorporation inside Flash, you'll want to output uncompressed audio and let Flash MX do the encoding. All audio samples inside a Flash movie have to use the same audio compression, so compressing the file before it's imported will generally entail another pass of compression on export.

*MP3*   MP3 is the best audio codec in Flash for higher bitrate applications, and it's the only authoring codec supported via Squeeze. Squeeze uses the excellent Fraunhoffer MP3 encoding library, and supports a good variety of data rates and sample rates. Flash MX itself has a very different interface to the same underlying codec. If you're encoding within Flash MX, make sure to set quality to Best.

You shouldn't use MP3 data rates above 160Kbps or at 8KHz—Flash can't always play those files back correctly.

*Speech*   Flash MX can also author to a Speech audio codec (a licensed version of the Nellymoser Asao codec). Like the other examples of its genre, Speech is suited to low-bitrate, low-fidelity speech content. You can only specify the sample rate of Speech—data rate is implicit, and output is always mono. MP3 does a fine job with speech and can provide better compression efficiency, so you'll generally only use the speech codec when targeting very low data rates, using sampling rates of either 5KHz or 8KHz.

Speech's data rate is an 8:1 reduction from the data rate of the uncompressed source. So 5KHz would be 10Kbps, and 8KHz would be 16Kbps. MP3 will often provide better quality even at those low data rates.

*ADPCM*   ADPCM is yet another Adaptive Pulse Code Modulation codec, like IMA. It offers good quality but only a little compression. It is very quick to decode, so it can make sense for short sounds like beeps and clicks. It shouldn't be used as the soundtrack for a video presentation due to its poor compression efficiency.

*RAW*   The RAW audio codec in Flash MX is straight-up uncompressed audio. It would be very unusual to use it.

# Chapter 22

# Miscellaneous Formats

This chapter is about the myriad older and special use formats. Some are excellent for specific, non-mainstream uses. Others you'll never encode to, but might still encounter in the wild.

## Bink

Bink, from Rad Game Tools, is the leading video and audio format for game developers. All of Bink's features are tightly focused on satisfying that market. Bink does a fine, if unspectacular, job playing cut scenes well, but its real strength is its extremely flexible SDK for incorporating into games video and audio as sprites, textures, or anything else for which a series of bitmaps can be used. Bink can be linked straight into an application, so it doesn't require a separate installation. The encoder is Windows only, but the decoder is available for Win32, Mac OS (Classic and X), Nintendo GameCube, Microsoft Xbox, and probably future consoles as well.

If all this has left you drooling, you might be the right target audience for Bink. If not (and that will be most of you), Bink isn't what you need.

### File format

The Bink file format includes both a video and an audio codec, wrapped in a format with the .bik extension. Neither are particularly revolutionary in compression efficiency—its video codec is in the ballpark of MPEG-4 Simple for compression efficiency. The audio codec is natively VBR, and a little more efficient than MP3.

### Encoder

Bink files can only be encoded with the free Bink encoder. It's Windows only, but to its credit supports both DirectShow and QuickTime for file reading. However, the DirectShow support only works for files up to 2GB (QuickTime support doesn't have this limit). The Bink encoder is remarkably slow compared to most modern codecs. It is optimized for multiprocessors, so a dual CPU or even quad CPU machine can help quite a bit.

Bink's video compression efficiency got a lot better with v1.2a (August 2001), and audio compression got a lot better with v1.5a (March 2002), so it may be worth reencoding files created prior to those versions.

The Bink encoder at first glance seems to have an overwhelming number of options, but in most cases there are only a few you'll use. If you need to preprocess, you'll generally want to preprocess in another application, like Cleaner or After Effects, which give you more control and visual feedback. You can adjust gamma and contrast, do deinterlacing, and so on in Bink, but the interface is very kludgy, and the codec itself is so slow you don't want to do any more trial and error than is absolutely required.

**22.1** The Bink encoder packs a lot of options into a single dialog.

In setting data rate, note the units used are bytes per second, not KBytes. These units indicate the total rate of audio and video data. A setting of about 250,000 is a good start for most content. You can also specify a peak data rate. A default of 3× the target data rate is typical, but it's better to test how high a peak your target platform can play back reliably, and specify that particular peak data rate. You can also specify keyframe sensitivity and "at least every" values. The defaults are good in most cases.

Bink has the unique feature of being able to set the number of preview frames. This does the initial processing of X-number of frames to scan into the future, but doesn't actually allocate the bandwidth for those frames. This lets it do a sort of 1-pass VBR. Your gut reaction may be to max this number up to 64. Strangely, bigger numbers often hurt quality. Good defaults are 8 for live-action video and 12 for synthetic, animated content. Higher values also use more memory on encode.

 **Note**

The Bink video codec is one of the few that can do real grayscale video. This can yield faster and better compression, as well as faster decode.

Also note the scaling compression modes, where you can do anamorphic compression (like 640×240 doubled to 640×480 on playback). This can be a good option when you have high-motion interlaced source where more than 240 lines would have been interpolated anyway.

Because encodes are so slow, it can be really frustrating to spend a couple of days encoding a file (really!) and discover that you got the settings wrong. Liberally use the "Preview" feature during encode to get a real-time playback preview of the frames you've encoded so far.

Bink audio is quality-limited VBR, so you can't actually specify a data rate. The default quality setting of 4 sounds great, and will be typically around 80–160Kbps for 44.1kHz stereo.

Once a file has been encoded, you can use the Analyze command to see a data rate graph of your file, which allows you to look for data rate spikes.

## Playback

Bink files are either played back via the Bink Player application or the SDK. Either way, you have a wide variety of tweaks available. These include full-screen playback, a wealth of different blitting modes including Y'CbCr overlays, scaling, and so on. Diagnostic tools are also available that can simulate playback with different speed drives, and provide useful run-time statistics.

## Business model

The Bink business model is unique, and a big part of its success. First, all the tools are free. You can incorporate video playback in your app for free, but the Bink logo will be displayed. To include video without the logo is $1,995 per title. SDK licenses are $5,000 and up. A full multi-platform, multi-title site license is $50,000.

# Java Media Framework

The Java Media Framework (JMF) isn't a format, but an API that Java applications can use to play back video in a variety of formats. The JMF is part of many different Java profiles, including the Java Game profile and a subset in MobileMedia. The JMF software is preinstalled on those platforms. Targeting other systems, like desktop computers, will require the JMF either be installed locally or included with the applet.

If all the above is gobbledygook to you, feel free to skip the rest of this section. You'll need to be working on a project being engineered in Java for the JMF to matter to you.

The JMF is available in two forms. The first is a pure Java version of the JMF that can run on any version of the JVM (Java Virtual Machine) from v1.1 on. This means JMF can play on the archaic versions of the JVM common on older Windows and Mac OS classic machines. Of course, native Java applications aren't as fast as those running in compiled code (although the gap is much smaller in modern JVMs). Thus, there are platform-accelerated versions of the JMF running native, compiled code. Sun has released "performance pack" versions for Solaris and Windows. Blackdown has done one for their port of Java to Linux. And Apple has announced that they will add a JMF compatibility layer to their QuickTime for Java implementation, which would use QuickTime to provide native code acceleration for the JMF on Mac.

The JMF is surprisingly flexible and complete. It can handle both file-based playback and RTP/RTSP. It can also capture (on Solaris and Windows) and compress video. The biggest complaint against the JMF is the weight it adds to a download. The pure Java is about 700K, although you can strip out codecs that aren't used in a particular project, so a version that only supports a particular format and video and audio codec will typically be around 300-400K. This will be cached like any other Web content, so the download will only need to occur once per session of video viewing. Also, systems using the Sun Java plug-in for Windows can automatically download the native code version of the JMF, and so will have faster performance and will only have to be downloaded once.

There are other pure Java video playback systems like Emblaze that have the advantage of only being a 50K download. However, they require you to buy another tool, and you must use their bad encoding tool. Because the JMF supports mainstream formats, you can use normal compression software, and the format is free.

These formats are supported for playback in all versions of the JMF:

| | | |
|---|---|---|
| **AIFF codecs** | Audio: | uncompressed |
| | Audio: | G.711 (U-law) |
| | Audio: | A-law |
| | Audio: | IMA ADPCM |
| **AVI codecs** | Audio: | uncompressed |
| | Audio: | DVI ADPCM compressed |
| | Audio: | G.711 (U-law) |
| | Audio: | A-law |
| | Audio: | GSM mono |
| | Video: | Cinepak |
| | Video: | JPEG |
| | Video: | None |

| | | |
|---|---|---|
| Flash 2 | | |
| GSM | | |
| MP2 | | |
| MP3 | | |
| QuickTime codecs | Audio: | None |
| | Audio | G.711 (U-law) |
| | Audio | A-law |
| | Audio: | GSM mono |
| | Audio | IMA4 ADPCM |
| | Audio | MP3 |
| | Video: | Cinepak |
| | Video: | H.261 |
| | Video: | H.263 |
| | Video: | JPEG |
| | Video: | None |
| RTP | Audio: | G.711 (U-law) 8kHz |
| | Audio: | GSM mono |
| | Audio: | G.723 mono |
| | Audio: | 4-bit mono DVI 8kHz |
| | Audio: | 4-bit mono DVI 11.025kHz |
| | Audio: | 4-bit mono DVI 22.05kHz |
| | Audio: | MPEG Layer I, II (MP2), III (MP3) |
| | Video: | JPEG |
| | Video: | H.263 |
| | Video: | MPEG-1 |

| WAV | Audio: | Uncompressed |
| --- | --- | --- |
| | Audio: | G.711 (U-law) |
| | Audio: | A-law |
| | Audio: | GSM mono |
| | Audio: | DVI ADPCM |
| | Audio: | MS ADPCM |

More formats are expected in the future, and codecs can be easily added. IBM has demonstrated but not yet released MPEG-4 support for the JMF.

## Authoring for the JMF

For delivering video content as a file for the JMF, I typically make QuickTime files, with H.263 video and MP3 audio. Making a straight file with Cleaner works just fine—you can even use Compressed Movie Headers. And remember to use H.263 in one of its three native resolutions: 128×96, 176×128, or 352×288 so the JMF won't have to scale. If decode efficiency is more important than compression efficiency, you could also do Cinepak video and IMA audio.

## Serving for the JMF

Delivering RTP to JMF clients is also easy. Any RTP server, like the QuickTime/Darwin Streaming Server or Kasenna MediaBase, will work fine. Note there's nothing specific to the JMF in the RTP stream—you can decode the same stream with both a JMF player and QuickTime. MPEG-1 is also a good option here, with even broader compatibility for playback, although somewhat lower compression efficiency than H.263 and MP3.

# Kinoma

Kinoma is a new company made up of former key members of Generic Media and Apple's QuickTime team. Their products are for video playback on mobile devices. For authoring, they provide Kinoma Producer ($29.99), for Windows, and MacOS 9.x and OS X. For playback, they offer the free Kinoma Player.

Kinoma Producer is based on Generic Media's gMovie Producer, which compressed video for black and white PalmOS devices. Kinoma licensed Cinepak Pro (!) from Compression Technologies, and used it as the basis for a color codec called Cinepak Mobile. This makes sense—Cinepak was originally designed to run on Motorola 68030 processors, an ancestor of the Dragonball processors used in all PalmOS handhelds until support for ARM was added in 2002. While RealVideo and Windows Media Video requires far too much processor power to decode on those slower machines, Cinepak fit the bill nicely.

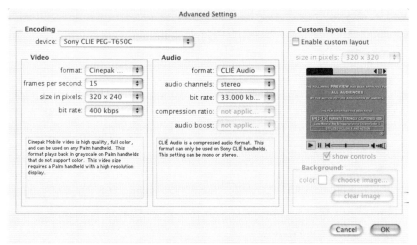

22.2   Kinoma Producer's Advanced Setting dialog. This is the default for Sony's CLIE PEG-T650C PalmOS handheld, one of the most powerful yet available.

22.3   A screen shot of the Kinoma player, as it would be seen on an old-school 160×160 PalmOS handheld. It can also run on 320×320 systems, and in color.

Kinoma Player is currently available for PalmOS 3.1. However, the screens on those machines vary from 160×160 black and white to 320×320 full color. Processor speeds vary as widely as well. It's best to optimize the compression for particular models, if possible. The lowest common denominator is quite low. Kinoma Producer includes presets for many models. Typically, the only thing you might tweak in the presets is the frame rate or number of channels.

The Kinoma Player can also play back gMovie encoded files.

## MNG

Multipart Network Graphics (MNG, pronounced "Ming") is the motion graphics counterpart to PNG. MNG is often described as being similar to Animated GIF, but it is much, much more capable.

MNG is a lossless RGB format, that can encode to a variety of bit depths from 32-bit on down. I find it most interesting as a replacement for codecs like Apple Animation for doing playback of synthetic graphics (where lossless compression can dramatically reduce file size). Compared to Animation, MNG offers much, much better compression efficiency; I've seen a 10× improvement in real-world projects.

MNG is patent and royalty free. There is also an open-source library for creating and playing back MNG files that many applications are based on. It's called Libmng.

So far, MNG support isn't particularly widespread, although that is changing rapidly. Playback is supported in a variety of players, including the Mozilla browser and its derivatives like Netscape 6+. Tarkvara is producing a MNG codec for QuickTime, currently in beta.

# Animated GIF

Animated GIF files were the first widespread Web video format. Unfortunately, they were remarkably lame, and have long since been supplanted except for a few tasks like banner advertisements for which ubiquitous playback is absolutely required.

An animated GIF is a GIF file (using the standard GIF89a format) that contains multiple images. Each has the normal limitations of GIF, which is a losslessly compressed indexed color format, with a maximum of 256 colors. Each frame has its own duration, and the movie can be set to play through just once, or it can be looped. Animated GIF doesn't have any interframe compression, nor is there any way to specify a data rate. Instead, data rate control consists of reducing the number of colors per image, and simplifying the image to make the lossless compression more efficient. The most difficult content to deliver with animated GIF is true-color images and dithering, both of which hurt compression efficiency. It's better to use graphics that have large areas of flat color of the sort you get with cel animation with animated GIF. Such images will compress extremely well.

Each frame can have either its own palette or use the same palette for the whole movie. Because the palette takes up some space, it's best to use a global palette assuming it works for your content.

Because there's no interframe compression, it's best to use an extremely small number of frames, with long durations. Looping can also help get more mileage out of a few frames.

You can fake a slight degree of interframe compression by using transparency and background images. For example, the first frame of a sequence can be a complex background image, and later frames can be made to play as an overlay on that background. Thus the background need only be transmitted first. Note you only have a 1-bit alpha channel. Also, each frame can be smaller than the size of the overall image; for example, you could embed a 160×120 image on top of a 320×240 background graphic.

If you look at professionally created animated GIF banner ads, you'll see they use fewer than 10 looping frames, and have only a few colors. The experience such banners can provide in under 10K can be very impressive. However, authoring these files requires a project designed from the ground up for them. Converting existing content to animated GIF, while possible, generally yields large, horrible looking files, and is almost never worth doing.

Of course, animated GIFs don't support audio.

All the major image editing applications like Photoshop, Flash MX, and LiveMotion can author animated GIF files, but video compression tools typically don't offer this ability.

# Advanced Authoring Format (AAF)

Advanced Authoring Format, a.k.a. AAF, has been on the verge of mattering for the better part of a decade. The goal was to create a new unified metadata format for authoring media, combining the best features of Avid's OMF and QuickTime. Microsoft and Avid have been AAF's main backers to date.

The promise of AAF is compelling—a full, object-oriented way to mark up and manage video content production. For example, you'd be able to look at the final edited file, and find out who the cinematographer was of each section. You'd also be able to get more reliable and complex online finishing from an offline edit than today's EDLs actually provide.

However, the number of applications that support AAF has yet to reach critical mass, though with v1.0 of the SDK being freely available, this could change. AAF showed signs of life at NAB 2002, so we may finally see mainstream products that can use it before too long.

AAF is an authoring format, not a delivery format. Even if AAF becomes popular for authoring, the AAF file will still be converted to a delivery format for final distribution.

# MP3

Although MP3 was originally designed for MPEG-1 (It's short for MPEG-1, Layer III audio), it is almost never seen in MPEG-1 files. However, it's quite popular both as a file format in its own right, and as a codec in formats like QuickTime, AVI, and Flash MX.

MP3 was created by Fraunhoffer. As part of an ISO standard, sample source code for MP3 encoding and decoding was released well before the licensing terms were announced. Many different software vendors created MP3 encoding and playback software while the technology's legal status was murky, and later had to pay retroactive license fees. Although it wasn't planned this way, it turned out to be an excellent way to launch the technology into ubiquity. However, this has left a certain legacy of confusion about when and where it is legal to use MP3, and whether or what fees need to be paid. Commercial "pay and download an MP3 file" sites definitely need to pay a fee for transactions. Commercial encoders and decoders need to do so as well.

MP3 was the first common, "good enough" format for encoding music—128Kbps for 44.1kHz stereo audio is quite listenable for casual listeners, and even picky folks like me have been known to dance to a 160Kbps MP3. Modern formats like AAC and WMA can offer much improved compression efficiency, but have a long way to go to be playable everywhere, as MP3 is. Heck, there's even an MP3 player for my NeXTStation Color Turbo!

Like other MPEG standards, file format and decoder behavior is precisely specified, but implementation of encoding is up to the vendor. Thus, many vendors have created MP3 encoders. The two most important are Fraunhoffer's own encoder (often called FhG), and LAME, both of which are used by many other applications.

## MP3 data rates

MP3 files are broken up into "frames," each 26ms long (or about 38fps). This frame rate isn't related to the video frame rate of any associated video. Each frame needs to be one of a limited number of data rates from 8 to 320Kbps (the minimum and maximum for a whole file as well). If there just isn't enough sound in a given frame to use up its assigned number of bits, leftovers can go into a "bit reservoir," for use by later frames that may need more than they're assigned.

*CBR* Originally, all MP3 files were CBR, where each frame had the same data rate. The nice thing about CBR files is you know how long a file is based upon its duration, and where you need to jump into the file to start playing it at a certain point. CBR is also universally compatible on playback.

*VBR* Each frame's header specifies that frame's size. The folks at Xing (now owned by RealNetworks) exploited this by allowing a file where data rate can vary frame by frame. These files are called VBR. All MP3 encoders I know of are 1-pass, and hence aren't 100 percent accurate at hitting the target data rate. Some don't even try to hit a target data rate. Encoders like Apple's iTunes can only raise the size of a given frame, not reduce it, so long silent sections of a file can use up a lot of bandwidth.

Still, VBR files sound a lot better bit for bit at lower data rates. However, very old MP3 decoders may not correctly support VBR files. For example, QuickTime 4.0 will crash trying to play back a VBR MP3 (although 4.1 and later does it without a hitch).

## MP3 stereo modes

An MP3 file can be mono, joint stereo, or normal stereo.

*Mono* Mono is, of course, a single channel of audio. As always, it's generally better to maintain sample rate than number of channels, so you'll normally encode any MP3 files that need to be less than 96Kbps as mono.

*Joint Stereo* In Joint Stereo mode, frequencies that are hard to spatially locate (at the high and low end of the audible frequency range) aren't encoded to both channels. Instead, the frequency is encoded in a single channel, and a directional hint is added to tell the decoder where to put the signal in the stereo field. Normally, this helps compression efficiency and hence quality.

However, changing high-frequency spatial information messes up Dolby Pro Logic routing information. While this isn't usually a problem with music source, it can be with movie and TV soundtracks.

*M/S encoding*   This is generally considered a subset of joint stereo. Like other Mid/Side-band solutions, instead of recording L and R as two different channels, L+R and L-R are encoded. Because most audio has a lot of information in common between the channels, L-R can be compressed a lot more aggressively, reducing data rate. Most encoders can switch between joint and M/S throughout the stream, using M/S when the two channels are similar, and joint when not.

*Normal Stereo*   In Normal Stereo, the two audio channels are encoded as separate tracks, with no information shared between them. However, the encoder is able to use different frame sizes for each, if one is a lot more complex than the other. This is the optimum mode for high bitrate archiving, or when you need to preserve Dolby Pro Logic information.

*Split Stereo*   In Split Stereo, the blocks in each channel are given exactly the same data rate, even if they have different relative complexity. This mode is also called Dual Channel. I haven't thought of anything this is good for yet.

# FhG

FhG is Fraunhoffer's licensable, professional MP3 encoder. For a long time it was by far the best MP3 encoder, although LAME's continual improvements have brought it into the ballpark with FhG. Most FhG-based encoders offer very few controls, providing choices of bitrate, sample rate, channel mode, and maybe a speed versus quality option—enough to suit the needs of most projects.

# LAME

LAME, originally the recursive acronym for "LAME Ain't an MP3 Encoder," started as an open-source patch to the available reference encoder. This reference encoder worked, but wasn't optimized for speed or quality, and so was enormously worse than the FhG encoder. However, lots of hackers cared about MP3, and so progressively enhanced the patch by adding new features to it. Eventually, for licensing reasons, they removed the remaining source code from the original reference encoder, and had what's effectively a clean room encoder.

LAME, which is freely distributed, doesn't license the MP3 patent. This limits the commercial work you can do with it, probably. There is some disagreement as to the legal status of using the encoder by itself, although it is clearly kosher to use a product based on LAME that pays the MP3 license fees.

## Chapter 22: Miscellaneous Formats

LAME is normally distributed as source code, which you can compile for your platform of choice (it supports virtually all modern operating systems). It doesn't come with a GUI, so most folks either just use the command line, a third-party interface, or an application that incorporates LAME support. Whichever way it's used, the standard parameters of LAME are available in most encoders. I'm going to discuss those most relevant to compression in the following sections.

*--abr (average bit rate)*   I just love the average bit rate command. This is classic, video style 1-pass VBR, where the file shoots for an average data rate while letting the size of each frame to go up or down as much as needed. When you encode with LAME in ABR mode, you get a histogram like the one below indicating the relative size of the different frames. Note the data rates of frames used run from 32Kbps to 320Kbps.

Table 22.1   A typical LAME encoding session. --abr 128 requests vbr with an average bitrate of 128. -q -0 means use the slowest, highest quality mode. The histogram shows how many frames use which frame sizes, with the specific number of frames of each type shown in the brackets.

```
[localhost:/Video] waggoner% lame --abr 128 -q 0 extazia.aiff
LAME version 3.92  (http://www.mp3dev.org/)
Using polyphase lowpass  filter, transition band: 15115 Hz - 15648 Hz
Encoding extazia.aiff to extazia.aiff.mp3
Encoding as 44.1 kHz average 128 Kbps j-stereo MPEG-1 Layer III (11x) qval=0
    Frame          | CPU time/estim | REAL time/estim | play/CPU |    ETA
   7843/7845 (100%)|    2:09/   2:09|    2:19/   2:19|  1.5805x|   0:00
 32 [    2] %3
 40 [    0] 3
 48 [    3] *3
 56 [    7] *3
 64 [    3] *3
 80 [  191] %****3
 96 [  378] %%%******3
112 [ 2896] %%%%%%%%%%%%%%%%%%%%%%%********************************3
128 [ 3034] %%%%%%%%%%%%%%%%%%%%%%%%%*********************************3
160 [ 1090] %%%%%%%%%%%%*************3
192 [  202] %%%**3
224 [   34] %3
256 [    5] %3
320 [    1] %3
average: 125.9 Kbps    LR: 3177 (40.49%)   MS: 4669 (59.51%)

Writing LAME Tag...done
```

*-v (variable bit rate)*   LAME's VBR mode is quality limited 1-pass encode, although you can set a minimum data rate. Even at the highest quality (0), the VBR algorithm can undershoot the bitrate needed to hear a constant acoustical quality, due to deficiencies in the psychoacoustic model. This is improving with more recent versions of LAME, but for high-quality music, I'd recommend you use a minimum data rate of at least 112. By default, LAME will go lower than the minimum for truly silent portions of audio.

*-q (quality)*   The quality mode is a speed-versus-quality control. I always encode with -q 0, the slowest, highest quality mode. It is still faster than real-time on my PowerBook G4 500 with typical projects.

*-m (mode)*   The -m flag sets the encoding mode of the file. You can do mono, normal stereo, joint stereo, automatic (which is normal or joint stereo, depending on data rate), or force m/s encoding for all frames. Generally, auto is just fine.

*--preset*   LAME also includes a number of high-quality presets for particular tasks. These include tuned psychoacoustics models for the particular tasks, like telephony, "CD quality," and voice. The voice preset is especially useful for encoding intelligible low-bitrate speech.

## MP3 Pro

The new enhanced version of MP3 is called MP3 Pro. MP3 Pro files are backward compatible with all MP3 players, but include extra information that an MP3 Pro decoder can use to enhance reproduction of higher frequencies. For example, a 64Kbps MP3 would normally encode at 22.05kHz. When encoded with MP3 Pro, perhaps 60Kbps would be spent on conventional MP3 data and the remaining four kilobits on Spectral Band Replication (SBR) data. An MP3 Pro decoder can use this extra information to synthesize the missing frequencies all the way up to 44.1kHz. A conventional MP3 decoder would just ignore the SBR, and play the file back at 22.05kHz. The SBR decode takes about three times the decoder power as normal MP3. This is unlikely to be a problem with modern computers, and any MP3 Pro playback device will be designed to handle it.

While MP3 Pro is definitely of higher quality than vanilla MP3 at data rates below 128Kbps, it can still sound a little artificial. Quality isn't fully competitive against all-new formats like AAC and WMA, so MP3 Pro is best used in cases where backward compatibility is required, but you anticipate an MP3 Pro decoder will be present.. So far, none of the mainstream architectures includes an MP3 Pro decoder, and you lose a little quality due to losing bandwidth to the SBR data, so there isn't yet any reason to use MP3 Pro over MP3 for most tasks.

## Ogg Vorbis

Ogg Vorbis was created as a patent-free, royalty-free, open-source alternative to MP3. Overall, Vorbis is now in the quality ballpark of good MP3 encoders, without any lingering patent issues. Playback is becoming common, although the only major player with support for it is RealPlayer. A third-party QuickTime component is in beta, however.

## Ogg Tarkin

Free software advocates often herald the coming of Ogg Tarkin, the sister video codec to Ogg Vorbis. However, Tarkin is a long, long way away—not only has no software been written yet, there isn't even a draft specification, nor even a consensus on the fundamental technologies the codec will use. So far, Tarkin is mainly a codec talking group. It may matter eventually, but probably not until the mid 'oughts. I only mention it here so you know not to worry about it yet.

## Theora

Because Tarkin is so far off, there is a new open source effort to make an Ogg video format called Theora. This combines the Ogg Vorbis audio codec with a tweaked version of On2's VP3 video codec. Work on the combined software is in the very early stages right now. It is being positioned as a royalty-free alternative to MPEG-4.

## Obsolete Formats

This industry has been around long enough to develop a lot of format cruft—old files created with old tools. Some of the most common of these are described in the following section. You're unlikely to ever make files in these formats, but the info here should help you recognize an old file type should you ever come across one, and give you some tips on how to play it or convert it to a modern format.

### Smacker

Smacker was the 8-bit predecessor to Bink. It's still used in games, such as Blizzard's StarCraft series, but almost all modern games want more than 8-bit indexed color graphics. Still, over 2,000 games were released using Smacker, and it was the first common technique for playing full-screen cut-scenes in DOS-based computer games (using the old DOS 320×200 full-screen standard). Today, Smacker plays back on DOS, Win32, Mac Classic, and Mac OS X.

Smacker's strengths are the same as Bink's—a very game-friendly SDK and install-free support for a variety of platforms. It's also a lot faster to decode than Bink—it predates Windows 95 and the Pentium! However, Smacker doesn't have nearly the same compression efficiency as Bink. You'll need data rates several times as high to even get in the same quality ballpark, and you'll still be suffering 8-bit resolution.

Smacker uses the same encoder application that Bink does, and has the same general set of controls. The only significant difference is that Smacker has palette support. Smacker can generate its own palettes, or you can provide them. You can change palettes as often as you wish, or use the same one over the entire file.

If you're actually targeting ancient machines or devices that display in 8-bit mode, you'll want to use palette rotation for palette changes. On some old systems, the 8-bit card will flash when you change a palette, showing the change in any pixel that is a color that has been changed. With palette rotation, you only change the colors that aren't being used. A palette rotation value of 128 means that each frame uses 128 out of the 256 possible colors. Thus, when a palette is using 0-127, 128-255 can be changed for the next frame, and vice versa.

The Bink Tools can read Smacker files, and convert them to other formats.

## FLI/FLC

The FLI (pronounced "fly") and FLC ("flick") formats were introduced by Autodesk in their Animator product many, many moons ago. They are both 8-bit indexed color formats. FLI was limited to 320×200 (the hoary DOS "full-screen" mode). FLC can use any resolution, and has somewhat better compression efficiency. Neither supported audio.

Needless to say, these were horrible formats that no one would ever want to use again. Should you encounter legacy files stored in either format, QuickTime can still read them both.

## IVF

IVF is the Indeo Video Format, originally introduced by Intel. IVF includes the Indeo 5 video codec and the 8:1 Indeo audio codec. IVF was an interesting early attempt to leverage wavelet banding in a video file. IVF was a progressive download format, but it front-loaded the lower wavelet bands earlier in the file. When the first five to 10 percent of the file had been transmitted, the audio could be heard, and low quality keyframes could be seen. As the rest of the file was downloaded, the quality and frame rate would improve.

The idea behind IVF was that you could quickly see whether or not the file was interesting enough to wait for the download.

Back in 1997, I did a pilot project for an advertising agency testing IVF's suitability for doing international transmission of dailies for TV commercials. While it was interesting, the advantages weren't strong enough compared to just using MPEG-1. Since then, the improvement in compression efficiency from other formats has proceeded apace, and IVF doesn't do anything competitive today.

Ligos acquired Indeo from Intel around 1999, and for a while pushed IVF pretty heavily. Ligos seems to have given up on that now, thankfully. The authoring tool itself never went out of beta.

The encoding tools and decoder are free, but are only available for Win32. I'm not going to describe how to use them to reduce the temptation for you to try this intriguing-sounding, but in fact, rather useless format.

## Vivo

Vivo was another early player in the Web video space, and was acquired by RealNetworks. Their tools are no longer available. The name lives on in Real's VivoActive PowerPlayer, which allows recording of streaming content to the local hard drive. RealPlayer can play Vivo files, but I don't know of any way to transcode them to another format.

## VXtreme

VXtreme was an early, proprietary, progressive download Web video format. The company was acquired by Microsoft, and members of the VXtreme team have played a central role in the development of Windows Media from NetShow 3.0 on. VXtreme authoring tools are no longer available.

## VDO

The last of the pioneering "V" companies, VDO was the only one not to get snapped up by Microsoft or Real. They provided two different codecs: VDOLive for UDP streaming and Lite VDO for progressive download. The company and the technology are long gone.

## Emblaze Video Pro

Emblaze Video Pro was an early attempt at a Java-only audio and video compression and delivery technology. It failed, largely because the playback requirements were such that it only worked on machines that came with Windows Media Player pre-installed. Emblaze also had poor audio sync, especially on slower machines.

Emblaze Video Pro is no longer available. The company Emblaze is now working on MPEG-4 video playback on mobile devices.

# Chapter 23

# Compression Tools

As you have probably gathered by now, quite a lot of compression tools are available these days. This chapter outlines the major ones. I'm not including MPEG-only applications here, because they don't typically handle nearly as much preprocessing and other aspects of the compression process. They're covered in the various MPEG chapters.

## What to Look for in a Compression Tool

There are a lot of factors to consider in choosing a compression tool and different factors will be more or less important depending on one's needs.

*Price*   Prices for workstation compression tools range from free to about $1,000. Bear in mind the compression tool itself might be a minor part of the cost for a full system, including computers, capture devices, video decks, codecs, and so on.

*Platforms*   We all have our favorite platforms to work on. The issues in picking the right platform are discussed in Chapter 24, "Workflow Optimization." With the exception of RealSystem Producer, which is available for some UNIX systems, the products that follow are only available for Mac OS Classic, Mac OS X, and/or Windows.

*Preprocessing*   Preprocessing is arguably the hardest part of a compression tool to get right, and there aren't any tools in the market that meet all my needs. One particular area of differentiation is support for inverse telecine. Some tools have excellent inverse telecine support, others none at all in that they treat telecined footage as any other interlaced video source, losing half the resolution of the source and requiring higher frame rates. Others have fragile implementations that don't work well with cadence breaks. Good inverse telecine is critical for folks that work with NTSC content originally shot on film.

Deinterlacing support also varies a lot among tools. Ideally, the software will support an adaptive deinterlacing system, so one of the two fields doesn't need to be thrown out wholesale.

Overall, good preprocessing requires controls for tweaking image settings, especially for luma. A surprising number of consumer tools don't include any luma control at all. Lastly, a good tool will intelligently support the many aspect ratios and pixel shapes of contemporary source. 360×240 should never be a default output.

*Input format support*   Not all applications can read all file formats. Each will support one or more architectures for reading files (at least one each for QuickTime, DirectShow, or VfW). Particularly notable problems are reading MPEG-2 in general, reading .avi on Mac with DirectShow-based Windows apps, and reading MPEG-1 with QuickTime.

The ability to read MPEG-2 source files requires a vendor to license patents from MPEG-LA (the MPEG Licensing Authority). It isn't expensive, but it is a factor. For example, these licensing fees are why QuickTime and DirectShow don't support reading MPEG-2 files out of the box, and so individual compression product vendors need to sign up for support.

QuickTime can read .avi and DirectShow .mov as long as the files in question use a codec available locally. Thus QuickTime can read .avi files that are uncompressed or use a DV codec. DirectShow apps are generally limited to reading QuickTime movies with just the None codec and much older CD-ROM codecs unless a third-party codec is installed.

At a minimum you want your software to be able to directly open files produced by any editing or capture software you want to use in your production pipeline. While it's possible to transcode between formats, this often entails a quality hit, and always eats up time.

*Output format support*   It may sound obvious, but your encoding tool should support the formats you care about, and offer the features you care about. Some more subtle features are only supported in a few encoding tools, like 2-pass VBR or MP3 audio in QuickTime. Because it has the deepest architecture, QuickTime support is the most variable among compression tools. Support for Windows Media and especially RealVideo is rather constant between platforms, due to having well documented, complete SDKs, and more focus.

Because MPEG encoders target a format, not a particular SDK, support for the different MPEG formats varies radically among products. Important areas of discrimination include speed and quality, 2-pass encoding modes, a preset for VideoCD White Book MPEG-1 compatibility, and support for more esoteric features like 4:2:2 color space.

Some tools may also support QuickTime Export components, which support a wide variety of other formats of varying utility, like FLC, AVI with QuickTime compatible codecs, and MPEG-4. Apple's awesomely fast MPEG-2 encoder from DVD Studio Pro will work if installed with any applications that support Export Components.

*Speed* Compression is one of the last tasks folks do on a computer that isn't real time, so speed matters. And speed varies a lot among different tools. Some of this may be in quality of algorithms (bicubic scaling is faster than sine, but doesn't look as good, and nearest neighbor is faster and worse looking yet). But of lot of what contributes to a system's speed is straight optimization. One very useful technique is Y'CbCr native color space processing instead of RGB; it is much faster and somewhat higher quality. ProCoder is the first compression tool to support this throughout.

Another important factor related to speed is whether or not a tool supports multiple processors. MP machines are rather inexpensive these days, so MP support is critical to a good performance tool.

The SIMD (Single Instruction Multiple Data) architectures like SSE in Intel and AMD chips, and AltiVec in Motorola's G4 processor can provide massive speed improvements when correctly supported. Video compression apps so far haven't found much use in the P4-only features of SSE2, which are much more oriented towards 3D rendering and scientific computing.

*Stability* With the scope of input and output formats a modern compression tool can address, there is the potential for a lot to go wrong. Without excellent QA and engineering, compression apps can have all kinds of problems. Failing stability is one of the primary reasons why Cleaner, once the compression tool of choice, has fallen out of favor with many users.

Stability for compression products also extends to correctness and compliance in compression. It isn't okay if the app doesn't crash, but the encoded files crash computers on playback! This kind of stability problem is especially pernicious, as it essentially requires you to test all your files completely to make sure they work correctly.

*Integration* Some editing tools support direct exporting to compression tools from their timeline. This eliminates the need to make an intermediate file, and can substantially improve workflow. Don't forget QuickTime reference movies can provide a lot of the benefits of direct rendering.

*Automation* For high volume encoding operations, some degree of automation is required. The easiest form is batch encoding, where multiple files can be queued up to be rendered together. Some applications also support scripting, through either the command line or DCOM on Windows and AppleScript on Mac.

On the high end, enterprise encoding systems can offer a complete automated workflow.

*Live capture support* Some tools can capture video and compress it in real time. Others can capture video from within the app, but still require offline compression after the capture. A few can even do live broadcasting of an incoming stream.

## Cleaner 5.1.2

Discreet's Cleaner was long the leading encoding tool. Originally from Terran Interactive, it was sold to Media 100 after they acquired Terran, and then was part of the sale of Media 100's software division to the Discreet division of Autodesk in 2001.

Cleaner was originally a Mac only, QuickTime only tool. Support for other formats was added in v3, and support for Windows came in v4. Version 5, the first built under Media 100 management, was something of a buggy nightmare, with lots of new but not well tested or integrated features. Support for new compression technologies has lagged since v5 was released; only a minimum of bug fixes and new features were made, often badly. (For example, the wrong buffer size variable was implemented in the Windows Media SDK and only fixed in 5.1.2 after over a year of users complaints.)

*Price*   Cleaner costs $599. Its Charger software upgrade for MPEG-1 and -2 encoding is $499, and the SuperCharger hardware upgrade for MPEG-2 encoding is $999.

*Platform support*   Cleaner runs on Windows 98 or later (it is longer available on NT), and Mac OS X and Classic. RealVideo encoding isn't currently supported on Mac OS X unless the application is run in Classic mode. Cleaner also has some wonderful features for the different formats, such as automatically generating HTML stubs and .asx/.ram metafiles with the correct path to the server.

*Input format support*   Cleaner reads files all files supported by QuickTime, and on Windows AVI through DirectShow, VfW, or QuickTime. Cleaner has its own MPEG-1 and MPEG-2 reader.

However, working with MPEG-2 source in Cleaner is exasperating due to its extreme slowness—it typically takes 10 seconds or more to change the position of the play head in the timeline with MPEG-2 source.

*Output format support*   Cleaner supports all the major delivery formats, including Windows Media, RealMedia, QuickTime, AVI, MPEG-1, MPEG-2, and MP3. Cleaner supports QuickTime with features unique to it and Squeeze, such as 2-pass VBR for Sorenson codecs and MP3 as an audio track. Cleaner can automatically make QuickTime Movie Alternates with a master movie and even a fallback, and set Autoplay modes so a file will automatically play full-screen when opened. Cleaner can support processing other tracks besides audio and video, so a Text track can be burned into the video frame or copied into the output without modification.

Cleaner supports the standard SDKs for Windows Media and RealMedia. It supports Windows Media's 2-pass VBR and CBR.

Cleaner's MPEG-1 and MPEG-2 output is licensed from PixelTools, and is infamous for being slow and buggy. The basic Cleaner has relatively good MPEG-1 support, and a very limited MPEG-2 exporter. The Charger software update adds 1-pass and 2-pass VBR modes for MPEG-1, and complete MPEG-2 support. Charger works reasonably well, although it's extremely slow compared to competitive tools. There is also a SuperCharger hardware upgrade that yields faster (but far from real time) MPEG-2 (it doesn't help MPEG-1). SuperCharger fails to live up to its advertised specifications (VBR doesn't work, Open GOP doesn't work, and so on). I don't recommend it, at least not unless Discreet substantially improves it. Cleaner export of all MPEG types is limited to 2GB files, except when it's through the Windows version of SuperCharger. Cleaner also has a 2GB limit on AVI file creation, although it can read larger files.

Cleaner also supports QuickTime Export components, which automatically show up in the file format list. However, because these show up automatically, newer components may not have been tested by Discreet, and may not work correctly. The MPEG-2 exporter in Apple DVD Studio Pro has been particularly problematic when used from Cleaner.

Future support for MPEG-4 in Cleaner has been announced for Cleaner 6 for Mac due late 2002. Cleaner 6 for Windows had not been announced as this book went to press.

*Preprocessing*   Cleaner was the first product to offer real compression-oriented preprocessing, including automatic aspect ratio correction, adaptive deinterlacing, and hands-off automatic, temporal noise reduction. However, a serious bug introduced in the inverse telecine algorithm in v5.02 and still present in v5.1 can cause the order of frames after a video cut or cadence break to be reversed. This causes a jarring jump in the motion.

Cleaner offers an excellent Dynamic Preview window, where the effects of the current filters can be previewed before rendering.

Cleaner's audio filters are lousy—you're better off using other audio tools like Digitdesign Pro-Tools, Sonic Foundry SoundForge, or Bias Peak for preprocessing.

*Speed*   For most formats, Cleaner's speed is substantially slower than competing products under Mac OS and single processor Windows machines. A bug prevents Cleaner 5.1 from using more than one processor on Windows, even when working with codecs that are MP-enabled. Cleaner uses some, but not much AltiVec and SSE optimization.

Where MPEG-1 and MPEG-2 source files and output are concerned, Cleaner is much slower than competitive tools. SuperCharger's MPEG-2 encoder makes the actual MPEG-2 codec real time, but Cleaner itself still must handle decoding and preprocessing of all the source files in software. This includes an unnecessary Y'CbCr>RGB>Y'CbCr conversion, even if no preprocessing is being applied. Cleaner 6 promises double the performance and will support Y'CbCr rendering.

*Stability* Cleaner was a stable product back at v3.x and before, but has become infamously buggy since. Large areas of functionality appear not to be even tested in the v5.x series, and any number of things that used to work fail. In general, features not available in presets may not work. Problems in versions up to v5.1.2 include both simple crashing to corruption of the installation (which requires a full reinstall) to corrupt files being output that won't work correctly on playback. Increased stability and improved testing is promised for Cleaner 6.

*Ease of use* Cleaner's hallmark is ease of use, and in that it remains unparalleled. The Cleaner UI is clean, with a well-thought-out structure of tabs with individual settings. The interface does an excellent job of keeping common element controls among formats looking the same, while not dumbing everything down to a lowest common denominator.

*Integration* Historically, many editing tools supported Cleaner integration, making Cleaner the default export module. Many editing apps shipped with Cleaner EZ, a stripped down version of Cleaner. Cleaner's install includes a Premiere plug-in.

*Automation* Automation is one of Cleaner's strong points. It can incorporate up to 3,000 files in a batch, all with unique settings. Cleaner also supports automatic FTPing of finished files (although it doesn't give different destinations for media and meta files).

For scripting, Cleaner supports AppleScript on Mac and both DCOM and command line on Windows. For high-end automation there's also the Cleaner Central automated encoding system.

*Live capture support* Cleaner ships with MotoDV, an otherwise discontinued DV capture application. It doesn't work under Mac OS X, and can't capture files longer than 2GB in length. You're almost always best off just using a different tool for capture.

# Tips

1. **Take advantage of mass production with metadata.**

Among Cleaner's strengths is its ability to encode a whole lot of disparate files at once. In Cleaner, the combination of an input file and an output is called a project, each of which is represented in one line in the Batch window. Each project also has an associated settings file, which can have modifications made to it via the Modifiers and Metadata control.

Let's look at an example where we have 10 input files, each of which are going to the same 10 output files. While it would be possible to just make 100 settings files, each of which would specify the metadata and the compression settings, this would be inefficient and difficult to maintain. Instead, we can set up each source with the correct metadata, and apply 10 settings files. First, import all of your files. Then set the metadata on each correctly. Once this is done, do a Select All, and go into Advanced Settings. Shift-click on all the settings you want to use (make sure

Untitled isn't still selected). Hit Apply. And *voilà*! Your 10 input settings and 10 output settings are now 100 projects! And, if the files need to be re-encoded to another format later, you just need to modify the settings file, and all 10 projects that use it will inherit the new settings.

You'll want to either set the file name of each with the Destination panel, or make sure you have "append settings to filename" on in preferences. The latter will output files in the format <source file name>_<setting name>.<extension>. Very handy, although Cleaner's 31-character limit means you'll need to keep the file and settings names short.

2. Skip the Wizard.

By default, Cleaner uses the Settings Wizard instead of the Advanced Settings dialog. The first thing you should do when installing Cleaner is to set it to use Advanced Settings by default. The Wizard doesn't make very good choices, and hides most of Cleaner's most important features from you. You're much better off jumping straight to learning the advanced settings dialog.

3. Turn off "Minimize Preview."

By default, Cleaner ships with the Minimize Preview item in preferences checked on. This forces the video frame in the output window to be drawn at lower resolution, only for every fourth frame, and disables preview of compressed output. Uncheck it, and you'll get a full resolution preview of every frame, and for QuickTime, RealVideo, and AVI, an A/B slider lets you see the frame before and after going through the codec. By watching the file as it encodes, you can often detect problems without having to wait for the entire file to render. There is a slight performance hit in this display, so just close the disclosure triangle for the preview when you're not looking at it.

4. Don't use Cleaner for audio processing.

While Cleaner includes a number of audio filters, all except the normalize filter and audio rate resampling are really bad. If you have source that requires audio sweetening, reprocess the audio earlier in the process.

5. Don't use Windows Media "Video Pre-Buffer Duration."

In one of the more embarrassing gaffes in recent Cleaner history, the wrong buffer size variable from the Windows Media SDK was implemented in Cleaner 5.0.2 and remains in v5.1.2 for Mac (but was fixed in v5.1.2 for Windows). Video Pre-Buffer Duration, instead of changing the buffer size of the encoding, just changes value in the file indicating buffer size without changing the underlying value. Leave the check box unchecked and the SDK will generate this value correctly—assigning any value can lead to a file that won't play correctly.

6. Make a dummy audio file for Windows Media.

One annoying limitation of the Windows Media SDK is that every file requires an audio track. Ideally, Cleaner would automatically generate some blank audio, but it doesn't. The old time workaround would be to append blank audio to the source file in QuickTime Player Pro or Premiere. However the "Enable Multipart sources" preferences option can help a lot. When this option is on, Cleaner will auto-associate any .wav file with the same file name as the video source—dragging the video file into Cleaner will automatically make the audio file the audio track for that project. Windows Media doesn't care how long the audio for the file is, so you can make a one-second-long audio file. And just copy it into the directory with the video file and rename it, and you're good to go.

Of course, because the audio is null, you'll want to use as low a data rate as possible—WMA 5Kbps. I hope a future version of Cleaner will support the "null" 0Kbps audio track.

7. Use High Quality frames for progressive download files.

For QuickTime files, Cleaner supports having either the first or the last frames be High Quality. This overrides the data rate control, and encodes them just as if they were encoded with a spatial quality value of 100. This can be great for progressive download files because you tend to look at both the first and last frames for quite a while. It doesn't do much for streaming files, of course. High Quality frames will bump up the data rate a little, but because it's two frames at most, the change won't be material except for the shortest files.

8. Encode MP3 as Highest Quality.

Until QuickTime 6 with AAC becomes mainstream, MP3 is the best general purpose audio codec for progressive download QuickTime movies, and Cleaner is one of the few products that supports encoding straight to MP3. However, by default, Cleaner sets the Speed versus Quality to "Normal." Make sure you always switch it to "Highest Quality (Slower)." It really does make a difference, especially at lower data rates.

# Canopus ProCoder 1.0.1

Canopus's ProCoder is the first application that has threatened Cleaner's dominance as a complete, standalone professional compression tool. Its v1.0 release is promising, although rife with rough edges. But, it's already fast and stable, providing a good platform for future enhancement.

*Platform support*   ProCoder requires Windows 98 or later, 2000, or XP preferred.

*Input format support*   ProCoder can read files with either DirectShow or QuickTime, and includes its own MPEG-1 and -2 parsers.

*Output format support* ProCoder supports output to QuickTime, Windows Media, RealVideo, AVI, MPEG-1, and MPEG-2. Its support for the non-MPEG formats is pretty generic, without any unique features. ProCoder includes Canopus's MPEG-1 and MPEG-2 codecs, which are extremely fast and provide excellent quality.

*Preprocessing* ProCoder has decent, but not exceptional, preprocessing. It is provides good, pre-render preview of the video, and a very nice side-by-side display. One delightful feature is audio previews, so you can hear how the filtering will sound. The initial release doesn't have inverse telecine, but this has been promised for a future, free update.

ProCoder does a good job of keeping aspect ratio and pixel shape straight, which is very important for working with MPEG.

*Speed* Speed is a major selling point of ProCoder. It leverages SSE and multiple processors heavily and works in native Y'CbCr color. ProCoder can also render multiple, simultaneous outputs from the same source in parallel—meaning source decode and preprocessing only has to happen once. On a typical dual processor Windows machine, ProCoder can be four times faster than Cleaner for typical projects, and many times faster yet for MPEG-1 and MPEG-2.

*Stability* ProCoder is extremely stable for a v1.0 product.

*Ease of use* ProCoder's biggest limitation is a somewhat clunky UI. It makes a false dichotomy between basic and advanced settings for each format, so you have to specifically go into an advanced mode to see most of the options. It'd be much simpler to have a single view available everywhere.

One nice feature is you can apply video and audio filters either to the source, or to a particular output. This lets you, say, get deinterlacing and luma adjustment correct on the source video, and then tweak gamma correction and scaling as appropriate for each output format.

ProCoder includes an encoding wizard that takes you through a step-by-step process of picking from a limited number of encoding options. It works fine, but I imagine anyone who cared enough to read this far in this book would just use the main UI.

*Integration* Canopus has a plug-in for Premiere, allowing exporting to ProCoder straight from the Premiere timeline.

*Automation* ProCoder 1.0 doesn't include automation features.

*Live capture support* ProCoder doesn't do live capture.

## Tips

**1. Render in Parallel.**

ProCoder separates preprocessing and compression into different passes. The preprocessing is only done once even if you have multiple outputs, making ProCoder a much more efficient tool for multi-format rendering. If you are going to make multiple versions of the same source, do them all together in a single project for maximum speed.

**2. Hint QuickTime movies for RTSP.**

ProCoder 1.0 doesn't support hinting QuickTime movies. If you want to RTSP stream a file encoded with ProCoder, you'll need to do an Export as Hinted Movie from QuickTime Player Pro.

**3. Let ProCoder transcode.**

ProCoder will do the "right" automatic preprocessing in most cases, so you don't need to sweat preprocessing when going to MPEG formats. ProCoder automatically deals with field order (switched when needed for interlaced output, automatically deinterlaced for progressive). It also will automatically crop 720 wide sources down to 704 before scaling to 352 for MPEG-1 output. And ProCoder can do darn good PAL/NTSC transcoding for software (although it isn't as good as dedicated hardware units). ProCoder will also automatically letterbox or crop (set in preferences) your video if the input and output aspect ratios don't match.

# Adobe Premiere 6.01

While it is primarily a video editing app, Adobe Premiere was the most-used compression tool before Cleaner—Premiere 4 predates the first version of Cleaner. However, Premiere's video compression features haven't evolved much since then. Instead, Premiere has added format-specific export modules for RealVideo and Windows Media. The native exporters are pretty much the standard options provided by those SDKs, as described in each of the formats' chapters. Cleaner and ProCoder both install Premiere export modules, so you can go straight to those applications from Premiere's timeline.

*Price* Premiere costs $549. It is also available in bundles with other Adobe products, and is bundled with video editing boards from a number of vendors.

*Platforms* Premiere runs under Windows 98 or NT or later, or Mac OS 9.x. Adobe has stated a future version will be available for Mac OS X.

*Input format support*   Premiere can read DirectShow or QuickTime files on Windows, and QuickTime files on Mac.

*Output format support*   The only standard formats Premiere's native Movie Export module supports are AVI and QuickTime (it also supports a mediocre Animated GIF export, and various numbered image formats such as FLC).

The export mode supports some old-school features, like automatic 8-bit palette generation (this only adds a palette to be used when the file is played back on an 8-bit computer display—it doesn't flatten the image down to an optimized palette).

*Preprocessing*   Premiere is mainly a video editing tool, and has a conventional set of filters. Most of its video filters are keyframeable. However, Premiere is not a finishing tool like its Adobe sibling After Effects, and lacks inverse telecine and the same kind of low-level control AE provides. However, it does have Adobe's wonderful levels-with-histogram control.

Premiere does a fine job working at arbitrary resolution, aspect ratios, and codecs, so you could edit an entire piece in 640×272 Huffyuv if you wanted. Thus other tools can be used to preprocess earlier in the compression process.

When exporting to QuickTime or AVI, you get access to the Special Processing control. This lets you specify some old-school preprocessing filters, including cropping, scaling, a blur, a simple deinterlace, on a decent resizer, and a gamma control. And when I say old school, I mean *old school*—this dialog has been essentially unchanged since before Movie Cleaner Pro 1.1. Still, it was used to create much of the CD-ROM video of the mid-90s, and is still useful for creating quick comps. Note the Gamma correction control is the reverse of most other tools, with values below 1 making the image brighter. Use 0.8 instead of 1.3 for Mac to Windows correction.

If you're exporting to RealVideo, you can also use the Real SDK's integrated inverse telecine, deinterlace, and scaling features.

Premiere does everything in 8-bit-per channel RGB.

*Speed*   Premiere is fairly fast. It supports SSE on Windows and dual processors on both Mac and Windows.

*Stability*   Premiere has been around for ages, and is well battle-tested. The application itself is very stable, although buggy drivers for third-party video editing cards are always a risk.

*Ease of use*   There has been a religious war among Premiere users over whether the UI of v4.x and earlier was the One True Premiere, or whether the UI introduced in v5.0 and refined in v6.0 deserves that title. I'm personally fine with the new model, even though Premiere 4.0 was the application I learned to edit on.

*Integration*  Premiere supports exporting directly to Cleaner and ProCoder straight from its timeline.

*Automation*  Premiere supports limited batch processing through its Batch Processing utility. This enables you to set up a list of project files for batch rendering. However, this only gives you access to the Export Movie option, and hence QuickTime and AVI files are the only significant formats available here. Premiere can't do batch encoding to RealVideo or Windows Media.

*Live capture support*  Premiere can capture video with the best of them (it is an editing app after all). It supports a wide variety of different capture cards and device control mechanisms. It can't compress in real time to a final file format though.

# Tips

1. Add keyframes.

Premiere is one of the few tools that supports manual keyframes. This is done with the Keyframes on Markers option. Just put a marker in the timeline at points where you want a keyframe, and if Keyframes on Markers is checked, those frames will be keyframes on output. Just hitting the * (asterisk) key while on the timeline will add an unnumbered marker at the current frame.

The Keyframes at Edits command automatically makes the first frame after an edit a keyframes. This would automatically happen in most cases, but if you have slow transitions or whatever, Keyframes at Edits ensures a keyframe will be created at those edit points.

2. Recompress when needed.

Premiere is great to work with when you need to make a change to an already compressed QuickTime or AVI file without recompressing parts that don't need to be changed. If you set Recompress to "Maintain Data Rate" and use output settings that match the input (resolution, frame rate, codec, data rate), only frames that have been changed, or were above the original target data rate, are recompressed. If Recompress isn't checked at all, only changed frames are recompressed.

3. Use enhanced rate conversion.

Premiere's default mode for resampling audio isn't very good. If you're changing the audio frequency in Premiere (say from 48kHz to 44.1kHz), make sure you have Enhanced Rate Conversion quality set to Best. The performance hit isn't noticeable on modern machines, and the quality gain is well worth it.

4. Use Better Resize.

If you're using the Export Movie control, make sure you have Better Resize turned on under Special Processing. This turns on Bicubic scaling; without it Premiere will use the atrocious Nearest Neighbor algorithm.

5. Optimize stills.

Optimize makes still frames into single frames of long duration instead of inserting multiple copies. This improves quality and saves on data rate.

6. Create a slide show with Frames only at Markers.

If you want to make a classic filmstrip-style movie, use Frames only at Markers. This will output a file that consists of only those frames at which you placed markers. This can provide excellent spatial quality at low data.

# Adobe After Effects 5.5

Adobe After Effects is a finishing tool for doing video compositing and effects. From that heritage, its real strength is as a preprocessing tool, although it has decent but not phenomenal support for encoding to QuickTime, AVI, Windows Media, and RealVideo.

*Price*   The Standard version of After Effects is $649, and the Production Bundle is $1,499. It'd be unusual to buy After Effects just for compression.

*Platform support*   After Effects is available for Mac OS 9.1 and OS X 10.1 or later, or Windows 98 and 2000 or later.

*Input format support*   After Effects can read QuickTime files on Mac and QuickTime and VfW on Windows. It also supports high-end formats like Cineon (high-resolution files output by telecine machines) and numbered still image files.

*Output format support*   After Effects can directly render to the compressed formats of QuickTime, AVI, RealVideo, and Windows Media under Windows. It can also support a variety of numbered image and high-end formats like Cineon. After Effects' QuickTime support is relatively limited compared to that of dedicated compression tools. RealVideo support is pretty complete except it has no 2-pass encoding modes. For most projects, you'd want to use After Effects to build an intermediate file for later compression to achieve optimal quality.

*Preprocessing*   The reason to use AE for compression is for its preprocessing. AE has a very broad range of features, and a rich ecosystem of plug-ins. While AE may be overkill for most projects with clean video source, nothing beats AE and its plug-ins for rescuing bad video. The one, glaring exception is that its inverse telecine can't work with source containing cadence breaks.

After Effects' audio support is weak—it doesn't even have a normalize filter.

*Speed*    After Effects can be quite fast if you run it in lower quality modes, but you wouldn't do that, would you? In 16-bit RGB mode things are somewhat slower yet, although they will have unrivaled quality. After Effects is MP- and SIMD-optimized.

*Stability*    AE is so stable you don't even think about its stability. I don't recall AE 5.x crashing ever.

*Integration*    A number of tools, including Media 100 and Final Cut Pro, integrate with After Effects, letting movies be exported to After Effects projects. However, this is mainly useful when using After Effects for finishing, not preprocessing.

*Automation*    After Effects supports Expressions, which allow you to associate functions to other functions. For example, so the color of an object can relate to its position relative to the camera. But on the rendering side, AE doesn't support scripting. That said, it does have a lot of robust rendering options: support for render farms including rendering on Linux clients; batch rendering via the Render Queue with any combination of sources and output formats; and it can even automatically split rendering among multiple copies of After Effects, which can be useful with long-form projects. This is done with the Watch Folder feature—you need to have at least one Production Bundle to set it up, but you can use any AE version to do rendering.

*Live capture support*    After Effects doesn't do live capture.

# Tips

### 1. Use 16-bit rendering.

There are two versions of After Effects, Standard and Production Bundle. The only important feature for compression in the Production Bundle is 16-bit rendering. Although After Effects is RGB only, when processing in 16-bit mode, it can do an excellent job of reducing the quality hit of color space conversions and multiple filters (although I still long for a Y'CbCr native version of AE). There is a speed hit, of course.

### 2. Render in Best Quality mode.

After Effects and many of its filters support Draft, Normal, and Best modes. Always output for compression in Best mode—it turns on antialiasing and other features that can make a smoother, easier-to-compress image. You can either set the quality flags to Best for the individual layers, or you can override everything to Best Quality in the Render Queue. I prefer to set all my layers to Best Quality while I'm working in the timeline to get as accurate a preview as possible.

3. Know the ins and outs of deinterlacing and inverse telecine.

After Effects can do a good adaptive deinterlacing and inverse telecine, but you have to jump through some hoops. These settings are specified in the Interpret Footage dialog. The first requirement is that you should know the field order of your source. After Effects can use that information to treat both fields as source, so you can do 59.94fps or 20fps output from 60i NTSC. If you are deinterlacing, make sure to have Motion Detect turned on— it's a good adaptive deinterlacer.

After Effects can do a fine job with inverse telecine, but it can be rather labor intensive, because it requires a specific cadence pattern. After Effects can "Guess 3:2 pull-down," by analyzing the first few frames of the video. However, this doesn't work if the first frames of the clip are black. In such cases, you're better off cropping the head of the file before importing it into After Effects, or just trying all the alternatives. If you're not sure what the field order is, that's 10 different variants you have to try. When troubleshooting field order, make sure you're looking at your source at 100 percent scale on the timeline. If you're working at a lower resolution, it can be very difficult to see interlacing problems.

23.1  After Effects' crucial Interpret Footage dialog, configured for Inverse Telecine.

Also, note the cadence needs to be the same for each clip. If you have a source with a cadence break, you'll have to break it up into multiple sections before taking it into AE. This generally isn't worth the effort.

4. Use the Levels filter for luma adjustment.

One of the joys of preprocessing with After Effects is the histogram in the Levels filter. This handy display allows you to tell at a glance whether your blacks are over-saturated or if your whites are blown out. Numeric entry allows you to quickly and easily set the black level to 16 and white to 235 for perfect video-to-RGB computer display range conversion (assuming your source footage is clean).

I wish other compression tools would incorporate this simple, powerful feature. Note the histogram is only shown for the current frame. Skip around the timeline to examine the values on other frames.

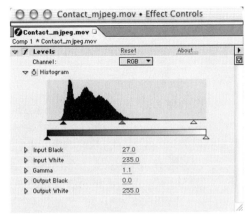

**23.2** After Effects' lovely Levels filter. It offers wonderful control and feedback over luma levels.

5. Use Color Finesse.

My favorite color-correction tool, Synthetic Aperture's Color Finesse, is only available as an After Effects plug-in. If you have serious color-correction work to do, check it out. Color Finesse is discussed in more detail in Chapter 7, "Preprocessing."

## VirtualDub 1.4.10

VirtualDub is a free software video compression tool. As free (as in speech) software, anyone can download and modify the source code of the application and distribute new versions, as long as they distribute the modified source code as well. Thus, there are a number of tweaked version of VirtualDub floating around, as well as a healthy plug-in ecosystem. However, VirtualDub is fundamentally a hobbyist tool, not aimed at professionals. This isn't always a bad thing—VirtualDub is wickedly fast, and is good at the kinds of things you'd want it to be good at if you did a lot of conversions of DVDs to other formats. However its UI is spartan at best.

*Platforms*   VirtualDub is Win32 only.

*Input format support*   VirtualDub uses VfW for reading AVI files, which means VfW AVI files that use DirectShow codecs won't work. VirtualDub also includes its own internal MPEG-1 (but not MPEG-2) file reader.

*Output format support*   VirtualDub only outputs to AVI. It can produce both v1.0 and v2.0 AVI, and can segment large files into multiple segments.

*Preprocessing*   Because of its open nature, a lot of folks have developed preprocessing algorithms for VirtualDub, thus it offers a fascinating but spotty bag of tricks. For example, VirtualDub has lots of different scaling options and a half dozen deinterlacing modes, but no inverse telecine or gamma adjustment.

*Speed*   VirtualDub is blazing fast, with excellent support for both MMX and SSE. It also supports native color space processing for optimal quality and speed. On a fast machine, VirtualDub can do real-time previews, of video as well as audio.

*Stability*   VirtualDub is extremely stable, and warning messages pop up when you're about to do things that might not be supported by certain codecs. Because it's open source, technically savvy users are able to track down and fix bugs themselves.

*Ease of use*   VirtualDub feels like it was written by programmers. Adding filters is a matter of picking Filters from a drop-down menu, hitting the Add filter button, and selecting and tweaking settings.

*Integration*   VirtualDub supports a Frame Server, which means it can make a dummy AVI file that other applications can open, but generate the frames in real time instead of having to make an entire intermediate file. This makes up somewhat for AVI's lack of QuickTime Reference Movies.

*Automation*   VirtualDub supports command-line scripting.

*Live Capture Support*   VirtualDub supports any VfW capture device. It can do a live compression during capture on a fast enough machine.

## Tip

1. Use Fast Recompress mode.

If you're going from one 4:2:0 codec to another, VirtualDub can do native color space processing, for perfect quality and tremendous speed.

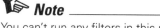
### Note

You can't run any filters in this mode, so it's just for transcoding.

# Sorenson Squeeze 2.0

Sorenson Media's reputation was built on being a codec vendor. Squeeze is their first standalone video compression product. Sorenson's marketing is somewhat confusing—there are two products listed in their line, but only one application. You can get Squeeze for QuickTime, which includes the Sorenson Video 3.1 Pro codec. You can also get Squeeze for Flash MX. However, if you have both, the single Squeeze application gives you all the options of both. A version with MPEG-4 export has also been announced.

*Price*  Squeeze costs $299 for the QuickTime version, $199 for the Flash MX version, or $379 for both. Squeeze for QuickTime also includes the Sorenson Video 3.1 Pro codec, which you can also use in other applications.

*Input format support*  Squeeze uses QuickTime to read files for both Mac and Windows, so it can read any formats supported by QuickTime. The Windows version can also use DirectShow to read AVI files. Squeeze can also capture live to a DV intermediate through a 1394/FireWire port on the computer. Compression itself happens off line, of course.

*Output format support*  Squeeze for QuickTime outputs a small subset of QuickTime—the SV3.1 video codec and the MP3, Qdesign Music 2, PureVoice, and IMA audio codecs. Squeeze is the only other tool besides Cleaner that supports 2-pass VBR with SV3.1 and using MP3 as an audio encoder for QuickTime files. The 2-pass VBR implementation is slightly better than the one in Cleaner.

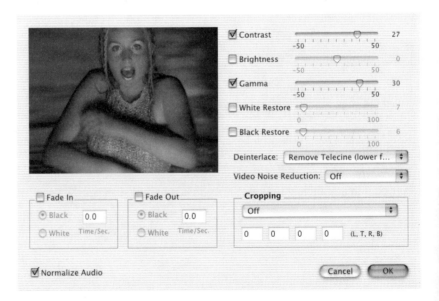

**23.3**  Squeeze's filter dialog. Squeeze has an excellent inverse telecine filter.

Squeeze for Flash MX has many unique, crucial features. You won't want to use anything else for mission-critical Flash work. See Chapter 21, "Flash MX," for more details.

*Preprocessing*   The first version of Squeeze was known for weak preprocessing, but it's much improved in v2.0. Still, Squeeze's preprocessing can be difficult to work with. You have standard filters like Contrast, Brightness, Gamma, White Restore, and Black Restore. However, these are all set with sliders, none of which provide numeric entry, making it difficult to precisely tune settings. The preview updates live as you change parameters, which is very nice. Squeeze has a very robust inverse telecine. My biggest problem with Squeeze's preprocessing is that you can't save filter settings!

Another annoyance is you can't scrub between frames while in the preprocessing window. Preprocessing takes place before scaling, so everything is automatically shown in a 320×240 window.

*Speed*   Squeeze isn't a speed demon. While the Squeeze application is not MP-enabled, it can use a second processor for the SV3.1 Pro codec on Mac, but not on Windows.

*Stability*   One advantage of having a very narrow feature set is that testing and bug fixing is much more easily accomplished, hence Squeeze is quite stable.

*Ease of use*   Overall, Squeeze is one of the easiest to use compression tools, although it still has a few rough edges. All the settings are provided in only a few windows and the feature set is constrained to those useful for its intended audience. I don't find Squeeze's settings very intuitive to use within my standard workflow. For each output file type, there are seven presets for different data rate targets. To tweak those, you need to double-click on the preset (Mac), or right-click

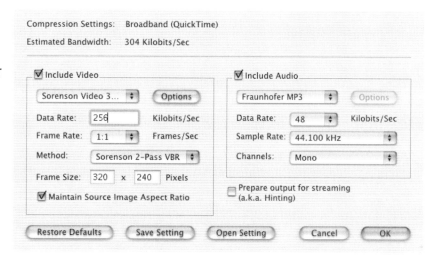

**23.4** Squeeze's QuickTime dialog. It supports many features that were formerly Cleaner only, like 2-pass VBR and MP3 audio.

on an item in the Output Files window (Windows). You can save modified presets, but I find the overall experience cumbersome.

Squeeze can only encode a single file at a time from its main UI. You can use Watch folders to do batch encoding, as described in the following section on Automation.

*Integration*   Squeeze doesn't have direct integration with any other products. But as long as an application can output a QuickTime-compatible file, you can use Squeeze in conjunction with it.

Squeeze nicely integrates with Sorenson's Vcast video storage and hosting service. You can directly render a file from Squeeze to your Vcast account.

*Automation*   Squeeze supports a Watch Folder, where any source file put in there is automatically processed based on the current settings. Alas, you can't have multiple folders with different settings. Squeeze doesn't support scripting on either Mac or Windows.

## Tip

1. Enjoy faster encoding with Preview off.

If you're doing a big batch, or a long file you aren't going to be babysitting, turn Preview off while rendering. It will speed things up a bit.

# HipFlics 1.1

HipFlics was created by venerable QuickTime vendor Totally Hip, best known for their indispensable LiveStage Pro and LiveSlideShow products. HipFlics was their first venture outside of interactive authoring, and has a frustrating combination of welcome innovation and critical shortcomings.

I've never worked with a product where I simultaneously longed for a 2.0 release while finding the current version effectively useless.

*Platform support*   HipFlics is available for Windows, Mac OS X, and Mac OS Classic.

*Input format support*   HipFlics exclusively uses QuickTime for import, and can read any QuickTime-compatible source.

*Output format support*   HipFlics only outputs QuickTime movies. It can use any available audio or video codec. HipFlics has the unique ability to use different codecs and settings on different sections of the video. However, these have to be set manually, and this feature isn't of practical use for the vast majority of projects.

***Preprocessing*** The most important thing to know about HipFlics' preprocessing is that it can't deinterlace. HipFlics relies on QuickTime for all its filters, so it lacks critical ones like deinterlacing, while providing not very useful options like lens flares.

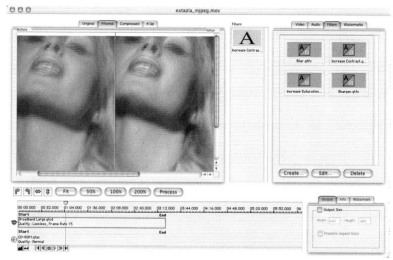

The frustration here is that HipFlics supports applying different filters to different parts of the file, in contrast to most compression tools that can only apply a global setting over an entire file. HipFlics also has an elegant side-by-side filter view, which lets you zoom in up to 200 percent, and do a synchronized pan of the image to closely compare before-and-after filters. But it still can't deinterlace.

**23.5** HipFlics' innovative UI. Still, no deinterlace filter is a hard one to get around.

***Speed*** HipFlics is pretty fast on a G4 Mac, but not on other platforms, as it relies on Apple's optimization of QuickTime.

***Stability*** HipFlics seems to be pretty stable, although it hasn't been extensively battle-tested by professional compressionists due to its limitations.

***Ease of use*** Overall, HipFlics is quite easy to use, and has some welcome innovations like the synchronized panes. I'm not fond of the drag-and-drop approach for filters, though, especially the way you can't visually see all the filters and their values applied to a given track.

***Integration*** HipFlics doesn't have any specific integration features.

***Automation*** HipFlics supports AppleScript on Mac, but doesn't have any scriptability under Windows. HipFlics can export Droplets that store a particular group of settings, and will compress any file(s) dropped on it.

# Tip

1. Wait for v2.0.

## Helix Producer

Helix Producer 9 is RealNetworks' video compression tool. It replaced RealSystem Producer. It's available in two versions: Basic and Plus.

*Price*   The Basic version is free, the Plus version is $199.95.

*Platform Support*   Helix Producer is available for Windows and Linux 2.2 for Intel. The older RealSystem Producer 8.5 is still available for Mac OS Classic and Solaris for SPARC.

*Input format support*   Helix Producer supports DirectShow on Windows and QuickTime on Mac. It now uses QuickTime for .mov support on Windows, but the initial version seems to have difficulty with some QuickTime files. Huffyuv AVI files also have some problems.

*Output format support*   Helix Producer only makes RealMedia files. It has full support for the Real SDK, but doesn't have any unique features compared to other tools that support the full SDK. It does RMVB progressive download files, which most of the other tools don't support yet.

*Preprocessing*   Helix Producer has decent deinterlacing and inverse telecine. Beyond scaling, cropping is your only real option—there aren't any image filters.

Helix Producer can extract a progressive frame from each field of video, à la After Effects, so you can do 59.94fps from NTSC to 50fps from PAL source.

*Speed*   Helix Producer is wickedly fast.

*Stability*   Helix Producer is tight and stable.

*Ease of use*   The old RealSystem Producer was very difficult to work with. Features that are file-specific, like inverse telecine, are set up as global preferences. Data rates can't be specified directly. Instead, you assign a setting to one of the eight default "audiences," like DSL 512K, and then specify the data rate and an audio format for the particular audience.

This all was much, much improved in Helix Producer Plus—although Helix Producer Basic doesn't allow customization of data rate bands.

*Integration*   Because the RealMedia SDK is so easy to adopt, most programs build in their own RealVideo support. Thus, no applications directly export to Helix Producer.

*Automation*   Helix Producer Basic doesn't support batch encoding but Plus does. However, it does support command-line encoding and you can use that to set up a batch for encoding. The command-line mode also lets you specify data rates directly, and otherwise overcome some of the limitations of the UI.

# Vegas Video 3.0a

*Platforms*   Vegas Video runs on Windows 98, Me, 2000, XP, or later.

*Input format support*   Vegas Video supports all DirectShow and standard QuickTime files. It doesn't work with Media 100 codec files, for some reason. Vegas can also import Windows Media files, a unique feature in editing applications.

*Output format support*   Vegas Video supports a plethora of output formats, including QuickTime, RealMedia, Windows Media, AVI, MPEG-1, MPEG-2, and even Ogg Vorbis.

For the most part, Vegas Video supplies all the standard SDK options without a lot of special tweaks; there's almost nothing missing. Two notable limitations: Sorenson Video 3.1 Pro's Bi-directional Prediction mode isn't supported, and is automatically turned off; and 2-pass VBR isn't an option for QuickTime.

One unique feature is that Vegas Video can edit and re-export already compressed Windows Media files!

*Preprocessing*   Vegas Video is more of a video editing app than a compression app. While it supports a number of filters, only a few are appropriate for preprocessing. It has an After Effects-style levels command, sans the enormously useful histogram. The levels command does include a very nice "Studio RGB to Computer RGB" filter. Alas, all its filters run in 8-bit RGB.

While Vegas Video itself doesn't have inverse telecine or other important preprocessing features, preprocessing can be used in exporting for SDKs (like Windows Media) that support it. Scaling is also handled by the SDKs. Because algorithms vary between platforms, this means that the output frames could be quite different among different platforms.

Letterboxing is supported in all formats.

Vegas Video mislabels the RealVideo Noise Filter settings as High Quality and Low Quality. The values are actually high and low amounts of noise reduction.

*Speed*   Vegas Video is pretty darn fast.

*Stability*   Vegas Video is stable.

*Ease of use*  Vegas Video's compression features feel somewhat bolted on—it's jarring to only be able to do inverse telecine with a single format!

*Integration*  Vegas Video, as a standalone editor, doesn't offer integration with other editing software.

*Automation*  Vegas Video doesn't support any automation features.

## Tips

1. **Always export with Best rendering quality.**

By default, the Render mode uses Good video rendering quality. Be sure to set it to Best.

2. **Make sure pixel aspect ratio is correct.**

Vegas Video doesn't automatically set the pixel aspect ratio correctly. Make sure you do this manually in the Video tab in the Custom Settings dialog.

# Windows Media Encoder 8/7.1

The current Windows Media Encoders are 8 for Windows XP and 7.1 for all pre-XP versions. There aren't any major differences between them.

*Platforms*  Windows Media Encoder requires Windows 98, 2000, or later.

*Input format support*  Windows Media Encoder imports any DirectShow compatible source.

*Output format support*  Windows Media Encoder only outputs WMA and WMV files. It doesn't currently support the 2-pass VBR and CBR modes in the current version.

*Preprocessing*  WME offers limited preprocessing: cropping, scaling, and deinterlacing/inverse telecine. It doesn't have any luma controls at all. Preprocessing in WME is made complicated because it doesn't have any kind of real-time preview.

*Speed*  Windows Media Encoder is extremely fast. It is the most MP-enabled encoder available, able to use up to four CPUs for video and one for each audio channel. It is highly optimized for SSE as well.

*Stability*  Windows Media Encoder is very stable.

*Ease of use*   Windows Media encoder is focused more on the novice and consumer user than the professional, whom Microsoft assumes will use a higher end, commercial encoding tool.

While Windows Media Encoder doesn't hide important options in preferences, it suffers from some of RealSystem Encoder's tendency to put "ease of use" features between you and the tool. For example, you can't directly specify a data rate, but you need to define a new audience that uses that data rate.

*Integration*   Because the Windows Media SDK provides access to the full functionality of the encoder, most applications support the full SDK and don't call the encoder.

*Automation*   The Windows Media Encoder doesn't directly support batch encoding. It does support command-line and DCOM automation, and command-line batch files can be used for batch encoding.

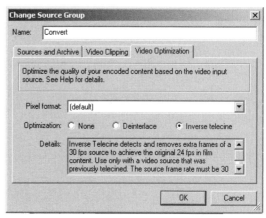

**23.6**   The well-hidden Video Optimization pane of Windows Media. It has a pretty good inverse telecine for live encoding.

## Tip

1. Find preprocessing.

Windows Media Encoder deftly hides its preprocessing features. First, you need to activate the Sources pane via the View menu. Then you right-click on the source file, and pick "Change." This gets you access to Clipping (cropping), and inverse telecine and deinterlacing.

#  Chapter 23: Compression Tools

# Chapter 24

# Workflow Optimization

## Why Workflow Matters

If you only have to compress an occasional file, it can be hard to understand the needs of medium- and high-volume production environments. On the high-end, there are installations that can produce dozens of hours of compressed content each day. Even a small shop can benefit from a well designed automated workflow. Processing just 20 30-second commercials to eight different file format/data rate combinations is 240 compressions, and making sure there aren't any significant errors in those 240 files or their associated HTML and meta files is a daunting task to tackle manually.

The bottom line is that a well-thought-out workflow makes the routine automatic, so you don't have to make decisions to which there is only one right answer. While optimum quality will always require some customization, you only want to have to make decisions that actually matter.

## Choosing an Encoding Platform

One central question compressionists face is deciding which platform or platforms to encode on. The real secret is, of course, is to pick the applications you want to run, and then pick the best platform to run them on.

### Mac OS

Back in the day, Mac OS was the dominant platform for video compression. PowerPC processors were faster than their Intel equivalents, and the Mac OS user interface was a lot easier to use. Since then, Motorola has been lagging behind Intel and AMD in the processor wars. And Microsoft has done a tremendous job improving Windows support for video authoring, and the current Windows XP Workstation is quite nice to work on. Still, given the heritage of the Mac in video authoring, it's the "native" platform for many video professionals. A number of unique features like AppleScript and ColorSync ensure the Mac will remain popular for video authoring for some time to come. If anything, the Mac has been gaining market share in video authoring, mainly due to Apple software. iMovie and iDVD have helped Apple machines be the best consumer video and DVD authoring platform. And Final Cut Pro has taken the professional world by storm.

As I'm writing this book, the fastest Mac systems run dual 1GHz G4 processors, and the fastest Windows machines run dual 2.2GHz Xeon (Intel's name for multiprocessor-enabled P4) processors. All else being equal, this should make a Windows machine much, much faster. However, all things aren't equal. First, the G4 gets a lot more done per GHz than a Xeon due to a fundamentally more efficient design. Second, the AltiVec SIMD (Single Instruction Multiple Data) architecture is substantially more efficient and easier to code for than Intel's SSE (Streaming SIMD Extensions, PIII) and SSE2 (P4) libraries. The net effect is that highly optimized software tends to be faster on the fastest Macs, and less optimized software is faster on the fastest Windows machines. Future processors from the different vendors may change this.

Of the major architectures, only Windows Media has radical differences between Mac and Windows for video compression. QuickTime is obviously kept up-to-date on Mac (and some Apple components, like live broadcasting, are Mac only). RealNetworks has traditionally kept software in-sync so far (although they still only support Classic Mac OS, and RealVideo 9 is Windows and Linux only so far) and there are a wide variety of MPEG-1 and MPEG-2 applications on the Mac.

## Mac OS 9 versus Mac OS X

The Mac universe is transitioning from the 15-year-plus-old Mac OS Classic (a.k.a. 9.x) operating system to the UNIX-derived Mac OS X. Overall, this in an excellent thing for compressionists. Mac OS X offers greater system stability, better dual processor support, better multitasking, and a number of other wonderful things for Mac users. All software compression applications I know of work just fine in the Classic emulation layer under Mac OS X, and some like Cleaner can actually run faster in Classic emulation than they did running on Mac OS 9. Many Mac compression tools, including Cleaner, Squeeze, and HipFlics are available as Mac OS X native applications. The Real libraries haven't been ported yet, so you'll need to run either as a classic application under OS X to encode to RealVideo until it is updated.

One important workflow advantage of OS X over Classic is that X can read and write files over 2GB over networks, something the AppleTalk implementation under 9 can't do out of the box. While there are workarounds, like using Thursby's DAVE software, having large file support is a wonderful thing.

### The ideal Mac encoding system

Because only Apple makes Macintoshes, it's easy to pick the best Mac encoding system—whatever the fastest machine is selling at the time. Cleaner, many QuickTime codecs, and other tools support dual processors quite effectively, so an MP machine adds a lot of value. Most encoding tools don't need much RAM. The 512MB RAM (the current shipping default) is fine for a machine that's used primarily as an encoding system.

Encoding systems aren't particularly sensitive to drive speed, so IDE drives are fine choices over much more expensive SCSI drives. Mac OS X supports RAID 0 out of the box, so getting two identical drives and making an array out of them can be a fine way to go.

The Xserve rack-mounted systems from Apple are well-suited to network compression—you can fit 21 dual 1GHz computers in a single 19" rack.

# Windows

Windows is the encoding platform used by a majority, although not by a huge margin in desktop compression. Windows' advantages are many. Just about any codec available anywhere is available for it. Windows hardware is relatively cheap, and available in a plethora of configurations, including in inexpensive rack-mount form. It's possible to stack a few dozen dual processor machines in a 19" rack for relatively little expense (although the 1U dual processor machines tend to use the slower PIII processor, giving them less bang per cubic inch than Apple's Xserve).

All the high-volume automated encoding systems run exclusively on Windows.

### Which version of Windows?

Although the differences are subtle, there are two basic flavors of Windows, the line derived from DOS and Windows 95, including Windows 98 and ME, and the line derived from Windows NT, including Windows 2000 and XP. Unless you're running some legacy hardware that requires the older operating systems, XP is unquestionably the best OS that Microsoft offers.

Personally, I find the default "Luna" UI of Windows XP really silly to look at, so I run it under the Windows Classic theme, which seems much calmer and more professional.

### The ideal Windows encoding system

Because so many vendors make parts and complete systems for Windows, there is a lot more room for optimization in building the ideal compression system. Both Intel and AMD make processors that run Windows XP. The current processor lines are the P4 from Intel and the Athlon from AMD. The biggest difference today is that the P4 supports the SSE2 SIMD architecture, while Athlon only supports SSE. However, the SSE2 adds support for double precision floating point numbers. If compression algorithms use floating point, they use single precision, so SSE2 doesn't do much for us. The fastest Athlons are quite a bit slower in clock speed than the fastest P4 machines, but the Athlon gets a lot more done per MHz, so in the real world, Athlons and P4s are fairly evenly matched.

Today, most compression software can use dual processors. Many QuickTime codecs are MP enabled. RealVideo encoding to SureStream can use one processor per SureStream, and Windows Media can use up to four for video. The best bang for the buck seems to be in dual processor systems. They're less than twice as much as a single processor, and nearly twice as fast. Quad and eight-CPU machines are much more expensive, and so the price:performance ratio isn't as good. However, if you need optimum latency or real-time encoding, and money is no object, use machines with more than two CPUs.

Dual P4 machines use the Xeon processor, which is *much* more expensive than the base P4. Athlon MP chips are quite a bit less expensive, so a dual Athlon machine is the price performance leader for compression.

Much talk has been made about Intel's IA-64/Itanium architecture. IA-64 is a radical reengineering of CPU architecture. However, initial versions simply haven't been competitive against existing solutions either in speed or price. Nor have there been any video compression apps available for these processors. The architecture is really designed more for running databases—SSE and SSE 2 on the P4 already accomplish a lot of what you'd want from an IA-64 chip.

AMD is also working on a 64-bit chip called Opteron, based on the x86-64 architecture. Unlike IA-64, x86-64 will be able to run normal x86 software with a recompilation—potentially a large advantage over IA-64. The main advantage of x86-64 over IA-32 (a.k.a. x86-32) is it'll support more than 4GB RAM—a feature that isn't important to compression.

## UNIX

UNIX and derivative systems (Linux, Solaris, and Irix) are quite popular in high-end media creation. The expensive, powerful workstations provide functionality and performance beyond anything available on Windows or Mac OS. And for network rendering, the speed and low cost of Linux is making it increasingly popular for render farms. The automation facilities of the UNIX shells are also very widely used in high-volume installations.

However, video compression tools are almost entirely lacking for these UNIX operating systems. Helix Producer is the only mainstream tool available on non-MacOS UNIX (Solaris and Linux).

The performance advantages of UNIX systems tend to be in memory bandwidth and other server-like operations, while video compression is almost entirely dependent on the speed of the processor.

Mac OS X is an interesting hybrid, in that it is a UNIX derivative, but also runs Mac software, in theory combining the best of both worlds. So far, the integration between compression apps and OS X's UNIX layers hasn't been very complete—none of the major compression apps can be controlled via the UNIX terminal, for example. Over time, integration between the two should improve.

### The ideal UNIX encoding system

The ideal UNIX system for video compression is the ideal Mac OS X encoding system for most uses. Barring that, you'd run Linux on the hardware from the ideal Windows encoding system. While Sun and SGI machines are wonderful servers, nothing yet beats the SIMD power of AltiVec or SSE for compression.

## Intermediate Files

One of the most powerful techniques to make compression faster and more manageable when outputting multiple files from the same source file is the use of intermediate files. An intermediate file is preprocessed, but without lossy compression. This lets you do the preprocessing once, making the final encode from the intermediate file. This saves a lot of time and ensures consistent quality. It also lets you do preprocessing and compression in different apps letting you use whatever tool is ideal for preprocessing and compression of a particular project. Intermediate files are also useful if your favorite preprocessing tool doesn't support a particular output format—make your intermediate files in a format that works with all the appropriate encoding tools.

For truly ambitious projects, you can also chain intermediate files, where a master intermediate is used to make further intermediates. For example, a 640×480 intermediate could be used to make a 320×240 intermediate. However, don't chain files that require too much processing into different output files, as this can create quantizing artifacts. For example, applying a gamma filter to an already gamma-corrected output can add noise.

A few tools, like Anystream and ProCoder, make intermediates internally when taking the same source to multiple outputs. But in most cases, you'll have to create intermediates manually.

### QuickTime intermediates

A QuickTime intermediate should generally use either PNG or Photo-JPEG video with uncompressed audio. PNG is lossless RGB, so it's the best option for intermediates that will be recompressed with an RGB-based tool like Cleaner or After Effects. Photo-JPEG is Y'CbCr, and hence is better for Y'CbCr tools like Canopus ProCoder. I normally run PNG in Best mode. This yields smaller files, although it's somewhat slower than the other options (all are completely lossless, of course). I use Photo-JPEG at 100 percent quality, which is almost mathematically lossless 4:2:0.

Of course, some applications including RealSystem Producer and Windows Media Encoder can't read QuickTime files. However, these are usually able to read a subset of QuickTime files via DirectShow, if there is a DirectShow codec available (make the files without a compressed movie header). Of the default set of AVI codecs, None (a.k.a. RGB uncompressed) is the only decent intermediate option. However, RGB uncompressed yields enormous files. Another option is to buy a DirectShow JPEG decoder, which will let most DirectShow applications read Photo-JPEG intermediates. PICVideo (www.jpg.com) has PicVideo Motion-JPEG that works well for reading QuickTime Photo- or Motion-JPEG files. Note some tools just won't recognize any .mov source files, so this workaround won't work in those cases.

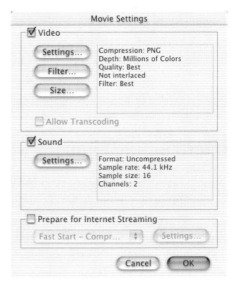

**24.1** A QuickTime Movie Export configured for a lossless RGB intermediate encode.

## AVI intermediates

By default, AVI doesn't have any good intermediate codecs. The only thing close is uncompressed RGB. However, my favorite intermediate codec, Huffyuv, is only available for AVI (at least so far). Huffyuv is lossless, and does both RGB and Y'CbCr 4:2:0. It's fast, free (as in speech and beer) and yields pretty good compression. More info about it is in Chapter 20, "AVI."

The flip side to .mov in DirectShow is .avi in QuickTime. If the same codec is available in both, no problem, but the lack of good default AVI intermediate codecs means you can't use QuickTime's good default ones. If you need to make an AVI intermediate in QuickTime, you're stuck with DV or RGB right now.

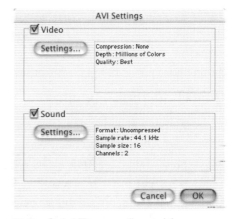

**24.2** QuickTime configured for a uncompressed RGB AVI export. This file will be much larger than the previous one.

## MPEG-1 intermediates

MPEG-1 also works for intermediates, because both QuickTime and DirectShow can decode the files. However, QuickTime can only decode MPEG-1 video, not audio, so MPEG-1 works best going from QuickTime to DirectShow. Use much higher audio and video data rates than normal—4000Kbps for 320×240 video and the full 384Kbps for Layer II audio.

## MPEG-2 intermediates

MPEG-2 intermediates aren't a really good solution yet, although that may be changing as MPEG-2 playback becomes common in future versions of the various architectures. At 15Mbps, an IBP MPEG-2 file is nearly lossless. The 50Mbps I-frame-only 4:2:2 variant is also becoming popular, although Cleaner is the only application that can read those so far (and only if they don't use 48Kbps audio, which crashes Cleaner 5.1.2).

***Note***

**How to be my friend.** So, after getting this far, do you want to curry favor with me? Simple: port Huffyuv to QuickTime for Mac and Windows! This would make it possible to both read and write Huffyuv QuickTime and AVI files in both DirectShow and VfW, which would give us a single file format that could be used in all encoding tools.

Huffyuv is free software licensed under the Gnu Public License (GPL). You're free to download the source codec and distribute modified versions as long as you also distribute the source codec as well under the GPL.

More info at http://math.berkeley.edu/~benrg/huffyuv.html

# Media Freshening

Another key goal for an automated encoding environment is to make it easy to update files to new file formats and codecs. Ideally, recompressing and uploading files from one format to another should be very simple to set up, including replacing media files automatically on the server. This means the system needs to know where the source files are, where the final files go, what settings were used before, and how to modify them.

Reattaching the source files to the batch can be the hardest part, especially if they aren't archived on a file server, but live on timecode stamped videotape.

# Multi-machine Rendering

No matter how fast your computer is, beyond a certain volume, rendering will need to take place on multiple machines. This makes things a lot more complicated. At the simplest level, you can put your source files on a file server, and load them over the network in the multiple encoding servers. On the high end, you can have centrally managed encoding.

## Latency versus volume

One critical question to answer when designing an automated encoding system is whether you want to optimize for low latency or high volume. A low-latency system minimizes the time between when a process starts and when the finished file is available. A high-volume system is designed to pump out as much content in a day as possible, no matter if a given file takes an hour or 10 to complete.

A number of decisions relate to this. For example, using intermediate files can help increase volume a lot because a given preprocessing pass needs to be made only once. But it increases latency, because the preprocessing pass must be done before the compression pass. If you want to optimize for low latency, run all the processes in parallel, with as few steps as possible.

## Networking

It used to be that a straightforward 100Base-T network was fast enough for rendering files over a network. However, as computers and compression software get faster, they consume more bandwidth. Most networks run on ethernet, be it the standard 100Base-T or the emerging, much faster Gigabit (1000Base-T). Some high-end installations use FibreChannel.

### Ethernet

For example, a 100Base-T network has a theoretical bandwidth of 100Kbps (although you're lucky if you get 80). DV runs at 25Mbps. So, you could get at most three real-time DV streams out of a 100Base-T connection. The cheap solution is to mix gigabit (1000Base-T) and 100Base-T ethernet. The file server itself plugs into the switch at gigabit, and the individual machines plug in at 100Base-T. You'd be able to serve dozens of DV streams in a switch (many will come off the server, but only one will be routed to each computer) because the different connections don't compete for bandwidth.

Most file servers and Network Attached Storage solutions are ethernet-based. They're cheap and work well. Their one limitation is that they don't work well for capturing video, especially with higher (greater than 25Mbps DV) data rates. Many capture systems won't even try to capture to a network volume. Thus, you generally have to take the extra step of copying the files from the capture machine to the server before compression can start.

### FibreChannel

FibreChannel is a rather different architecture. Unlike ethernet, FibreChannel gives you shared storage that looks like a local drive to different machines. This means you can copy straight to the shared storage device without having to make a copy. Saving that step can help reduce latency quite a bit. A system is sometimes called a SAN, for Storage Area Network.

## Scripting

In building an automated encoding system, you can either roll your own, or buy one off the shelf. Buying an off-the-shelf system can save huge amounts of time. However, even the off-the-shelf systems tend to require a fair amount of customization to work in the real world. They're obviously also designed for general use. If you have a very unusual or very specific project, it may make more sense to build a custom system based on scripting.

## AppleScript

AppleScript is the scripting environment native to Mac OS. It's been around for years, and is well supported and documented. AppleScript is definitely the easiest of the scripting environments to write for and to debug. However, it is Mac-only. It also requires that applications explicitly have AppleScript support built in. QuickTime Player has excellent support for AppleScript, supporting everything QuickTime Player can do through the normal UI, and a lot more. Cleaner and Hip Flics also have decent AppleScript support.

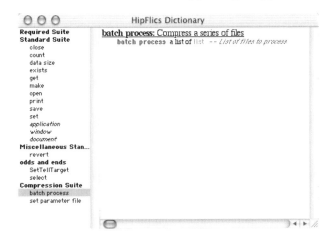

**24.3** The HipFlics AppleScript Dictionary. It doesn't have a lot of options, but can make for easy batch encoding of files.

One neat thing about AppleScript is that the Script Editor application can list the AppleScript commands available to any application, instead of requiring explicit documentation.

## DCOM

Microsoft's Distributed Component Object Model is the high-end scripting model supported under Windows. Among those who know about such things, AppleScript is considered a lot simpler for small projects. However, DCOM is thought to be superior for implementing industrial-level applications. There aren't a lot of apps that support DCOM—though Cleaner and Windows Media Encoder do.

## Shell

The old school scripting model is shell scripting, a.k.a. command line scripting. It originated in UNIX, but can also be used via the command line in Windows, and the Terminal window in Mac OS X. And, of course, applications can use it to communicate with each other.

Shell scripting can work well, as it's simple to set up small and repetitive projects with. However, you'd be unlikely to do shell scripting for a project if either AppleScript or DCOM is available.

Most Windows apps support command line options (although often with only a subset of the application's functionality), but there aren't many that do so on the Mac.

# Workgroup Encoding Products

## Anystream Agility

The Agility product line from Anystream was the first enterprise encoding system. It's still the most powerful and complete solution available. Anystream was spun off from ICE, and the Agility line was based on the BlueICE Ultra board that once powered Media Cleaner PowerSuite and provided After Effects acceleration.

The Agility line's strength is its depth. Depending on the model, Agility products can incorporate live Webcasting, automated capture off satellite feeds, insertion of advertising into a stream, and so on. Agility has excellent Y'CbCr preprocessing, including a good inverse telecine.

## Telestream FlipFactory

FlipFactory, from Telestream, provides a very different encoding model than the other tools mentioned so far. With FlipFactory the central server "flips" media, transcoding files among different formats. It doesn't handle integrated capture. It does handle on-the-fly transcoding in some models, so you don't even have to define what file formats you want to stream—you dial in what you want, and FlipFactory delivers it immediately.

## Cleaner Central

Disclosure: I was the original product manager for Cleaner Central many moons ago. Cleaner Central is Discreet's entrée into the enterprise encoding market. Cleaner Central's selling point is that it's based on Cleaner, and hence supports good preprocessing, 2-pass VBR for most formats, including QuickTime, and a familiar settings file format. Thus, you can actually reuse Cleaner settings, making it possible to automate an existing Cleaner workflow. Cleaner Central is also the cheapest of the volume encoding systems—$4,995 for the server and first encoding node, and

$1,495 for each encoding node. However you'll also need to have SQL Server. The encoding nodes include Charger for full MPEG-1 and MPEG-2 support.

However, the low cost of Cleaner Central's encoding software is somewhat offset by the general slowness of the Cleaner engine. For typical projects, a single Agility node should be able to process quite a bit more content. To get the same capacity, you'd need to buy more nodes, and hence more computers, than the other solutions.

Cleaner Central has unparalleled support for QuickTime among the Enterprise solutions, with 2-pass VBR, automatic reference movie creation, and MP3 audio tracks. It is also much more stable than the basic Cleaner product—hopefully they'll be able to port back the bug fixes from Cleaner Central to the original version.

# Chapter 25

# Tutorials

This chapter takes everything discussed in the previous 24 chapters and shows you how to apply it to three real-world projects. We'll consider content, communication goals, and audience as we approach each project and I'll show you how those considerations affect specific questions as they arise.

The CD-ROM[1] included with this book includes short snippets of each project's source files, along with sample output files. If you'd like to work with the full-length content, you can order the companion DVD-ROM with complete project files from Customflix; keep an eye on my Web site (www.benwaggoner.com) for information.

## Tutorial 1: NASA Footage for Web Streaming

In this project, we're going to take a longish DV source file and encode it for real-time streaming in multiple formats and data rates.[2]

### Content

The source is a 9:40-long NTSC DV clip assembled from a variety of sources, some progressive, some interlaced. The video is 16:9 letterboxed in a 4:3 window. It's provided on the DVD ROM as a DV .avi file, which is readable by both QuickTime- and DirectShow-based apps (although the APIs used to read the files may cause it to behave differently on different platforms).

The source is fairly noisy. It appears to have gone through an analog composite video connection at some point, and was probably transferred from analog tape. While a lot of the noise will average out when the file is scaled down to lower resolutions, we definitely want to apply some kind

---

1. Due to the lead time required to produce the CD-ROM, the Tutorials had to be frozen a couple of weeks before the rest of the book. At times, the CD refers to versions of software older than the ones described in the book.

2. The source file was provided by eMotion Studios.

of noise reduction. The video itself has fairly high black levels. Bear in mind different DV codecs may or may not expand the 16-235 range to 0-255 automatically (the Apple DV codec will).

The overall complexity of the video varies a lot. Much of the content is simple interview footage shot on a static background. There are also sections in which the video contains complex motion graphics and is sped-up. Finally, there are very grainy, very high-motion archival sequences of rocket launches. Varied content such as this is well-suited to 2-pass encoding.

Audio is mainly voice with background music, with a few sound effects and rocket blast-offs. Using voice codecs would probably work on these clips, as all the important content is speech.

**25.1** A sample frame from the NASA footage, note the letterboxing and left and right blanking.

## Communication goals

Imagine this file is being encoded for NASA's Web site. The goal is to get the general public interested in the International Space Station. Our goal is to leave them with a positive feeling about the project. A clip that buffers a lot, provides low quality, or is otherwise a glitchy experience would definitely detract from the experience and erode the effortless sense of high tech we wish to provide.

## Audience

Space exploration naturally attracts a high-tech audience. However, a clip like this would also be of interest to schools, politicos, and so on—an audience of which a majority would have broadband connections, but many would still be using modems. Because we're talking about a school audience, you can assume there will be more than an average number of Mac OS clients, and that older computers will be in the mix.

For the most part, this isn't an audience that is going to be so interested in the clip that they'd be willing to stand for long buffering times.

## Our approach

The balancing act is to be able to deliver decent quality to a wide variety of machines and connection speeds. If the clip were shorter, I would strongly consider progressive download, but at almost 10 minutes in duration, a slight mismatch between data rate and connection speed could cause many minutes of buffering.

Given that we're targeting older and academic machines, we can't count on any one media player being available. So we'll target RealVideo, Windows Media, and QuickTime to make sure there will be a format that works for each viewer. Given these files will be streamed to the public Internet, we'll definitely want to use a multiple bitrate solution. Because Windows Media's Intelligent Streaming and RealVideo's SureStream use a fixed resolution, we'll prepare modem data rate and broadband versions with different resolutions, and we'll use QuickTime Movie Alternates to link to different files through a single link. Because Mac users are likely to pick QuickTime as their viewing format, we'll create two different sets of QuickTime movies, one for Mac and the other Windows gamma. Also, because real-time streaming to modems via QuickTime is so lousy, we're going to use progressive download for modem rates, even though we can use RTSP streaming for higher data rates.

## Preprocessing

Preprocessing for these clips is pretty straightforward. First, we'll crop the 16:9 source to remove out all letterboxing. Next, we'll apply adaptive deinterlacing. We're going to increase contrast to get blacks to black for the title cards. Then we'll scale down the resolution to our target. Because we're aiming at relatively low data rates and have somewhat challenging footage, let's target 320×176 for the broadband resolutions, and 192×112 for modem resolutions (both are divisible by 16, although this isn't strictly required for the formats we're targeting). We could probably push up the broadband resolutions a little higher if we wanted to be aggressive, but for this project's communication goals, a reliably functional experience is better than a file that provides cutting-edge quality with greater fragility.

In cropping the 16:9 source, our original 480 lines were cropped down to 360 lines, and half of the fields were thrown away on frames where adaptive deinterlace isn't able to extract. This leaves us with only 180 reliable lines of source. Using a resolution higher than 320×176 could introduce scaling artifacts on the vertical axis.

One complexity: Apple's DV codec applies a gamma correction when decoding to RGB on Mac, but not on Windows. Thus, if QuickTime is used to decode the file (which it would be on a

Mac), the RGB values in the middle of the 0-255 range would be quite a bit lower than if the file were decoded on Windows.

## In After Effects

Preprocessing for this clip is rather easy in After Effects. The first step with any After Effects project is to set up the Interpret Footage dialog correctly. Because this is DV source, it is Lower Field First. We definitely want to turn Motion Detect (Best Quality Only) on, to get adaptive deinterlacing. Make sure Pixel Aspect Ratio is set to the DV 4:3 option.

Next, make compositions. Though we want 320×176 and 192×112 video, I prefer to figure out my filter settings at full resolution, so I can see all the lines and spot any interlacing problems or whatever. So, start with a 640×352 Composition, and modify it later.

The easiest way to crop in After Effects is to scale the source video track to be larger than the frame.

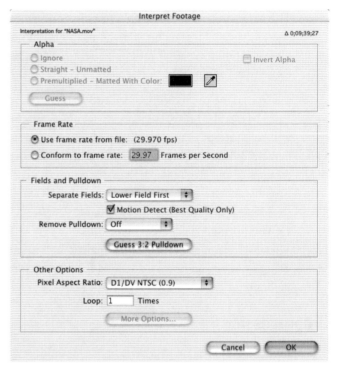

**25.2** After Effects' Interpret Footage dialog, configured for adaptive deinterlacing and DV 4:3 source.

Scaling up our video to 101 percent nicely eliminates the black edges on almost all shots. However, it's important to skip around to different parts of the video to make sure some don't use significantly different framing. For example, there's a trouble spot at 00:54:03, where there's more black edge at the top and bottom of the screen. This same problem occurs in other archival shots of rocket launches. To address this, raise the scaling to 103.5 percent, which removes all black edges on all the shots in our tutorial footage.

Next, increase contrast. Because this file uses the QuickTime DV codec, it expands the 16-235 range of DV to 0-255 on decode to RGB, the same range used by After Effects. However, due to the noise and other analog artifacts, the black levels on title cards are still quite high. By increasing the Input Black point, we increase contrast. Looking at the histogram on a frame that should be video black, you'll see a bar instead of just a line. This is because the noise in the video means pixels that should all have the same value actually vary quite a bit. So, we raise the black point to

**25.3** This frame and other archival shots have thicker black borders at the top and bottom of the screen.

**See also**
For the color version of Figure 25.3, see Figure V in the color section.

the highest level in that black band. This will make all of the black background mathematical zero. However, while this increases contrast, it also decreases brightness.

To fix that, you can do a corresponding luma range reduction, although there aren't any frames of white to calibrate to. We'll be less aggressive about that, so as not to blow the image out. We'll also do platform-specific gamma adjustment (bearing in mind the Apple codec is already doing its own on Mac). If encoding on Mac for Windows, we'll use 1.3, and on Windows for Mac, 0.8.

After Effects doesn't have good noise reduction. The best we can do is a narrow gaussian blur—0.5 pixels helps reduce single-pixel grain without making the image notably soft.

After Effects doesn't have an audio normalization filter, so you'll have to apply normalization in an external audio application such as BIAS Peak or Sonic Foundry SoundForge.

### In Cleaner 5.1.1

Cleaner is a whiz at this kind of preprocessing. In the project window, make sure the source is 4:3 interlaced, and set the crop rectangle's constraint to 16:9. Then manually select the active region. Cleaner provides fast updating while scrubbing, making it easy to find areas where the crop box needs tweaking.

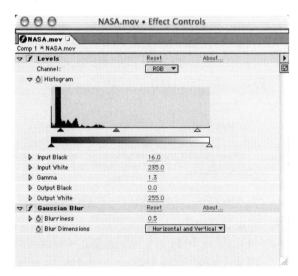

**25.4** Our filters, as configured in After Effects for Mac OS X.

In the Image tab, define the output resolution we want. For Deinterlace, select Auto (which means Odd will be applied, because Interlaced is checked in the Project window), with Adaptive on. Because this video is quite noisy, use adaptive deinterlacing at all resolutions, even though we're only using 180 lines of source. We'll also use the Adaptive Noise Reduction option. While most of this clip would benefit from Temporal Processing, the small text throughout the clip is badly blurred by this filter, so we have to turn Temporal Processing off. In the Adjust tab, tweak gamma by +30 if encoding on Mac for Windows, and by −30 you're working on Windows for Mac. Then increase contrast to the point where blacks are really black. However, this will make the whites too blown out, so we should use a combination of reducing brightness and increasing contrast. Cleaner's Dynamic Preview mode is quite handy in finding the ideal settings.

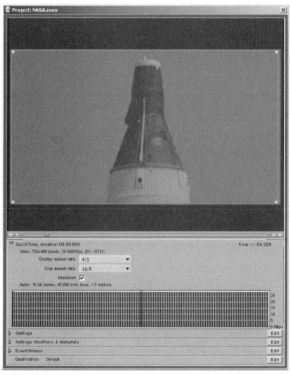

**25.5** Cleaner's Project window, where settings specific to a given output file are made, including manual cropping.

For audio, set Normalization to 90 percent.

### In ProCoder

ProCoder makes this preprocessing a breeze. Import the footage, and go into Advanced mode to apply filters.

First, apply the Adaptive Deinterlace filter. ProCoder's default settings are fine for this.

Next, add a crop filter. ProCoder has a nice visual crop filter in the "Cropping Dialog" mode. However, it doesn't include a movie controller, so you'll have to continually go back to the Video Filter window to change the frame being looked at.

Lastly, adjust contrast and brightness with the Color Correction filter.

None of ProCoder's noise reduction filters are particularly well-suited to this kind of clip, alas. The best bet is the Circular Blur, which can be configured to a 0.5 pixel size.

Finally, apply the Normalize audio filter; set it to "Normalize Peak" at −3dB.

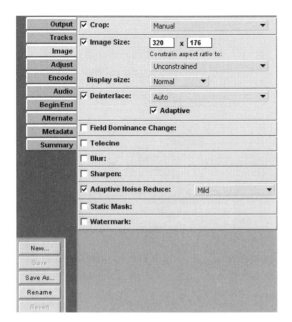 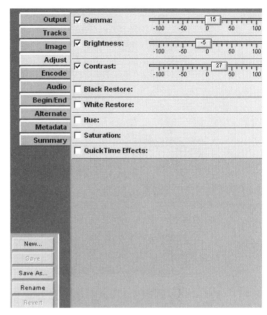

**25.6** The Image and Adjust tabs of Cleaner, configured for our preprocessing settings.

### In RealProducer Plus 8.5

One of the frustrations of RealProducer is that it puts many features that should be setting-specific in its global Preferences. In the Video Filter tab, set the Noise Filter on Low, Resize on High Quality, Inverse Telecine off, and Deinterlace Filter off (because RealProducer processes NTSC DV as 360×240, single-field).

In Video Settings, set up the crop. Alas, RealProducer only shows the first frame of the video, which in this case is black. So you have to just know what the right settings are. However, RealProducer (like many other DirectShow apps) thinks of DV as being 360×240, not 720×480. Thus, you need to correct for both the lower resolution, and the fact it's measuring numbers in non-square pixels.

More irritatingly, RealProducer doesn't give values for the right or bottom, it just shows you the size of the window, and you have to calculate the right values from there. Ugh. Anyway, the only way to preview the settings to make sure they work is to encode the file. Turn off 2-pass encoding, so you don't have to wait all the way through the first pass, and watch the output window for glitches. After some trial and error, I wind up with Left=12, Width=340, Top=35, and Height=168. To add to the aggravation, you can't even directly input the numeric values. Instead, you have to use the up/down arrows. After all this, you can't save your combined settings, so keep careful notes of settings if you want to be able to replicate results later.

There isn't any contrast, brightness, or gamma adjustment in RealProducer, so you'll just have to live with what you have. If possible, you may prefer to use After Effects for preprocessing, create

## 404  Chapter 25: **Tutorials**

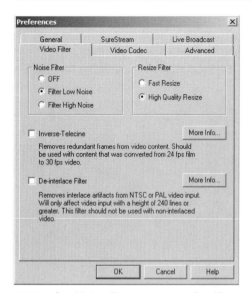

**25.7** Our Video Filters set up in RealProducer's preferences.

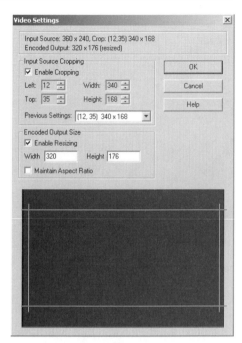

**25.8** The laborious crop dialog of RealProducer 8.5. I really wish there was a way to actually change the frame.

an intermediate file with it, and only use RealProducer for the final encode, given all the above annoyances and limitations.

### In Windows Media Encoder

Windows Media Encoder requires a profile for most encoding settings, so you'll have to create a new profile that specifies the 16:9 target resolutions. WME has even more anemic preprocessing features than RealProducer, but they're easier to use.

Like RealProducer, Windows Media thinks that DV source is 360×240. Thus there's no need to deinterlace, but you'll need to use proportional cropping values. To input the cropping settings, go to Video Clipping (hidden as a pane in the Change window, accessible by right-clicking on the source in the not-shown-by-default Sources window). Alas, you get no visual preview, so as with RealProducer, you'll have to go the trial and error route.

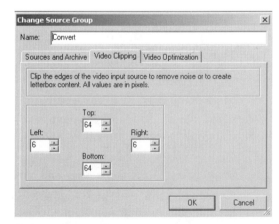

**25.9** Windows Media Encoder 7.1's Video Clipping window.

## In Squeeze

Squeeze 2.0 has preprocessing that's much improved compared to what was in v1. However, it still can be a little clunky. Note Squeeze does a proper Y'CbCr decode, and so you'll need to do manual contrast and gamma adjustment. In the Filters window, tune in appropriate contrast, brightness, and gamma settings. Either draw a crop box or manually enter crop values. One irritation of Squeeze is that you can't scrub in the Filter window, so you need to keep going back to the main window to change the frame and go back into Filter mode to make sure the settings work for a given frame. This can be especially irksome when trying to find ideal crop settings with a clip like this; still, it's a lot better than RealProducer and Windows Media Encoder. Alas, you can't save Filter settings to a file.

Leave Adaptive Deinterlacing to its default setting, and turn on audio normalization.

# Compression

We're targeting three different formats for this project: RealVideo, Windows Media, and QuickTime. For RealVideo and Windows Media, we'll create a modem and broadband MBR file for each. For QuickTime, we'll make individual files at different data rates in both Mac and Windows gamma.

### Data rate targets

Our audience is a mix of broadband and modem users. For broadband users, expect clusters of connection speeds around 500Kbps and 256Kbps (typical DSL speeds). We also want to pick a rate that will work with the popular-in-Europe Dual ISDN connections. Because you should always leave a little headroom, we'll target total data rates of 450Kbps, 220Kbps, and 100Kbps for the broadband clips.

For the modem data rate targets, we'll also use three data rates. On the high end, a total of 44 is about what you can expect out of a good connection. For our low end, we'll use 22, which should work in the case of a 56K modem that has to fallback to 28K (a common occurrence). And we'll stick another data rate in the middle, at 34K.

Data rate targets for broadband: 450, 220, 100.

Data rate targets for modem: 44, 34, 22.

### Encoding to RealMedia

lets us vary our settings for both audio and video, so we can use progressively higher audio data rates. This lets us find the optimum balance at each data rate.

RealSystem's 2-pass VBR significantly improves compression efficiency with this kind of varied content, with more effect the bigger the buffer size. We'll increase the buffer from the default of 7 to 10 seconds for the broadband clip, and to the maximum of 25 seconds for the modem clip. While this will increase buffering latency, we need every last bit of quality we can squeeze out at modem data rates.

I always prefer to use RealVideo in Smoothest Motion mode—most people find artifacts less distracting than frame dropping.

|  | Target data rate | Video data rate | Audio codec | Frame rate |
| --- | --- | --- | --- | --- |
| Modem | 28Kbps | 15 | 6.5 – Voice | 5 |
| Modem | 56Kbps (low) | 25 | 8.5 – Voice | 6 |
| Modem | 56Kbps (high) | 32 | 12 – RA8 stereo | 10 |
| Broadband | Dual ISDN | 80 | 20 – RA8 stereo | 15 |
| Broadband | Low broadband | 188 | 32 – RA8 stereo | 30 |
| Broadband | High broadband | 406 | 44 – RA8 stereo high | 30 |

### Encoding to Windows Media

Encoding to Windows Media is quite similar to RealVideo, with the big exception being you can only have one audio setting per file.

You can encode these files with Cleaner on Mac, but you'll have to use the WM7 versions of the codecs, and can't access Intelligent Streaming. Because you can only do a single audio track per file, I'd just do the middle data rate for each band (so 220 for broadband and 34 for modem), knowing that users with slow connections will have problems and users with fast connections won't get the best possible experience. For this reason, I don't encode in Mac OS for professional quality Windows Media. However, this doesn't mean you can't do encode *on* Mac OS. Let me explain. I create a preprocessed None codec AVI file in Cleaner or After Effects, and do the final encode in Connectix's VirtualPC. Performance is surprisingly good because only codecs are being applied, without preprocessing.

Ideally, you'd encode all these files with the 2-pass CBR algorithm. However, this isn't available in any commercial tools yet—Windows Media Encoder itself only supports 1-pass CBR, and Cleaner does all the modes but 2-pass CBR. I hope this will have changed by the time this book is published (if not, check out Microsoft's command-line only Windows Media Encoding Utility). Specify a buffer size of 10 seconds for the broadband clip and 25 for the modem clip, to match the RealVideo settings used in the previous example.

# Tutorial 1: NASA Footage for Web Streaming

|  | Target data rate | Video data rate | Audio codec | Frame rate |
|---|---|---|---|---|
| Modem | 28Kbps | 14 | WMA 11.025kHz, 8Kbps mono | 5 |
|  | 56Kbps (low) | 26 |  | 7.5 |
|  | 56Kbps (high) | 36 |  | 10 |
| Broadband | Dual ISDN | 68 | WMA 44.1kHz, 32Kbps mono | 15 |
|  | Low broadband | 188 |  | 30 |
|  | High broadband | 418 |  | 30 |

## Encoding to QuickTime

QuickTime's unique advantages and limitations mean the optimal setup for the QuickTime files is radically different from those of the other formats. Because there is no MBR system, we need to make an explicit file for each target bandwidth. Also, lacking MBR, making an RTSP file for modem use isn't really worth doing—a file with a low enough data rate to play everywhere would offer really low quality for most users. With QuickTime, it'd be much better to deliver the modem data rate files as progressive download. There will be some buffering time, but it's really the only way to deliver adequate quality.

Although QuickTime's Movie Alternates don't offer true MBR, they have a number of unique features. Because each QuickTime Alternate is a different file, we can set them to vary quite a bit: different audio codecs, streaming versus progressive download, different Mac and PC gamma.

Cleaner provides excellent support for automatic authoring of movie alternates. If you use other tools, you'll need to use MakeRefMovie after the fact, which adds a semi-laborious extra step.

For the progressive 56K clips, encode the video at 40Kbps with 2-pass VBR. For the audio codec, encode with MP3 at 16Kbps. This will yield a total data rate of 56Kbps. Let's figure out the latency assuming a real-world connection speed of 48Kbps and a clip duration of 9 minutes. Using the formula in Chapter 11, "Web Video," on page 158 we get

$$96 \, \text{seconds} = \left(\frac{56 \text{Kbps}}{48 \text{Kbps}} - 1\right) \times 580 \, \text{seconds}$$

So, an average buffering time is about a minute and a half on a fast modem connection, and worse with slower connections. While this isn't ideal, we can at least deliver decent quality at this data rate.

Set the modem clip to play back at the full 320×176, though it's encoded at 192×112. This means we can embed the master movie into a Web page, with a fixed 320×176 movie size. Lastly, use a compressed movie header.

For the streaming clips' video, use SV3.1 Pro in its 1-pass VBR mode (ideal for streaming). For audio, we'll use QDesign Music 2 Pro (not available under Mac OS X unless you're running your app in Classic mode.). We'll have to use somewhat higher data rates for audio than we might like to keep phasing to a minimum when encoding voice with background music. We'll also encode all files to mono to improve quality. Speech will still be a little mechanical sounding, but the QDesign codec does a great job of reproducing the bass rumble of the rocket launch.

We're also going to use B-frames for all clips. This is especially important for the streaming clips because the QuickTime Streaming Server can drop B-frames under bandwidth stress, and provide the closest thing to a true MBR that QuickTime can manage. B-frames also improve overall compression efficiency, so we'll use them to squeeze out that extra last bit of quality. The big drawback of B-frames is they cause a 1-frame delay in video relative to audio, throwing sync off, especially at lower data rates. To fix this, shift the audio by a single frame. For the streaming tracks, this will require rehinting as well. We'll cover that procedure next.

When the final files are complete, you'll need to make a master movie (unless you're using Cleaner). We'll walk through that process as well.

| Target data rate | Video data rate | Audio codec | Frame rate |
|---|---|---|---|
| 56Kbps | 40 | MP3 11.025kHz, 16Kbps | 7.5 |
| Dual ISDN | 80 (1-pass VBR) | QDM2 22.05kHz, 20Kbps mono | 10 |
| Low broadband | 180 (1-pass VBR) | QDM2 44.1kHz, 40Kbps mono | 15 |
| High broadband | 394 (1-pass VBR) | QDM2 44.1kHz, 56Kbps mono | 29.97 |

### Compression in After Effects

While After Effects can encode to RealVideo and QuickTime, it doesn't do a good job with either. For QuickTime, it lacks support for hinting, MP3 audio, and so on. After Effects can't encode Windows Media at all (although this has been announced for a future release). Instead, we'll just make intermediates to encode in another application.

We made our original composition at 640×352. We could duplicate our previous settings and filters and modify them to get our lower resolutions. But AE gives us an easier way via nested compositions. Simply make new compositions at the target resolutions, and drag the first composition into the smaller ones' time lines. Adjust scaling to fit—50 percent for 320×176, and 33 percent for

192×112, and we're ready to render. Among other things, this technique allows you to tweak filters in the original encode, and have the changes ripple through the child compositions.

If encoding with QuickTime-based tools, I'd use the PNG codec for RGB-based applications and Photo-JPEG for Y'CbCr-based applications like ProCoder. If encoding with AVI tools, I'd use Huffyuv in either RGB or YUV (as it labels its Y'CbCr) mode as appropriate (make sure it's installed on all your encoding machines). If encoding in AE on Mac for future compression in Windows Media Encoder (on a Windows-based computer or via Virtual PC), use None, but beware the resulting gigantic files. Somebody port Huffyuv to QuickTime, stat!

## RealProducer

In Video Codec Global Preferences, we want 2-Pass Encoding, VBR, and Loss Production set to On. Between encodes of the broadband and modem clips, you'll also have to change the buffer size from 10 seconds to 25 seconds in the Global Preferences menu.

Our target data rates aren't quite what RealPlayer normally uses. So, you'll have to change the Target Audience settings definitions for the various settings, and tweak them accordingly. The audio content is Voice with Background Music, so set the audio codec's setting for Voice with Background Music to the appropriate option.

One thing you'll notice after all this tweaking: RealProducer is wicked fast. On my dual PIII machine, RealProducer renders the project in 16 minutes, where Cleaner takes over three hours. Some of that is because the DV codec isn't doing a full-resolution decode of the source, and is doing much less preprocessing.

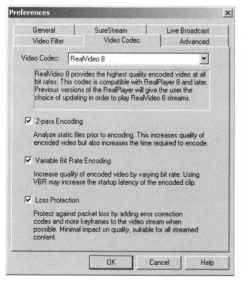
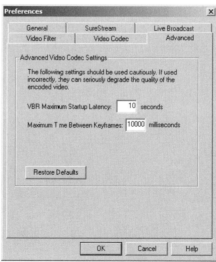

**25.10** We need to set these values in RealProducer Preferences (changing the VBR Maximum Startup Latency between the modem and broadband encodes).

## Fixing Sorenson B-frame Sync

Sorenson Video 3.1 Pro's B-frames option offers advantages in compression efficiency and RTSP bandwidth scalability. Alas, it delays the video two frames relative to audio, which throws the video and audio out of sync. While this should be a trivial thing to fix in a compression tool, none offer the ability (even Sorenson's own tool, Squeeze, doesn't allow you to fix sync), so you have to fix it after the fact. This is a rather lame state of affairs, but the advantages of B-frames, especially when RTSP streaming, make the hassle worth the effort.

There are two ways to correct sync: via QuickTime Player Pro or through Roger Howard's ReSync AppleScript. The QuickTime Player option has the advantage of working on Windows (although Howard promises a RealBasic version of his tool that will eventually run under Windows).

Let's walk through the QuickTime method. Load the movie. Then Extract Tracks to make a new movie with just an audio track (you don't need to save it). Pick Select All, and then Copy, loading the entire audio track into the clipboard. Next, Delete Tracks to remove the old audio track, leaving the original file with just video. Now, simply press the right arrow key twice. This moves you forward two frames in the movie. Pick Add, which inserts the audio at that point in the movie, two frames delayed from where it started. And the audio is in sync!

Skip to the end of the movie, and see if there is a white frame of video. If so, trim it out. At the very last frame of the file, hold down the shift key, and hit the left arrow. This will select the last frame of the movie. Then hit Delete, and the spurious last frame is gone. Sometimes there is more than one spurious frame, so you might have to repeat this procedure. Now you can save your movie. You Save As to a file name that already exists in the same directory. I generally Save As with the correct name, but I save the file to the desktop. Then I quit QuickTime Player (not saving any of the other movies), and copy the new file over the old one. By replacing the file with a new one of the precisely same filename, URLs are maintained. Note if the file is for streaming, Export as a Hinted Movie instead of doing a Save As.

To use the ReSync AppleScript (you can download it at http://www.binaryboy.org/resync/), you simply drag the file onto the script, and you're good to go.

# Tutorial 1: NASA Footage for Web Streaming 411

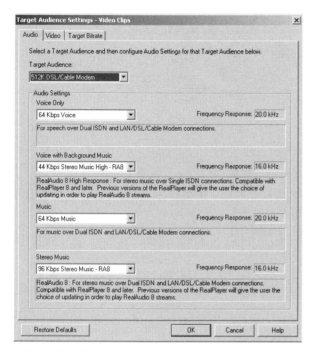

**25.11** RealProducer 8.5 forces us to change the settings for each data rate globally. We can tweak audio, video, and bitrate.

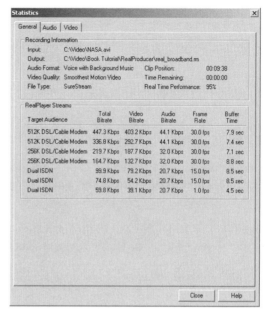

**25.12** RealProducer creates a very useful statistical report for the encoded file. Note how additional data rates that we didn't specify have been added in the output.

In the final encode, RealProducer adds extra SureStream bands below what was requested. Currently, there isn't any way avoid that, but the only real drawback is that the file is larger.

## Windows Media Encoder

Windows Media Encoder lets you either set up a session via a Wizard (the default), or tweak files by setting properties. I normally create base setting with the Wizard and tweak from there.

Like RealProducer, Windows Media Encoder requires you to jump through a lot of hoops to use custom settings. First, you make a custom profile. Then you define custom Target Audiences with the appropriate data rates. A peculiarity of Windows Media is that the "effective" data rate is always a little higher than you requested. So, to get a real 100Kbps stream, you need to request 93Kbps, because another 7Kbps will be added. This happens both in Windows Media Encoder and other applications based on the Windows Media SDK.

**412** Chapter 25: **Tutorials**

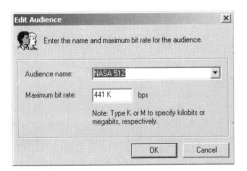

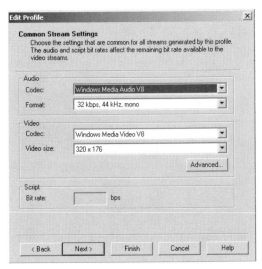

**25.13**  Setting up a target audience in Windows Media Encoder. A roundabout way to set a data rate! *(above)*

**25.14**  The Common Stream Settings, where options common to all streams such as resolution, are set. *(right)*

Windows Media Encoder lets you specify frame rate, keyframe rate, and image quality versus frame rate for each data rate.

Windows Media Encoder lets you save your session and its settings as a .wme file; a very welcome feature.

### Sorenson Squeeze

Squeeze is much easier to use than either Cleaner or AE, although the UI can be a little odd. To target our four different data rates listed previously, click the icons at the top of the screen. This will apply four defaults. You can then modify to achieve the settings you want by right/option clicking on each.

Compared to the complexity of Cleaner, Squeeze has a very simple compression interface. You can tune data rate, audio codecs, and so on. You can also save settings for later use, although to access a saved setting, you first need to apply one of the generic settings, and load the saved settings with the Load button in the settings dialog.

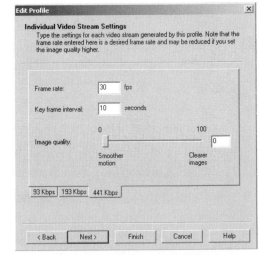

**25.15**  Individual Stream Settings, where frame rate and other parameters that can vary by stream are specified.

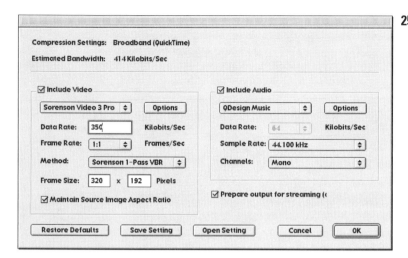

**25.16** Squeeze's codec setting dialog. Nice and simple. Because Squeeze does proper Y'CbCr processing, we don't have to second-guess the codec's luma processing.

### Cleaner

Cleaner is the Big Kahuna of compression here—it support all the formats, with most of the features you'll need. For QuickTime, Cleaner does it all, including excellent support for auto-generating the reference movie, including streaming links. For this project, the output will include folders to put on the Web and streaming server, as well as a ReadMe file that describes the project's structure. Note you need to have Set Server Path set to On in order to get streaming files to reference correctly.

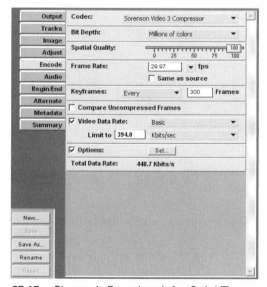

**25.17** Cleaner's Encode tab for QuickTime.

**25.18** Cleaner's Alternates tab—support for QuickTime Alternates is one of Cleaner's standout features.

**25.19** The folder structure of output files, nicely laid out into multiple directories.

Cleaner's RealVideo support is complete. No tricks here. Just directly specify the data rates and streams that you want.

Cleaner's Windows Media support doesn't include the streaming-optimized 2-pass CBR mode, or the ability to set frame rate or keyframe rate on a per-stream basis. And on the Mac, you don't have access to Intelligent Streaming or the V8 codecs. It's best to encode Windows Media under the Windows OS.

### MakeRefMovie

If you authored your QuickTime files in a tool other than Cleaner, you'll need to create the reference movies after the fact. Apple's free MakeRefMovie is the simplest tool for this (although you might use Peter Hodde's XMLtoRefMovie for automated, high-volume projects). MakeRefMovie is available for Mac OS 9.x, OS X, and Windows, but the Windows version doesn't yet support the data rate options introduced with QuickTime 5. Because these options include everything between Dual ISDN and T1 rates, you'll need to use MakeRefMovie on Mac for this project.

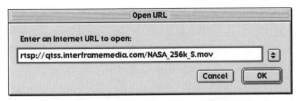

**25.20** Using Open URL in New Player to make a streaming reference.

There's a trick to embedding streaming links in a file with MakeRefMovie. You can't do this directly. Instead, you make a movie that contains a streaming track that points to the file you want. In QuickTime Player, use the Open URL in New Player command, and type in the streaming URL you want (it'll start with rtsp:, not http: or ftp:), then save that file. This file contains just the link to the streaming file. Note the connection doesn't have to be active—you just need to know what the URL is going to be. Now you can build the master movie with those files included.

**25.21** Setting up our movies in MakeRefMovie. While functional, its interface doesn't draw very well under Mac OS X.

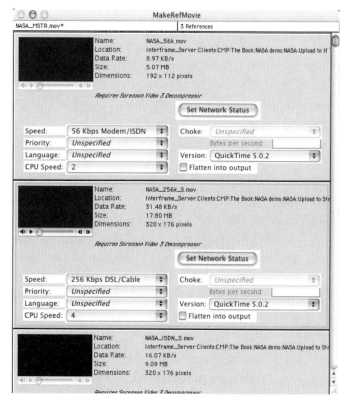

MakeRefMovie is easy from here on. Drag all the files you want to link into the window, and set the properties for each. Cleaner and MakeRefMovie support different options. MakeRefMovie can't discriminate based on OS, so you can't use it to specifically target Mac and Windows gamma. However, MakeRef does support minor version numbers of QuickTime, so you can discriminate between v4.0 and v4.1 (v4.1 was the first version that didn't crash with VBR MP3), and v5.0x and v5.0.2 (v5.0.2 was the first version bundled with Sorenson Video 3).

# Tutorial 2: Music Video for Progressive Download

This project involves encoding a music video for high-bandwidth progressive download.[3] Imagine the file is being used by a record company to promote a recording artist. The techniques described here can apply to other kinds of short, high-action content such as movie trailers.

---

3. If the file looks familiar, it's because I used the same clip in my Cleaner CD-ROM tutorial published by Safer Seas.

## Content

Our source is an NTSC music video from Extazia, a one-hit wonder in Denmark. It was originally shot on film, but was edited in NTSC and had effects added in video without any attempt to maintain the 3:2 cadence, so it's going to be a real stress test for inverse telecine. Tools that can't correct for cadence breaks will have difficulty with this project.

The source is a square-pixel, 640×480, Motion-JPEG QuickTime file. It'll work well under QuickTime, but a third-party JPEG codec will be required to use it on a Windows machine.

The video was shot full-frame, and typical of 640×480 captures, it doesn't have any edge blanking. Because you're outputting for higher resolutions, you don't need to do any cropping at all.

The file starts with a few frames of black. Because it is progressive download, we want to give the user something to look at during buffering, so we'll want to trim out the first few frames. Following those opening frames, there aren't any other black frames, so we don't have to specifically try to get black areas down to 0. Just adding some contrast to expand to the 0–255 range is enough.

## Communication goals

As a musical promo piece, the quality of the video and especially audio are paramount. The video itself must be compelling. It's better to have a high-quality baseline experience than to have real-time performance.

**25.22** A progressive frame of the source video, showing some composite noise artifacts.

### See also

See Figure W on page XV of the color section.

## Audience

For pop music, expect a wide-ranging audience, with all kinds of computers, connection speeds, and levels of media savvy. You can assume a substantial college audience for this kind of content, all with broadband connections. Most users will be running Windows, with relatively few Macs. Fortunately, because we're dealing with a clip that's under four minutes, even large mismatches between data rate and connection speed won't yield buffering times that are too high.

## Our approach

Given the project is a short clip with high audience quality requirements, progressive download is the way to go. An advantage of progressive download is the user can view the file multiple times after downloading without using server bandwidth, to save on hosting costs.

We'll make one file each in QuickTime, Windows Media, Flash MX, and MPEG-1 (as a fallback for people who don't want to or can't download players). Shoot for 500Kbps for all four clips. For 56K modem users with a real-world connection speed of 48Kbps, that will give us a total buffering time of:

$$1930 \sec onds = \left(\frac{500 \text{Kbps}}{48 \text{Kbps}} - 1\right) \times 205 \sec onds$$

That yields about 30 minutes total buffering time for modem users. It's long, but audiences are clearly willing to wait that long for a music video or movie trailer by an artist they care about. And, of course, performance will be nearly real time over any kind of broadband connection. You could consider providing low data rate versions of each clip, at perhaps 100Kbps, but the audio/visual experience wouldn't be high enough to meet our marketing goals. In this case, it would be better not to target a modem-bound audience than to provide them low quality media.

For the QuickTime, Windows Media, and Flash MX clips, 500Kbps is enough to consider resolutions of more than 320×240. Let's target 400×300 with both formats. Because MPEG-1 has lower compression efficiency, use a maximum 320×240 for that format.

## Preprocessing

The source is 640×480, so no aspect ratio correction is required. It was originally shot on film, so ideally you'd use inverse telecine to restore a full 640×480, 23.976fps image. However, this video is rife with video speed effects and cadence breaks, so only a strong inverse telecine algorithm will be able to tackle it. You'll have to do a normal deinterlace with other tools, which has two big drawbacks: you'll potentially lose half the lines of the source image, and you'll have to increase the frame rate by 25 percent to 29.97, reducing the number of bits per frame and the smoothness of motion, and increasing processor requirements for decode.

The video has composite noise artifacts throughout. This would be a great clip for my theoretical chroma-channel-only noise reduction algorithm; alas, it only exists in my dreams.

Don't forget to normalize the audio.

### Preprocessing in After Effects

Because After Effects can't deal with cadence breaks in film source, treat the source as interlaced. Turning Motion Detect on will improve quality.

Scaling is just a matter of setting the right size for the project, and picking the right scaling value to fill the screen.

Setting the input frame in AE is a little tricky. First, drag the left-hand edge of the layer's control to the first frame. I find the easiest way to do this is to position the playhead precisely at the first frame with an image on it, and then drag the track's handle so its edge aligns perfectly with it. This procedure is frame-accurate at the highest zoom level. Then you position the playhead back at the start of the movie, and drag the new, shorter layer to align with the start of the movie.

### Preprocessing in Cleaner

Cleaner historically had the most robust inverse telecine in the business. However, a bug introduced in v5.0.2 and maintained through v5.1.2 causes the order of frames to be reversed after a video cut in some cases. Thus, the second frame would be played, then the first, then the third. This causes the video to "jump" at cuts. The problem doesn't occur at every cut, and is subtle in slow-moving content. So, while earlier versions of Cleaner were able to process this file just fine, more recent versions have trouble with it. Still, the advantages of higher resolution source and lower frame rate make using Cleaner for preprocessing more than worthwhile. There are only a handful of really obvious jumps in this video, and we'll just have to live with them until Discreet fixes this bug.

For Cleaner, scaling is dead simple—just type in 400×300.

You can set the new in-point of the video in the Project window by scrubbing to the new Start frame, and picking Set In-Point from the Edit window.

Lastly, set the audio volume control to Normalize, setting a peak of 90 percent.

### Preprocessing in ProCoder

ProCoder 1.0 lacks an inverse telecine filter, but Canopus has promised one in a future free upgrade. Today, you have to use the Adaptive Deinterlace filter set to Field and Frame mode. Unfortunately, the threshold command doesn't offer enough control to get perfect deinterlacing—some field errors still creep through. I hope this will be a non-issue by time this book is out.

**25.23** Preprocessing in Squeeze on a Mac for Windows delivery.

**See also**
Figure X on color page XVI.

Set an in-point in the Source's Advanced—Setup window. Simply scrub to the first frame, and hit the In button in the Trimming box.

The audio normalization filter set to Normalize Peak at –3dB is perfect for our purposes.

### Preprocessing in Squeeze

Squeeze is the only tool that's able to get the inverse telecine on this clip right—whoo-hoo! Set Squeeze to Inverse Telecine—Upper Field Dominant, and you're good to go. (See Figure 25.23.) Alas, Squeeze can't set the video's start frame. Make a trimmed Reference Movie in QuickTime Player Pro, and use that as the source. Also turn on Squeeze's Audio Normalize filter.

### Preprocessing in MPEG Power Professional

Heuris MPEG Power Professional 2.5 has a limited, but appropriate set of preprocessing filters. For our MPEG-1 needs, these include scaling and a fine inverse telecine. In the Video 1 tab, turn Inverse Telecine on and set display resolution at 320×240. I also like to use the Global Filters on Auto. This applies content-specific filters to the video to make it easier to encode. This rarely hurts quality, and definitely reduces artifacts at low data rates. In Video 2, set aspect ratio to 1.0, and fps to 23.976.

MPEG Power Professional doesn't offer any way in its GUI to set a start frame, although this can be done by manually editing the ECL file. The simplest approach is to trim the video in Quick-Time Player, and Save As a reference movie with the edit.

MPEG Power Professional doesn't do audio normalization.

**420** Chapter 25: **Tutorials**

## Compression

We're doing progressive download for all clips, and only one data rate each for QuickTime, Windows Media, and MPEG-1. This makes the project much simpler than the NASA example in Tutorial 1.

Because the music is central to the experience with these clips, we want to spend a significantly higher proportion of bandwidth on audio than we did in the first Tutorial. We'll want to use at least 32kHz, and ideally 44.1kHz stereo to maintain pristine audio. Because the audio was mixed for broadcast, there shouldn't be any audio above 32kHz anyway.

### Encoding to QuickTime

For QuickTime, use Sorenson Video 3 for the video. SV3.1 has the best compression efficiency at these data rates, and has the ideal-for-progressive 2-pass VBR mode available in both Squeeze and Cleaner. As with the first Tutorial, use B-frames for their compression efficiency, and compensate for the sync shift as described in the first Tutorial. Turn off Streaming, of course. There's no point in spending bits on loss recovery with progressive download, where packets aren't dropped. You can leave Image Smoothing on, but it will automatically be turned off when being played back on anything less than a G4 or PIII at this resolution. With 2-pass VBR, a peak of 800Kbps is a good choice. More than that wouldn't help quality much, but would raise processor requirements quite a bit.

For audio, use MP3, available in both Squeeze and Cleaner (although it's not in most other QuickTime compression tools). MP3 requires a relatively high data rate for a clean 44.1kHz stereo. Target 96Kbps, which leaves 404Kbps for video.

The file itself should not be hinted (no streaming), and should have a compressed movie header (to make it smaller).

**25.24** Preprocessing settings in MPEG Power Professional. Not much to do here—most of it is handled by the application behind the scenes.

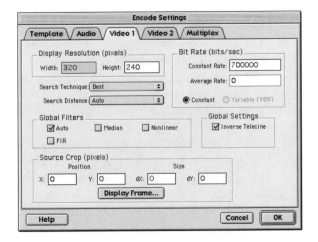

## Encoding to Flash MX

We want to make a self-contained .swf file, playable in the Flash MX player. For this, we'll use the Spark Pro codec in Squeeze for Flash MX. The UI for Spark Pro follows Sorenson Video 3 closely, so the structure is about the same. We will turn on Image Smoothing in the codec because this file is meant to play on desktop computers, not mobile devices. MP3 is the only audio codec we can use for this. As always, we'll encode with 2-pass VBR for optimal quality.

## Encoding to Windows Media

Windows Media now includes an excellent 2-pass VBR mode for progressive download files. However, this feature is just now being rolled out in mainstream encoding tools. It is supported in Cleaner, but, oddly, not in Windows Media Encoder yet.

There aren't any real tricks to working with the Windows Media file. Due to the superior compression efficiency of Windows Media Audio 8, you can allocate just 64Kbps to audio, leaving 436Kbps for video.

## Encoding to MPEG-1

Due to the lesser compression efficiency of MPEG-1, you'll need to use a smaller resolution and a higher data rate than with the other formats.

For video, use 320×240 square-pixel at 23.976fps. Different tools let us set different speed versus quality controls. Set these to the maximum in all cases. And if you have a 2-pass VBR mode available, use it as well. Because MPEG-1 is so easy to encode, a peak of 2,000 should work fine. Video will target a data rate of 700Kbps—much below that, and quality would be noxious. You'll also want to use Open GOPs for better compression efficiency, and a GOP size of 15 instead of the normal GOP of 12 for film, also for increased efficiency.

For audio, use Layer 2 192Kbps, 44.1kHz stereo. This will sound about as good as the 96Kbps MP3 in QuickTime.

## Encoding with Squeeze

Squeeze handles both QuickTime and Flash MX output. With the exception of the audio sync shift due to B-frames, Squeeze does the job with QuickTime perfectly. Just tune in the data rates and codecs, and the file comes out just right. The only subtle problem is that Squeeze doesn't have 23.976 as an output frame rate, just 24. However, with a clip of this short length that frame rate mismatch won't cause any significant problems.

The Spark Pro codec used for Flash MX follows the UI for QuickTime very closely. We can use almost exactly the same settings. However, 96Kbps isn't an option for MP3 here, so we'll have to use 80Kbps. Because of this, we'll raise the data rate to 420 to fill up our allotted bandwidth.

**25.25** The Spark Pro dialog looks almost like a QuickTime file.

## Encoding with Cleaner

Cleaner is the only tool that supports all three target formats. Like Squeeze, the QuickTime setting is trivial—apply the settings as discussed previously. The only trick is to make sure you have MP3 set to the Highest Quality (Slower) mode. You could use MP3 VBR, but the implementation in the current v5.1.1 of Cleaner is buggy, and can yield crashes and unpredictable output data rates. However, a good VBR implementation, à la LAME would improve quality somewhat. You could encode the audio in LAME, and then paste it in with QuickTime Player Pro. If you go that route, use this option:

```
lame Extazia.aiff --abr 96 -q 0
```

**25.26** Cleaner's MPEG-1 Encode tab. Having Charger installed gives the 2-pass VBR option.

This specifies a VBR that targets an average data rate of 96Kbps, in the slowest, highest quality encoding mode. You'll need to use QuickTime Player Pro to extract the audio into an .aiff file first, of course.

Cleaner's Windows Media on Windows support is just as easy. Because this project is progressive download, use Windows Media Video v8 in 2-pass VBR mode. Note the Video Quality control is ignored in 2-pass VBR, although Cleaner still draws it. On a Mac, the file will need to be encoded with the v7 codecs, and without 2-pass VBR. Still, the gap between Mac and Windows encodes isn't as big with progressive download, as Intelligent Streaming isn't an issue.

For MPEG-1, Cleaner also does well (although with very slow render times). Make sure the Image tab is set to 320×240 at a 1.0000 display ratio, with the Encoding Speed set to Slower (Higher Quality). If you have the MPEG Charger add-in installed, you can and should use 2-pass VBR.

### Encoding with ProCoder

ProCoder 1.0 supports QuickTime and Windows Media, but not 2-pass VBR, MP3, or the other important features we want to use. (Canopus has announced improved support for those formats in future versions.)

ProCoder has world-class MPEG-1, including 2-pass VBR. However, it is limited in its minimum video data rate—it can't go below 900Kbps (also due to be addressed in a future release). Because ProCoder can't currently do inverse telecine (another soon-to-be-addressed issue), use a slightly higher data rate to compensate for having to use 29.97fps. That said, quality in the Mastering Quality mode is superlative. There are a ton of esoteric options that relate to packet headers, but these can be safely ignored.

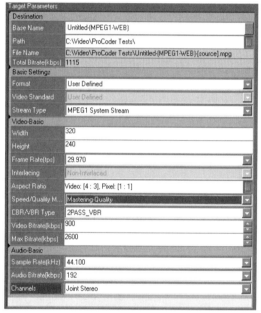

**25.27** ProCoder's MPEG-1 settings for this project. We don't need any of the myriad extra options in Advanced for this project.

### Encoding with Heuris MPEG Power Professional

Heuris MPEG Power Professional (MPP) is an excellent MPEG-1 encoding solution, although many important options are hidden under a somewhat esoteric interface.

In the Video 1 tab, set resolution and data rate as Normal. Note MPP can't do VBR with MPEG-1, although it can with MPEG-2. The Search Technique parameter should be set to Best, and Search Distance to Auto. While it might be tempting to pick Extended for Search Distance, this can actually hurt quality with some video, so it's best to just let MPP pick the right settings.

In Video 2, set GOP size as appropriate. Check the innocuous–looking Auto Analyze checkbox. It's critical to achieving optimum quality. Auto Analyze enables an initial analysis pass to help determine appropriate values for the other options set previously, do better I-frame injection, and make other quality improvements. Auto Analyze provides a lot of the benefits of a traditional 2-pass encode at a fraction of the speed hit.

In Multiplex, make sure both Multiplex Target and Initial Clock Reference are set to Auto.

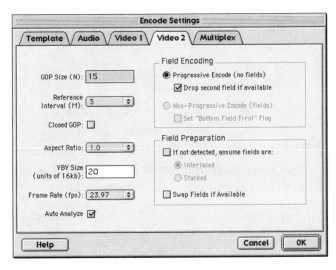

**25.28** The Video 2 tab in MPEG Power Professional (see Figure 25.24 for the Video 1 settings). The Auto Analyze check button looks unassuming, but is critical for maximum quality.

# Tutorial 3: Animation for Video Game Cut Scene

In our final tutorial, we're going to compress computer animation for a video game cut scene with multiple target delivery platforms. This process would work well for most high data rate CD-ROM and kiosk applications. For file format, we'll use Bink, which has excellent playback support on Windows, Mac (OS 9.x and OS X), Microsoft XBox, and Nintendo GameCube.

## Content

The source is progressive, computer-generated, 24fps RGB. It's provided on the disc as a QuickTime PNG movie in lossless RGB, rendered with Mac gamma. Because it was computer-generated, gradients are smooth and there's absolutely no noise. However, there are sequences of very fast motion.

## Communication goals

Our goal for the project is simple and lofty—we want the playback experience to look as much like the source as possible. With Bink projects, typically the main limitation is file size, not decode performance. This is a particular problem on disc-based games, where most of the game is already taken up by game art and the playback engine, often leaving only a limited amount of space for video. Or, for a game that uses a lot of video, there is a battle between minutes of video

**25.29** A sample frame from the Myst trailer.

(Official credits: Myst 3: Exile trailer produced by Presto Studios, Inc. published by Ubi Soft Entertainment, S.A. © 2001 Ubi Soft Entertainment, S.A. & Cyan, Inc. All rights reserved.)

**See also**
See Figure Y on color page XVI.

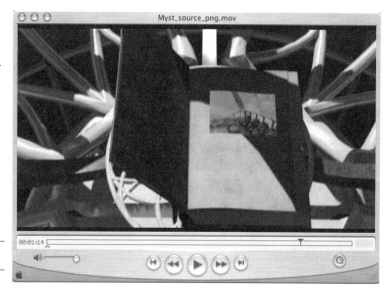

and the quality of that video (as anyone who tried to do a full install of *Diablo 2* knows all too well).

## Audience

Let's assume this project's video will be played inside a computer game. Our target delivery platforms are an 800MHz PIII Windows machine, a 500MHz G3 Mac (running either Mac OS X or 9.x), a Nintendo GameCube, and an Xbox (one great thing about consoles is they represent a uniform target platform).

## Our approach

For any given game, you would only encode to a single format, which in this case is Bink. Bink offers merely decent compression efficiency. But it has unrivaled support for consoles, excellent decode performance, and a very rich API.

If you wanted to target just Mac and Windows machines, you could encode in QuickTime with the VP3 video and IMA audio codecs or .AVI with DivX 5 video (with 2-pass VBR) and MP3 audio. While either of those options would offer superior compression efficiency, they both require a lot more decode horsepower. Bink's deep API and excellent decode performance makes it possible to do things like preload levels or even install the game in the background while the video is playing.

## Preprocessing

There is very little image preprocessing required for this file. For the non-Mac file, you need to apply a standard Mac-to-Windows gamma correction. Also, crop out those six pixels of black at the top and bottom, leaving a 640×348 image.

Although the Bink encoder isn't a preprocessing powerhouse, it's more than capable of doing the cropping and gamma correction required by this project.

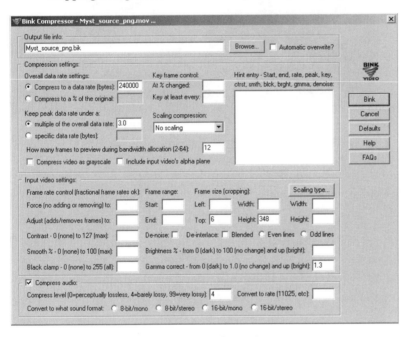

**25.30** The Bink UI, as we have it configured.

The Bink UI is rather cluttered. One common source of confusion is that there are two boxes labeled "Width" and "Height." Cropping uses the ones on the left (next to "Left" and "Top"). The others, under "Scaling Type" are used to set the scaling of the video (which is processed *before* cropping). Because we want to crop 6 pixels off the top and bottom, enter a Top value of 6, and a Height value of 348 (360-6-6).

Use the standard 1.3 value for gamma correction.

The audio is already normalized, so you don't need to do any more audio processing. As you can see, computer-rendered sources make preprocessing trivial.

## Compression

We're targeting a total data file size of 20MB. Like QuickTime, you set data rate in Bink as the total KBps from which it subtracts the audio data rate to determine video data rate. So, to figure our target data rate, divide 20000K by the clip's 90 second duration, to get 222K. However, Because there are some longer sequences of black frames or very simple graphics that will use much less than our requested data rate, we know we can raise the data rate slightly. Let's shoot for 240K, and enter 240000 in the Bytes field.

Bink's default peak data rate of 3.0 is a good default for most projects. Set frame allocation to 12, the recommended default for rendered sources (we'd use 8 if it were mainly a video piece).

Lastly, leave audio at its default compression level of 4. This produces good compression efficiency and nearly audibly lossless quality. I wouldn't want to encode a classical music concert at this setting, but it will be more than enough to handle the narration-plus-music soundtrack. Although Bink Audio is a quality-limited VBR, with a level of 4, you'll generally see an output data rate in the ballpark of 128Kbps.

And that's it! After compressing the file, you'll get a file size of 19,975KBps, only 25K shy of our target file size.

You can analyze the file with Bink's excellent Video Analysis tool. The blue line is at the theoretical peak speed of a 2X CD-ROM. As you can see in the figure, most of the file falls under that line, although there are occasions where it spikes up to 800KBps. However, the target machines all shipped with at least 16X CD-ROM drives, so you don't need to be concerned about this at all.

The output quality looks good, but there are visible artifacts. If we had encoded to targets of 350K instead of 240K, it would have been more impressive, but we would have missed our file size target dramatically.

One odd effect of the 12-frame look-ahead is that the data rate actually goes *up* before easy sections of the video. Because the codec keeps the data rate average over 12 frames, the data rate starts going up 12 frames before the cut to black.

## Delivery

Normally, the application author would use the Bink SDK to control the video. However, we can test it within this Bink application, and can even save the movie and playback settings as a stand-alone, executable application. Here's how:

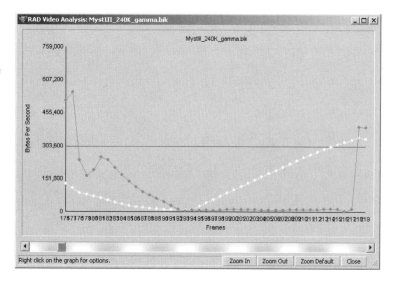

25.31 The Bink Video Analysis window. Quite useful for finding difficult-to-compress data rate spikes. Here we see how the video rate drops around a section of total black.

Go into the Advanced Play dialog. For settings, I like "Clear outside window to black," which draws letterboxing bars. Set Full Screen resolution to 640×480, into which 640×348 fits perfectly. I like color depth set to "Try True Color." This will use the 8-bit-per-channel 32-bit and 24-bit modes by default, but will fall back to the 5/6-bit-per-channel 16-bit mode if needed. For Blitting Style, pick Choose Blitting Style Automatically. This will normally default to DirectDraw to Primary Surface, using the YUV overlay if available, falling back to lower quality or speed modes if needed. Set Window Style to "No title bar or border." Turn on "Hide Windows mouse cursor in the Bink window." To get statistics on playback (useful for troubleshooting playback problems) check "Show a playback summary."

Now hit play, and you'll get a great, full-screen cut-scene experience. You can hit the Spacebar to pause and play, or Escape to quit. To make a standalone executable, use the "Make EXE" command. Because you're probably using the unlicensed version of the encoder from the CD-ROM or the Rad Game Tools Web site, you'll get the Bink logo at the end of playback in either mode.

Some game vendors market their games by distributing cut scenes this way. I'd normally use a format with better compression efficiency for Web delivery.

# Glossary

*1-pass encoding*  a technique that encodes a file as it is read.

*2-pass encoding*  a technique in which a file is first analyzed, and then compressed in a second pass. 2-pass encoding is slower but yields higher quality than 1-pass.

*4:1:1*  the color space used in NTSC DV format video. It uses one chroma sample per 4×1 block of pixels.

*4:2:0*  the color space used by most delivery codecs. It uses one chroma sample per 2×2 block of pixels.

*4:2:2*  the color space used by many authoring formats. It uses one chroma sample per 1×2 block of pixels, thus never sharing a chroma sample between fields.

*5.1 Dolby Digital*  a surround-sound format developed by Dolby Labs. "5.1" refers to five discrete audio channels, each going to its own speaker, plus a single low-frequency channel that goes to a sub-woofer.

*AAC (Advanced Audio Coding)*  an audio codec originally designed for use in MPEG-2, and standard in MPEG-4.

*AAF (Advanced Authoring Format)*  a protocol and file format developed jointly by Microsoft, Avid, and others to define metadata defining certain aspects of a production in a manner that can be read by any AAF-compliant application. Not yet used in any products.

*ACELP (Adaptive Code-Excited Linear Prediction)*  a low bitrate speech codec.

*adaptive deinterlacing*  the technique in which some data from the discarded field is used to construct a higher quality progressive output.

*ADPCM (Adaptive Differential Pulse Code Modulation)*  an audio compression technique that encodes the difference between samples, not the value of the voltage itself. The basis of the IMA codec

*alpha channel (also "key channel")*  an additional channel of grayscale information, that can be used to key such things as transparency. 32-bit video normally includes an 8-bit alpha channel.

*AltiVec*  a SIMD architecture built into Motorola G4 CPUs that dramatically improves the performance of media processing with software that is optimized for it. See *SIMD*.

*animated GIF*  a very early Web video format. Essentially, a series of 8-bit or fewer GIF images bundled into a single file. It offers poor compression efficiency for natural images, but can do very well with flat colors. Often used for Web banner advertisements.

*API (Application Programming Interface)*  a set of programming "hooks" that provide a means for developers to write programs that connect to core functionality within someone else's application.

*artifact*  a visible defect in a compressed image. Typical artifacts are ringing or blockiness.

*ASF (Advanced Streaming Format)*  the file format of Windows Media.

*.asf*  the file extension originally used for Windows Media files. See *.wmv* and *.wma*.

*aspect ratio*  the ratio of length to height of pictures. Standard television is 4:3, widescreen, video 16:9. Feature films often use 1.85:1, 2.35:1, and other values.

*.asx*  the original file extension for Windows Media streaming metafiles that reside on Web servers. See *.wax* and *.wvx*.

*ATSC (Advanced Television Standards Committee)*  the group tasked with developing the technical standards of digital television. Also used to refer to the U.S. standard for digital television. www.atsc.org

*AVI (Audio Video Interleave)*  a format developed by Microsoft that originally allowed audio and video to be combined in a single file that could play off a CD-ROM. Technically obsolete, but still widely used.

*banding*  the visible distortion caused by insufficient bit depth. So-called because it manifests itself as visible bands in what should be smooth color or tonal gradients.

*bartending*  the monitoring of the progress bar on a compression machine. While immensely aggravating on a deadline, with the right attitude bartending can make compression a contemplative experience. Bartending is an excellent time for zen meditation, email, and rereading this book.

*bit*  (short for Binary digiT) a bit can have one of two possible values—one or zero. A byte is a group of eight bits. Two bytes (16 bits) grouped together are called a *word*. Among audio people, it's common to hear phrases such as "16-bit word" and "20-bit word." In these phrases, the term "word" describes a single sample, no matter how many bits are used to measure it.

*block*  the basic unit of DCT and other kinds of compression. Typical codecs use 8×8 blocks.

*blocking*  the distortion in which different macroblocks exhibit different errors, so the edge between them becomes visible.

*byte*  a group of eight bits.

*CBR (Constant Bit Rate)*  an encoding method in which the bitrate remains constant across the length of the file being encoded. Note that while audio codecs are often truly CBR, the size of video frames will vary somewhat.

*CCD (Charge Coupled Device)*  the light-sensitive elements used to convert light to a corresponding electrical charge in digital imaging systems such as scanners and digital video cameras.

*CELP (Code Excited Linear Prediction)*  a common low bitrate speech codec in MPEG-4.

*chroma*  the nonlinear color component of a video signal, not independent of luma.

*chroma subsampling (or color subsampling)*  the process of reducing color resolution by taking fewer color samples than luminance samples. 4:2:0 is a Y'CbCr subsampled color space.

# Glossary

*codec (COmpressor, DECompressor)* the technology used to compress and play back video and audio.

*color space* the various ways color is described mathematically. RGB and Y'CbCr 4:2:0 are color spaces.

*compression efficiency* refers to how well a codec preserves the quality of the source at a given bitrate. Codecs with better compression efficiency can deliver better quality at a given data rate or lower data rates at a given quality.

*DCT (Discrete Cosine Transform)* a mathematical technique in which a series of numbers is turned into coefficients for a series of cosines. In compression, it's widely used as the first stage of encoding digital video. DCT operations work on pixel blocks usually in multiples of 8 (normally, 8×8).

*deinterlacing* the process of converting interlaced video to progressive.

*DirectShow* the current Windows API for encoding AVI files and playing back all kinds of media files. Largely replaced VfW. Can have compatibility problems with some VfW video codecs.

*dither* a method to reduce banding by randomly distributing the error between the source pixel values, and the pixel values possible in the output color space. Dithering can improve quality before compression, but makes compression much more difficult.

*DRM (Digital Rights Management)* the technology used to control access to technology to licensed users. Different varieties are provided by many different vendors.

*.divx* the extension for an AVI file using MPEG-4 for video and typically MP3 for audio. Typically authored with the DivX codec.

*DVD* (often referred to as Digital Video Disc or Digital Versatile Disc, but without any official meaning for the acronym) the name of a series of technologies based around a high-density 5-inch optical disc format.

*DVD-ROM (DVD-Right Once, read Many)* one of several DVD formats.

*DVD-R (DVD-Recordable)* one of several DVD formats, and the most compatible with the widest range of DVD players for recording.

*DVD-RW (DVD-ReWriteable)* one of several Digital Versatile Disc formats.

*DVD+RW (DVD+ReWriteable)* one of several Digital Versatile Disc formats.

*ECL (Edit Control Lists)* used to store information such as where to inject I-frames, what filters to run, and inverse telecine cadence in MPEG encoders.

*field* the odd and even lines of a frame of interlaced video are each one field. Typically referred to as "top and bottom," with top being every other line from the first line, and bottom being every other line from the second.

*FireWire* the Apple trade name for IEEE 1394. (see *IEEE 1394, i-Link*.)

*frame rate* a measurement of frames per second in an image. NTSC default is 29.97, PAL is 25, and film is 24.

*fps (Frames Per Second)* the unit of measurement for frame rate.

*full screen* traditionally 640×480—from the standard resolution of the Mac II computer.

*gamma* the name for exponential value used to control the relationship between stored non-linear luma and displayed linear luma. Changing gamma values changes the brightness in the middle of the luma range, but not the ends. Different display technologies (Macintosh and Windows monitors, film, and so on) have different gammas, and video must be encoded to target its eventual display gamma.

*GOP (Group of Pictures)* in MPEG video, an I-frame followed by P- and B-frames. A closed GOP is a self-contained unit.

*HD (High Definition)* an image format that is higher resolution than SD (Standard Definition) television. Images of 720 lines of resolution or more are usually considered HD.

*HSV (Hue, Saturation, Value); sometimes HSL (Hue, Saturation, Luminance)* a color space system. It is used within content creation applications, but not for storage.

*I-frames* the MPEG terminology for a keyframe. An I-frame is self-contained, and starts every GOP.

*IEEE 1394* a serial digital interface standard capable of running at speeds of up to 800Mbps or more.

*i-Link* the Sony trade name for IEEE 1394. (see *IEEE 1394*, *FireWire*.)

*interframe encoding* the technique in which consecutive frames of video are compared so redundant elements can be removed. Interframe encoding is core to all delivery codecs.

*interlace* a method of capturing and displaying video in which each video frame consists of two fields, referred to as upper and lower. As each frame is scanned onto a display such as a television screen, first one field then the other is shown. The second field consists of scan lines that fall in between the first field's scan lines, hence the term 'interlaced.' The technique makes fast motion appear smoother and reduces flicker. The opposite of interlace is progressive scan, in which each line of video is scanned onto the display in successive order.

*inverse telecine* the reverse of the 3:2 pull-down process, in which excess frames that were created to generate 60-field per second video are removed.

*IPMP (Intellectual Property Management and Protection)*

*IRE* a unit of measurement defined as 1 percent of the video range between blanking to peak white.

*ISMA (Internet Stream Media Alliance)* the industry organization formed "to accelerate the adoption of open standards for streaming rich media—video, audio, and associated data —over the Internet." Its members include: Apple, Cisco, IBM, Kassena, Philips, and Sun.

## Glossary

*ISO (International Organization for Standards—yes, that doesn't match)* an international body governing standards. Among others, they are responsible for the MPEG formats.

*ITU-R BT.601 (formerly CCIR 601)* the standard that defines the parameters of encoded digital video signals. Typically thought of as 4:2:2, sampled at 13.5MHz, with 720 luminance samples per active line digitized at 8- or 10-bits. ITU stands for International Telecommunications Union. R is for Radio spectrum, and BT is for Broadcast Television.

*JPEG (Joint Photographic Experts Group)* an organization who creates digital still image technologies. Also, the JPEG, which uses DCT, and is very common on the internet.

*JPEG 2000* a wavelet-based still image codec from the creators of JPEG.

*JMF (Java Media Framework)* the API-enabling audio, video, and other time-based media to be added to Java applications.

*JVM (Java Virtual Machine)* the core of Sun's popular Java language. The "virtual machine" refers to processor architecture emulation that allows Java applets to be run on a variety of platforms, regardless of CPU.

*KBps (kiloBytes per second)*

*Kbps (kilobits per second)*

*keyframe* in compression, a self-contained frame, which doesn't require a reference to any other frame. Immediate access in a compressed file is only possible to a keyframe. Called I-frame in MPEG. Also, a term from cel animation in which a single image plays an important (i.e., "key") role in defining the action that follows. It has also come to refer to a group of parameters that define a set of actions that might change over time, as the program interpolates between one snapshot of the values and another.

*LFE (Low Frequency E----)* the .1 low-frequency channel in Dolby Digital 5.1 surround systems. The 'E' in the acronym can mean a number of different things depending on its use: Extension, Effects, or Enhancement.

*lossless* a type of codec that preserves all of the information contained within the original file. Depending on the source, lossless compression can result in dramatic space savings or none at all.

*lossy* a type of codec that discards some data contained in the original file during compression. Lossy compression offers much smaller data rates than lossless.

*luminance (luma)* a video component pertaining to brightness, referred to as Y in YUV and Y'CbCr.

*macroblock* a group of four (usually) 8×8 pixel blocks (yielding a 16×16 block) used for motion estimation during encoding.

*MP@ML (MainProfile@MainLevel)* in MPEG, a 4:2:0 profile that covers broadcast television formats up to 720×486 at 30fps (NTSC) or 720×576 at 25fps (PAL), with data rates raging from 2Mbps to 9Mbps.

*MP@HL (MainProfile@HighLevel)* in MPEG-2, a 4:2:2 profile at 720×486 (NTSC) or

720×576 (PAL), with data rates up to 50Mbps.

*MBR (Multiple Bit Rate)* a method of encoding in which files scale to match the available bandwidth.

*M-JPEG (Motion-JPEG)* a 4:2:2 codec that stores each field in its own bitmap used for authoring, prevalent in the mid-90s.

*MMX (MultiMedia eXtensions)* an early SIMD instruction set added to Intel's Pentium processors.

*MNG (Multipart Network Graphics)* the motion graphics counterpart to PNG. a lossless RGB format that can encode to a variety of bit depths from 32-bit on down.

*Moore's Law* (named for Intel founder Gordon Moore) in its modern use, it states that the performance of CPUs will double every 18 months at the same price.

*motion estimation* a technique used to calculate if and where elements of an image have moved between one frame and the next. Motion estimation is a big differentiator between codecs, and the major consumer of CPU cycles during compression.

*motion search* the name given algorithms used in motion estimation.

*Motion-JPEG* see *M-JPEG*.

*MPEG (Moving Pictures Experts Group)* a group working under the auspices of the ISO formed to develop international compression/decompression standards.

*MPEG-1* the original MPEG format. It is used in Video CD and a variety of multimedia tasks.

*MPEG-2* an enhanced version of MPEG that added support for interlaced frames, as well as enhanced compression. It is used for DVDs, digital satellite video transmission, and digital cable.

*MPEG-4* a major new version of MPEG with much enhanced support for streaming, low bitrates, and compression efficiency.

*MPEG-7* not a video format, MPEG-7 is a metadata solution for video, somewhat analogous to AAF. It is not yet supported in any products.

*MPEG-21* a forthcoming format for rich, interactive media delivery. Still some years away from being used in real products.

*.mov* the standard QuickTime file extension.

*OHCI 1394* a particular kind of IEEE 1394 card for Windows. All modern cards are OHCI, but older ones like the Digital Origin Lynx cards aren't OHCI, and won't work with DirectShow applications.

*PCI (Peripheral Component Interconnect)* a high-speed bus for connecting peripherals to a CPU developed by Intel in 1993. It is used for capture cards in both Mac OS and Windows computers.

*perceptual noise shaping (PNS)* an encoding scheme that reduces redundancies and irrelevancies within audio signals.

*pixel doubled*  a technique used to scale an image up by repeating each pixel, thus a 160×120 image that is pixel doubled would become a 320×240 image.

*premultiply*  a technique in which opacity is removed from an alpha channel. The term is derived from the math involved in the process—the opacity is multiplied by the value of the color against which an alpha channel is composited. Premultiplied alpha channels are opaque.

*progressive download*  a technique in which a file is transmitted and will begin playing back before the entire file has been sent.

*progressive scan*  a method of capturing and displaying video in which the signal is displayed in consecutive scan lines, as opposed to interlaced. Video meant for playback on computers is almost always progressive. Also, it has become common to use the letter "p" to indicate progressive scan content. For example: 24p (24fps, progressive) or 1080p (1,080 lines of resolution, progressive).

*progressive GIF*  a GIF file that will be displayed in greater and greater detail, as the file is received.

*progressive JPEG*  a JPEG file that will be displayed in greater and greater detail, as the file is received.

*QT*  Apple QuickTime.

*QTVR*  Apple QuickTime VR, as in Virtual Reality. Apple's 360 degree panoramic image technology.

*.qti*  the QuickTime Image file extension.

*quarter screen*  canonically 320×240, or half the height and width of the 640×480 resolution used by the first generation of color Macintosh computers.

*QuickTime*  the first and most complete video authoring, distribution, and playback architecture. From Apple Computer.

*raster*  the pattern of horizontal scan lines that make up a video picture.

*RealMedia*  the native streaming media type used with RealNetworks products.

*Red Book*  the spec for CD audio, first published in 1982.

*resolution*  the height and width of an image.

*RGB (Red, Green, Blue)*  the color space of computer displays.

*ringing*  the compression errors that occur around sharp edges.

*RLE (Run Length Encoding)*  the compression technique that depends on long strings of identical pixels, and as many as possible identical lines between frames. Files sizes are reduced by replacing those pixel strings with simple code that instructs the decoder to repeat x-color, y-many times. RLE encoding is mostly used on black and white line art.

*.ra*  the older file extension for RealAudio, supplanted by the .rm extension.

*.rm*  the RealMedia file extension.

*RTSP (Real Time Streaming Protocol)*  the standard client-server control protocol for streaming.

*safe area* the portion of the image area guaranteed to be displayed on all televisions. The most common of these are title safe and action safe.

*SDTV (Standard Definition TV)* 720×486 pixels at 29.976fps (NTSC) or 720×576 at 25fps (PAL).

*SDI (Serial Digital Interface)* a lossless, digital interconnection format used in high-end facilities for video capture from DVCPRO50, Digital Betacam, and similar systems.

*SDK (Software Development Kit)* a set of developer's programming tools for writing code that takes advantage of an application's API.

*SIMD (Single Instruction Multiple Data)* the processor architecture that simultaneously performs a single operation on multiple items of data. SIMD can radically improve speed of codecs and other media processing for software that uses it. (See *AltiVec*, *MMX*, *SSE*, and *SSE2*.)

*SMIL (Synchronized Multimedia Integration Language)* (pronounced "smile") a programming language for authoring interactive presentations.

*spatial compression* the technique that reduces file size by reducing redundancy within a frame.

*SSE (Single SIMD Extensions)* the group of 70 instructions added to the Pentium III CPU to improve media processing performance over Intel's initial SIMD offering, MMX.

*SSE2 (Single SIMD Extensions 2)* the group of 144 instructions added to the Pentium 4 CPU to further enhance performance. These extensions are of less utility in video compression.

*statistical multiplexing* an inter-channel VBR used in cable and satellite delivery systems. Because the total bandwidth of cable and satellite systems is fixed, and any given channel's bandwidth can vary within that overall bandwidth limit, massive hardware statistical multiplexors simultaneously analyze all channels at once, dynamically adjusting the bandwidth among them to offer the highest average quality possible.

*streaming* real-time transmission of video and/or audio data. Streaming requires a particular kind of server software.

*subsampling* the process of reducing spatial resolution by taking samples that cover larger areas than the original samples or of reducing temporal resolutions by taking samples that cover more time than the original samples.

*telecine* the process of converting film to video. In NTSC, each film frame, which is essentially a progressive scan, is converted to fields, with the first frame becoming three fields, the next frame two fields, then three, then two, and repeating. This is called 3:2 pull-down. Reversing this process is called *inverse telecine*. In PAL, telecine simply speeds the film up to 25 fps and it is captured progressive.

*temporal compression* the technique that reduces file size by reducing redundancy between frames. A critical feature of all delivery codecs.

*UDF (Universal Disc Format)* the standard describing a practical, recordable, random access file system. Typically used on DVD media.

*UDP (User Datagram Protocol)* a connectionless transport protocol that runs on top of IP networks. UDP offers only light error checking. It is used by all streaming formats.

*UI (user interface)* the means by which a user interacts with software-hardware or hardware-based devices.

*VBR (Variable Bit Rate)* an encoding technique that allows the bitrate to fluctuate according to the complexity of the content being encoded.

*VBV (Video Buffering Verifier)* the portion of the MPEG-1 specification defining a minimum buffer size to allow MPEG-1 to work on memory-limited devices.

*vector* a method of storing information about graphical elements such as lines and curves as data describing lengths and angles. Vector graphics have the advantage of taking relatively little space yet they can be scaled to any size without the usual distortion associated with scaling an image of fixed resolution too large.

*vector quantization* a video compression technique used in Cinepak and Sorenson Video 1 and 2.

*Velocity Engine* Apple's trademark for the AltiVec architecture used in G4 processors.

*VfW (Video for Windows)* the original Windows API for authoring and playing back AVI files. Largely replaced by DirectShow, although some applications still use VfW.

*VLB* an outdated high-speed bus used in Intel 486-based computers. It has been supplanted by the PCI bus on Pentium-class systems.

*VOB files (Video Object files)* the DVD Video files that hold video and audio information.

**wavelet compression** a compression technique in which a signal is converted into a series of frequency bands. It is processor-intensive, and so is considered better suited to still images than video.

*.wax* the Windows Media file extension for streaming audio metafiles that reside on Web servers.

*White Book* the document written by Sony, Philips, and JVC that extended the Red Book audio CD format to include digital video. The result is commonly referred to as Video CD.

*WMA (Windows Media Audio)* the name of the default audio codec in Windows Media.

*.wma* the file extension used for audio-only Windows Media files.

*WMV (Windows Media Video)* the name of the default video codec in Windows Media.

*.wmv* the file extension used for Windows Media Video files.

*word* see *bit*.

*.wvx* the Windows Media file extension for streaming video metafiles that reside on Web servers.

*Y'CbCr* the luminance and color difference signals in digital video. Note that isn't a single "curly" quote mark after the Y, but a straight single quote ( ' ).

*YUV* when used properly, Y refers to luminance, and U and V refer to subcarrier modulation axes in NTSC color coding in which U stands for blue minus luminance and V for red minus luminance. However, YUV has become shorthand for any luma-plus-subsampled chroma color spaces. When referring to files that reside in a computer, Y'CbCr is the proper term and is what this book uses.

*YUV-9* the obsolete 9-bit per pixel color space that uses one chroma sample per 4×4 block of pixels.

*YUV-12* another name for 4:2:0 color space. It uses 12 bits per pixel.

*YUV overlay* the special hardware in a video card that allows it to take Y'CbCr data directly, and convert it to RGB inside the card. This is faster and of higher quality than requiring the computer to handle the conversion itself. All modern video cards include a YUV overlay.

# Index

## Symbols
.asf 430
.asx 430
.divx 431
.mov 434
.qti 435
.ra 435
.rm 435
.wax 437
.wma 437
.wmv 437
.wvx 438
μ-Law 238, 335

## Numerics
16-bit color 34
1-pass encoding 128–129, 429
 vs. 2-pass 128
24-bit color 35
24p 72
2GB limit 93
2-pass encoding 129, 429
2-pass VBR 149
30-bit color 37
3ivx 225–226
4:1:1 31, 429
4:2:0 29–31, 136, 429
4:2:2 30, 136, 429
4:4:4 30
48-bit color 37
5.1 Dolby Digital 429
8-bit color 33
8-bit grayscale 34
8mm 81

## A
AAC 196, 234
AAC-LC 234–235
AAF 351, 429
ABR mode 354
AC-3 supported channels 288
ACE 312
ACELP 429
ACELP.net 255
ActiveMovie 323
ACT-L2 222–223
Adaptec Toast Titanium 284
adaptive deinterlacing 429
Adobe Premiere 6.01
 *See also Premiere* 368
ADPCM 57, 342, 429
Advanced Audio Coding (AAC)
 289, 429
Advanced Authoring Format
 *See AAF*
Advanced Streaming Format
 *See ASF*
Advanced Television Standard
 Committee
 *See ATSC*
AES/EBU 89
After Effects 74–75, 208, 371–
 374, 400–401
 and AVI 324
 Color Finesse plug-in 112
 inverse telecine filter 104
 Motion Detect mode 102
 pan and scan tools 100
 preprocessing 418
 safe area boundaries 98
 scaling 105
 tips 372, 374
Agility 394
Akamai 165
A-Law 238, 336
aliases 24
aliasing 24
 noise 24
alpha channels 231, 328, 429
AltiVec 429
animated GIF 350, 429
animation 76, 231
 tutorial 424
antialiasing 74
Anystream 389
Anystream Agility 394
API 429
Apple DVD Studio Pro 298
Apple iDVD 298
AppleScript 393, 410
applications
 custom 164
artifacts 430
ARTS 308
ASF 239–240, 430
aspect ratio 72, 105–106, 108,
 122, 430
 MPEG-1 274–275
 MPEG-2 290
aspect ratios 275
ASX 162
ATSC 79, 293, 430
 formats 295
audio 56
 compression 56, 120
 connections 87, 89
 preprocessing 119
audio codecs
 *See Chapter 9*
 general purpose vs. speech
  141
Audio Video Interleave
 *See AVI*
AVI 177, 181, 321–336, 430
 *See Chapter 20*
 audio/video interleave 322
 authoring codecs 333
 authoring tools 323

**439**

# 440  Index

delivering files in 322
delivery codecs 325
intermediates 390
progressive download 185
AviSynth 324

## B

background
    detail 76
        reducing 73
balanced audio 88
balanced mediocrity 65, 134
    testing for 65
banding 430
bands 47
bartending 2, 430
Betacam 81–82
Betacam SP 82
Betacam SX 82
B-frames 53–55, 126, 137, 205, 408, 410
    and MPEG-1 272
    DivX 329
    *See also bi-directional frames*
bi-directional frames 53–54, 126
    *See also B-frames*
BIFS 171
Bink 183, 343–345, 424, 426, 428
    Audio 427
    audio 345
    business model 345
    data rates 344, 426
    encoder 344
    playback 345
    and Smacker files 357
bit depth 122, 136, 145
    and audio codecs 142
bit-buffers 58
bitrate accuracy 59
bits 20, 430
black levels 114
blanking areas 95–97
blitting 428

block refresh 133, 139
    and RTSP streaming 133
blocking 46, 430
brightness 26, 114, 116
broadcast standards 77
broadcasting
    live 160–161
bytes 430

## C

Canopus ProCoder 1.0
    *See ProCoder*
capture 77–92
    *See Chapter 6*
    codecs 92
    optimizing 92
    resolution 89
Carlos, Wendy 18
casts 154
Cb and Cr channels 122
CBR 127–128, 143–144, 146, 430
    and MP3 352
    vs. VBR 127, 143, 146
CCD 430
CELP (Code Excited Linear Prediction) 430
channels 28
Charge Coupled Device
    *See CCD*
Charger 282, 297, 422
    and MPEG-2 297
    MPEG-1 282
chroma 430
    adjustment 116
    blocks 45
    subsampling 430
chrominance 7
CIE color charts 9
Cinepak 228, 331
    and Kinoma 348
    grayscale mode 34
    keyframes 125
    tips 228, 331
Cinepak Pro 229
    tips 229
clamping 32

Cleaner 211, 282, 297, 323, 362–366, 394–395, 401, 413–414, 422–423
    Adaptive Deinterlace 102
    and MPEG-2 297
    and Windows Media 249
    deinterlace filters 104
    encoding 422
    Flat encoding mode 128
    MPEG-1 282
    noise reduction 110
    preprocessing 418
    and QuickTime 363
    scaling 105
    tips 364–366
    vs. Premiere 368
Cleaner Central 394–395
ClearVideo 332
CMYK 31
coaxial connections 85
codebooks 40–41
codecs 431
    audio
        *See Chapter 9*
        channels 142
        general purpose vs. speech 141
        settings 141
    choosing 135
    options 180
    speech 58
    video
        *See Chapter 8*
        settings 121
coding
    progressive vs. interlaced 44
colocation 165
color 5, 10–11
    correction 112–113, 117
    indexed 33, 123
    perception 11
    safe 71
    spaces 28–31, 122, 136, 431
        CMYK 31
        conversion 38
        RGB 28
        Y'CbCr 29

Index **441**

subsampling 430
U and V 32
Color Finesse plug-in 112, 117, 374
colorblindness 8
companion DVD-ROM 3
compatibility
    codecs 61
    OS 175
Component codec 233
composite noise 111
    artifacts 416
compression 19, 40–61, 67
    *See Chapter 3*
    audio 56, 120
    data 40–41
    efficiency 126–127, 133, 135, 431
    fractal 48
    lossless 41
    lossy 41
    spatial 42
        methods 42
    speed 131
    sub-band 57
    temporal 49
    wavelet 46
compression tools
    *See Chapter 23*
cones 7–8, 16
connection speed 156
connections 85
    audio 87–89
contrast 114
    values 115
Corona Windows Media 9 codecs 180
cropping 95, 97, 99
    amounts 98
cross-fades 76
Customflix 397
cuts 66
    rapid 75
    straight 76

## D

data rates 55, 91–92, 126–130, 145, 427
    audio codecs 143
    Bink 344
    buffered vs. whole-file 128
    control 54–55, 126–129, 204
    mode 138
    multiple 127
    quality-limited 129
    spikes 427
DCOM 393
DCT 42–46, 48, 431
DeBabelizer 34, 232
deblocking 133–134, 252
deinterlacing 101–103, 105, 431
    After Effects 373
delivery 67
delta frames 52, 124
    *See also P-frames*
depth of field
    controlling 73
deringing 133–134, 252
detail
    background 73
device control 90
digital cable
    and MPEG-2 293
Digital Rights Management
    *See DRM*
digital television
    and MPEG-2 294
Digital Theater Systems
    *See DTS*
digital video workflow
    *See Chapter 4*
digitization 66, 77–92
    *See Chapter 6*
Director 150, 152–154, 208, 325
    and AVI authoring 325
    and QuickTime 153
    formats 153
    interface 150
    optimizing for video 154
    and QuickTime 209
    rich media apps 164
DirectShow 150, 323, 390, 431
    and Bink 343
disc-based playback 181
Discreet CineStream
    pan and scan tools 100
discrete cosine transformation
    *See DCT*
dither 35, 431
DivX 319, 328–330
Dolby Digital 287–288
dpi 20
drive space 92
drive speed 92
DRM 243–245, 311, 431
    MPEG-4 311
    Windows Media 243–244
dropframe timecode 91
DTS 288
DTV 79
Dual Channel 353
DV 333
    codecs 230
DV25 82
DVCAM 83
DVCPRO 84
DVD 289, 291, 293, 431
    audio 292
    enhanced 292
    multi-angle 292
    progressive 290–291
DVD+RW 431
DVD-R 431
DVD-ROM 292, 431
    MPEG-2 292
DVD-RW 431
DVD-Video 182

## E

ears
    anatomy 16
    how they work 15
ECL 431
edge blanking 95–97
Emblaze Video Pro 358

eMotion Studios 397
encoding
    platforms 385
    speed 144, 146
    stereo 145
Envivo Authoring Studio 171
error resilience 133, 139
ethernet 392
EventStream 172

## F

fallback movie 210
Fast Fourier Transform
    See FFT
FFT 43
FGS 308
FhG 353
    See also Fraunhoffer
FibreChannel 392–393
field 431
files
    intermediate 389
filters 110, 114, 134
    black restore 100
    contrast 116
    inverse telecine 104
    lowpass 111
    noise reduction 110
    postprocessing 133, 139
Final Cut Pro
    color correction 112, 117
    hue adjustment 118
FireWire 86, 431
Flash 172
Flash MX 152, 172, 177, 181, 183, 337–342
    See Chapter 21
    audio codecs 342
    authoring 338
    encoding 421
    progressive download 184
Flask MPEG 319
flattening 35
FLC 357
FLI 357
FlipFactory 394

formats
    obsolete 356
fovea 8, 10
fps 432
fractal compression 48
fractal transform 49
frame dropping 55, 131, 138
    avoiding 131
frame rate 118, 123, 137, 432
    interpolation 133–134
    mixing 124
    and sync 76
Fraunhoffer 236, 335, 342, 351–353
full screen 432

## G

gamma 27, 38, 112, 115–116, 432
    adjustment 115
    correction 426
    values 115
GammaToggle 112
gear 71
GIF 351
    animated 350
Giles, Darren xii
Gleick, James 48
Global Filters 281
GOP 273, 292, 432
    Open 273, 292
Gouras, Peter 18
Graphics codec 232
Group of Pictures
    See GOP
Guile, Steven 169

## H

H.263 223, 340, 348
HD 286, 294, 432
HDTV 79
hearing 13–18
Helix Producer 259–262, 380–381
    and RA Surround 268
Helix Producer Plus 262

Helix Server 241, 257, 260, 301
Heuris MPEG Power Professional (MPP) 281, 298, 423
    preprocessing 419
Hi8 81
HiFi 80
high-end gear 71
hint tracks 132
hinting 132, 139, 218
    modes 132
    ProCoder 368
    QuickTime 191
HipFlics 206, 378–379, 393
hits 81
hosting 164–165
HSV 432
HTML+Time 170
hue 117
    adjustment 118
Huffman encoding 40
Huffyuv 324, 333–334, 390–391

## I

iDVD 298
IEEE 1394 431–432
I-frames 50, 125, 432
    and MPEG-1 272
    See also keyframes
i-Link 432
IMA 237
IMA ADPCM 335
Image 341
image smoothing 341
iMovie 284
in-browser video 162
Indeo 3.2 230, 332
    tips 230
Indeo 4.4 331
Indeo 4:2:0 334
Indeo 5 226–228, 326–328, 357
    tips 228, 328
Indeo Raw 334
indexed color 33
in-house hosting 165

## Index  443

installed base 178
interactive video 167
interframe encoding 432
interlace 432
interlaced coding 44
interlaced video 286
intermediate files 389
inverse telecine 101, 103–104, 432
    After Effects 373
    algorithm 417
    support 359
    Windows Media Encoder 383
IPMP 432
IRE 432
iShell 164
ISMA 432
ISMA standards 309
ISO 299, 433
Iterated Function System 49
ITU-R BT.601 433
IVF 357–358
IVI 327

## J

jacks 88
Java Media Framework
    *See JMF*
JMF 345–348, 433
    authoring 348
    formats supported 346
    serving for 348
JPEG 433
JPEG 2000 433
JPEG2000 193
JVM 433

## K

key channels 429
keyframes 50, 52, 55, 123–124, 127, 137, 326, 433
    and MPEG-1 272
    and Premiere 370
    flashing 55, 124
    inserting 125
    rate 124, 137
    sensitivity 125, 137
kilobits per second (Kbps) 126, 433
kilobytes per second (KBps) 126, 433
Kinoma 348–349
kioskmode
    QuickTime 201

## L

LaBarge, Ralph 298
LAME 353–355
latency 161
latency vs. volume 392
letterboxing 98–101, 291
levels
    matching 88
LFE 433
light 5–6
    visible 5
lighting 72
Linux
    and MPEG-1 279
live streaming
    and Windows Media 242
lossless compression 41, 433
lossy compression 41, 433
lowpass filters 111
luma 29, 36, 45, 86, 111–112, 432–433
    adjustment 112
        After Effects 374
    processing 413
    values 114
luminance 7, 10, 34, 36, 433
    perceiving 10

## M

Mac OS 385–387
    encoding on 385
    and gamma 115
Mac OS Classic 386
Mac OS X 386
    and Cleaner 364
MACE 238
macroblocks 45, 433
Macromedia Director
    *See Director*
Macromind Director 1
MakeRefMovie 211–212, 215, 414–415
mask smoothing 220
masking 17, 219
    information 220
Master Movie 209
MBR 434
McGowan, John 336
media freshening 391
metafiles 162
MetaSound 256
MetaVoice 256
Microcosm 233
Microcosm codec 37
Microsoft Producer
    and Powerpoint 168
Microsoft RLE 334
MIME 163
miniDV 83
mipmapping 76
M-JPEG
    *See Motion-JPEG*
MMX 434
MNG 349–350, 434
Moore, Gordon 434
Moore's Law 434
motion 74
    blur 74–76
    estimation 75, 434
    perception 12
    search 132, 138, 434
    tracking 111
Motion-JPEG 231, 333, 390, 434
MovieShop 208
MP3 236, 276, 334, 342, 351–353, 355, 366
    data rates 352
    stereo modes 352
MP3 Pro 355
MPEG 434

MPEG-1 125–126, 132, 152, 180, 271–284, 423
    *See Chapter 17*
    aspect ratios 274
    audio 274
    authoring 280
        tools 281
    disc-based playback 182
    encoding 421
    for CD-ROM 279
    for VideoCD 276
    intermediates 391
    playback 279
    progressive download 184, 279
    streaming 280
    video codec 271
MPEG-2 125–126, 132, 285–298
    *See Chapter 18*
    aspect ratio 290
    audio 287
    authoring 296
    and CD-ROM 295
    disc-based playback 182
    DVD video 289
    DVD-ROM 292
    encoding tools 296
    intermediates 391
MPEG-4 46, 177, 181, 183, 222, 299–319
    *See Chapter 19*
    applications 313
    architecture 299
    audio 234
    audio codecs 303–309
    authoring 317
    format 301
    licensing 317
    progressive download 184
    QuickTime 196
    servers 301
    variants 319
    video codecs 302
    wireless standards 311

MPP
    *See Heuris MPEG Power Professional*
MS MPEG 319
multicasting 159–160
Myst 3
    Exile 425

# N

National Television Systems Committee
    *See NTSC*
N-bit 312
networking 392
noise
    composite 111
noise reduction 110–111, 114, 120
non-dropframe 91
None 334–335
normalization 120
NTSC 77–79
Nyquist frequency 23–25
Nyquist theorem 22–23

# O

Ogg Tarkin 356
Ogg Vorbis 335, 356
OHCI 1394 242, 434
on-demand streaming
    and Windows Media 241
OpenDML 321
optical audio 89
OS compatibility 175–176, 178

# P

PAL 78–79
PalmOS 348–349
    and Kinoma 349
PCI 434
PDAs 242–243
    playback 225
pels 20
perceptual noise shaping (PNS) 434

Perry Mason effect 85
perspective 12–13
P-frames 52, 54
    and MPEG-1 272
Philips WebCine 317–318
Photo-JPEG 389–390
pixel doubled 435
pixels 20, 109–110
    non-square 20
    shape 108, 136
platforms
    encoding 385
    issues 385–389
        *See Chapter 13*
playback
    anamorphic 212
    disc-based 181
playback performance 135
playback scalability 341
plugs 88
PNG 232–233
    codec 232
PocketPC 242–243
portable devices
    and Windows Media 242
postprocessing
    filters 133
postproduction 66
PowerPoint 168
    and Windows Media 250
Premiere 105, 207, 323, 368–371
    and WM 250
    Keyframes on Markers feature 125
    tips 370
premultiply 435
preplanning 70
    tips 70
preprocessing 66, 399, 417, 426
    audio 119
ProCoder 283, 297, 366–368, 423
    encoding 423
    preprocessing 402, 418
    tips 368

Producer
    and Windows Media 250
*Producing Great Sound for Digital Video* 119
progressive coding 44
progressive download 157–158, 183–184, 435
    Cleaner 211
    and data rates 138
    RealVideo 259
    tutorial 415
    and Windows Media 241
progressive GIF 435
progressive JPEG 435
progressive scan 72, 104, 435
    mode 104
psychoacoustics 17, 274
Pulse Code Modulation (PCM) 287
PureVoice 185, 236
PVAuthor 305, 318

## Q

QDesign Music 237
QDesign Music 2 180, 234–235
QT
    *See QuickTime*
QTVR 435
quantization 19, 25–26
    and bit depth 26
    errors 37
    levels 33
    value 46
    vector 48
quarter screen 435
Quesa project 194
QuickTime 19, 129–130, 152, 176, 178, 189–238, 279, 435
    *See Chapter 14*
    alternate movies 209
    and AVI 323
    audio 191
    authoring tools 205
    B-frames 126
    Broadcaster 199

delivery codecs 215
and Director 153
disc-based playback 181
encoding 420
and frame rates 119, 124
history of 190
intermediates 389
and JMF 346
Movie Export 390
and MPEG-1 playback 279
MPEG-4 196
progressive download 157, 184
rich media support 186
skin 164
spatial quality 130
streaming 192
QuickTime Player Pro 205

## R

raster 435
RAW audio codec 342
RealAudio 267–269
    codecs 266
RealMedia 263, 435
RealNetworks 257
RealOne 176–179, 258–259, 261
    Player 258–259
RealPresenter 261
RealProducer 403–404, 411
    preferences 409
RealSystem 257–269
    *See Chapter 16*
real-time streaming 159, 183, 185–186, 397
RealVideo 130, 134, 176, 179–180, 259–260, 264–265
    codecs 264
    disc-based playback 182
    encoding tools 261
    progressive download 184, 259
recompression 53
Red Book 435
reference files 162

resolution 108, 110, 136, 435
    constraints 108
    and video codecs 121
RGB 10, 28, 32–33, 36–38, 435
    and YUV conversion 38
    component 86
rich media 167–168, 170, 186, 259, 315
    and MPEG-4 315
    authoring 169
    support 186–187
ringing 46, 435
RLE 42, 334, 435
rods 7–8, 16
Rose, Jay 91, 119
RTSP 435
RTSP streaming 218
Rudiak-Gould, Ben 324
run-length encoding
    *See RLE*

## S

S/PDIF 88
safe areas 72, 96–97, 436
sample rates
    audio 142
sampling 19–20, 22
    rate 20, 145
    space 20
    time 20
satellite
    and MPEG-2 293
saturation 117
    adjustment 117
SBR data 355
scalability 286, 308, 328
    playback 341
scaling 105–108
    bicubic 105
    bilinear 105
scripting 393–394
SD TV 436
SDI 37
SDK 436

SDTV 79
SECAM 79
Serial Digital Interface (SDI) 87, 436
Shannon limit 42
Shannon, Claude 39, 61
sharpening 111
shell scripting 394
shutter speed 73
SIMD 361, 436
skins 163–164, 173, 194
sliding window 128
Smacker 183, 356–357
    and Bink 356
    vs. Bink 357
SMIL 169–170, 436
Sonic Foundry 249
Sonic Foundry SoundForge 401
Sorenson Squeeze 339, 376–378, 412
    *See also Squeeze*
SoundForge 401
space
    perception 12
Spark 337
    and Flash MX 337
Spark Pro 339–340, 422
Spark video codec 340–341
spatial compression 436
spatial quality 130, 138
    threshold 130
Speech audio codec 342
sprites 194
Squeeze 206, 338–340, 342, 376–378, 412–413, 419
    and Flash MX 377
    and QuickTime 377
    encoding 421
    interface 339
    inverse telecine filter 376
    and mixed source 104
    preprocessing 405, 419
    tips 378
SSE 436
SSE2 436
statistical multiplexing 436

stenciling 220
stereo encoding mode 145
Stitcher 249
Streambox 222–223
streaming 397, 436
Studio Pro 298
subsampling 436
SuperCharger 297
    and MPEG-2 297
SureStream 185, 258–260, 262, 266, 405
S-VHS 80
S-Video 85
sync
    correcting 410
Synthetic Aperture 112–113, 117, 374

## T

tape formats 80–85
tearing
    and VHS 96
telecine 103, 436
telecined video 103–104
Telestream FlipFactory 394
temporal compression 437
temporal quality 130
temporal scalability 265
Terran Interactive 2
Theora 356
timbre 14
timecode 90–91
    non-continuous 90
TMPGEnc encoder 283
Toast 151, 276, 284
    Titanium 151, 284
trails
    and block refresh 133
transparency 328
tripods 73
TrueMotion 332
tutorials 3
    *See Chapter 25*

## U

UDF 437
UDP 437
UI 437
unbalanced audio 87
UNIX 388–389
    and MPEG-1 279
    encoding system 389

## V

VBR 127, 144, 146, 437
    additive 144
    and MP3 352
    LAME 355
    subtractive 144
    vs. CBR 127, 143, 146
VBV 273, 437
VDO 358
vector 437
vector quantization (VQ) 48, 437
vectorscopes 117
Vegas Video 381–382
    tips 382
Velocity Engine 437
VfW 321, 437
VHS 80
video
    composite 85
    in-browser 162
    interlaced 286
    telecined 103
Video 1 332
video codecs
    *See Chapter 8*
Video Finesse plug-in 113
VideoCD 182, 276, 278
    formats 276
    players 278
    preprocessing 277
videoconferencing 161, 186, 223, 338, 340
VirtualDub 110, 324, 374–375
    tips 375
visible light spectrum 5

vision 5–13
Vivo 358
VLB 437
VOB files 290, 437
VoxWare MetaSound 256
VP3 224–225, 325–326
VQ 48
VRML 315
VXtreme 358

## W

watermarks 220–221
wavelet compression 46, 48, 437
web streaming
    tutorial 397
Web video 155–165
    See Chapter 11
    integrating 162
    types 157
white
    perceiving 11
White Book 276, 284, 360, 437
Windows
    encoding on 387
Windows Media 130, 152, 176, 179–180, 239–256, 405
    See Chapter 15
    audio codecs 254–256
    data rate control 129
    disc-based playback 181
    embedding in a Web page 243
    encoding tools 246–254
    and portable devices 242
    and PowerPoint 250
    progressive download 184, 241
    streaming 241–242
    video codecs 251
Windows Media Encoder 382–383, 404, 411–412
    encoding 421
    preprocessing 404
    tips 383
Windows Media Player 240
Windows Media Screen 253
Windows Media Video 252
Wired Sprites 170–171, 190, 194
wireless standards 311
WMA 437
WMV 322, 437
word 438
workflow
    See Chapter 4
    See Chapter 24

Worrall, David 18

## X

Xtras 150, 182, 325

## Y

Y'CbCr 26, 29–30, 36, 86, 361, 389, 409, 438
    Agility 394
    and After Effects 372
    and Squeeze 405
    AVI 390
    and Cleaner 363
    color space 29, 36
    component 86
    and connections 85
    processing 36
    to RGB 38
YUV color space 30–31
    and Y'CbCr 438
YUV overlay 438
YUV-12 438
YUV-9 122, 327, 334, 438

## Z

ZyGoVideo 180, 225

# The Authority on Digital Video Technology
# DV MEDIA GROUP

The **DV Media Group** provides video professionals technical insight, inspiration, and knowledge to improve their skills, productivity, and make informed buying decisions.

DV Expo West
December 9-13, 2002
Los Angeles Convention Center, CA

**DV**
PRINT

**DV.com**
ONLINE

**DV expo**
EVENTS

www.DV.com

# The Ben Waggoner Compression Style Guide to Cleaner

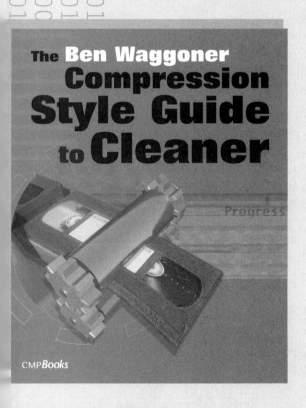

Get better results faster with this essential best-practices guide! It features point-by-point descriptions of every feature and control of Discreet Cleaner—it even documents bugs and their workarounds. Learn how to choose an appropriate codec, how to select optimal settings, and how to use Cleaner with many third-party codecs and formats not covered in the Cleaner manual.

ISBN 1-57820-212-4
$24.95
64 pages, PDF format

**Distributed exclusively through www.ebooktech.com**

CMP Books

## Audio Postproduction for Digital Video

### by Jay Rose

**Perform professional audio editing, sound effects work, processing, and mixing on your desktop.** You'll save time and solve common problems using these "cookbook recipes" and platform-independent tutorials. Discover the basics of audio theory, set up your post studio, and walk through every aspect of postproduction. The audio CD features tutorial tracks, demos, and diagnostics. 304pp, ISBN 1-57820-116-0, $44.95

## DVD Authoring & Production

### by Ralph LaBarge

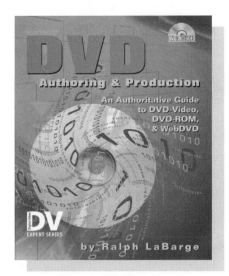

**Save time and money with the latest tools and techniques for publishing content in any DVD format—and get sage advice on tried-and-true marketing and distribution methods.** Organize, execute, and sell your projects efficiently with a complete understanding of the theory *and* the practice. Includes a product guide and the author's own *StarGaze* DVD. 477pp, ISBN 1-57820-082-2, $54.95

Find CMP*Books* in your local bookstore.

e-mail: cmp@rushorder.com
www.cmpbooks.com

Order direct 800-500-6875
fax 408-848-5784

# Final Cut Pro 3 Editing Workshop

Second edition

by Tom Wolsky

**Master the art and technique of editing with Final Cut Pro 3 on Mac OS X.** Thirteen tutorial lessons cover the complete range of tasks—from installing the application to color correction and outputting. All of the ingredients, including raw video footage and tips & tricks, blend to give you a working knowledge of the principles taught in film schools. CD-ROM and KeyGuide™ included, 583pp, ISBN 1-57820-118-7, $49.95

# Creating Motion Graphics with After Effects

Volume 1, Second edition

by Trish & Chris Meyer

**Create compelling motion graphics with After Effects v5.5.** Boasting the same color-packed presentation, real-world explanations, and example projects that made the first edition a bestseller, this first of two volumes features a chapter to get new users started and a focus on the core features of the application. CD-ROM included, 432pp, ISBN 1-57820-114-4, $54.95

Find CMP*Books* in your local bookstore.

e-mail: cmp@rushorder.com
www.cmpbooks.com

Order direct 800-500-6875
fax 408-848-5784

# What's on the CD-ROM?

The companion CD-ROM for *Compression for Great Digital Video* includes project files and sample media for the three tutorials in Chapter 25—plus settings for encoding them in various applications.

Software is not included on the CD-ROM because it changes too rapidly. To download the latest and greatest installers for the formats and demos for the various applications, go to:

www.benwaggoner.com/bookupdates.html

Visit the same site for corrections, additions, offers, and other updates for the book. Happy compressing!